T5-CVD -633

The Chemistry
of Catalytic
Hydrocarbon
Conversions

The Chemistry of Catalytic Hydrocarbon Conversions

HERMAN PINES

Department of Chemistry
Northwestern University
Evanston, Illinois

1981

ACADEMIC PRESS
A Subsidiary of Harcourt Brace Jovanovich, Publishers
New York   London   Toronto   Sydney   San Francisco

ACADEMIC PRESS, INC.
111 Fifth Avenue, New York, New York 10003

United Kingdom Edition published by
ACADEMIC PRESS, INC. (LONDON) LTD.
24/28 Oval Road, London NW1 7DX

Library of Congress Cataloging in Publication Data

Pines, Herman, Date.
 The chemistry of catalytic hydrocarbon conversions.

 Includes bibliographies and index.
 1. Catalysis. 2. Hydrocarbons. I. Title.
QD505.P55   547.1'395   80-1778
ISBN 0-12-557160-7

PRINTED IN THE UNITED STATES OF AMERICA

81 82 83 84   9 8 7 6 5 4 3 2 1

To Dorothy, Judy, Jeff, David, and Debbie

Contents

3
Heterogeneous Hydrogenation

4
Dehydrogenation and Cyclodehydrogenation (Aromatization)

5
Oxidation

6
Homogeneous Catalysis by Transition Metal Organometallic Compounds

7
Metathesis of Unsaturated Hydrocarbons

8
Synthesis of Liquid Hydrocarbons (Synthetic Fuels)

Preface

The catalytic conversion of hydrocarbons is the key method for the production of both high octane gasoline and petrochemicals. A thorough understanding of the chemistry of reactions accompanying these conversions is of the utmost importance for developing new catalytic processes and for improving existing ones. Standard textbooks of organic chemistry pay only scant attention to this conceptually and industrially important field of chemistry.

This book, with few exceptions, is confined to a description of the various aspects of catalytic conversions of hydrocarbons. However, in view of the recent interest in synthetic fuels, it contains a description of the catalytic synthesis of hydrocarbons from carbon monoxide and hydrogen and from methanol.

The material is arranged according to the type of catalyst used in the various reactions. The first chapter covers the acid-catalyzed reactions of hydrocarbons. Acid catalysts are widely applied in a multitude of catalytic reactions and are the backbone of many major petrochemical processes. Because of its length and vastness of material, this chapter is divided into eight sections, and the references are also grouped by these same section divisions. In the remaining seven chapters references are not separated. References with an asterisk contain either broad citations of pertinent literature or a review article.

In writing the book an attempt was made to present mechanistic interpretations of the reactions, and, whenever possible, examples of industrial relevance were included.

The author wishes to express his gratitude to Professor Charles D. Hurd who read the entire manuscript. His many valuable suggestions are greatly appreciated. The various chapters of the manuscript were examined by experts in their respective fields. Drs. Fred Basolo, R. L.

Burwell, Jr., W. O. Haag, G. W. Keulks, Gordon Langlois, Jr., D. A. McCaulay, D. F. Shriver, Samuel Siegel, W. H. M. Sachtler, and W. M. Stalick. Their critical comments and constructive suggestions are gratefully acknowledged.

This book would probably not have been written but for the urging and nudging from Professors John B. Butt and W. Keith Hall. I thank them now for it.

Ms. Vonita Curbow deserves special recognition for the difficult task of typing the manuscript from the handwritten copy and for making india ink drawings of some of the chemical structures. All this she has performed skillfully and cheerfully.

It is the author's hope that this book will be of value not only to all those active in the field of catalytic hydrocarbon conversions but that it will also be used as an auxiliary text for seniors and graduate students in organic chemistry and related areas.

Herman Pines

Terminology and Abbreviations

For general abbreviations and terminology of catalytic systems consult "Manual of Symbols and Terminology for Physicochemical Quantities and Units—Appendix II. Part II. Heterogeneous Catalysis," *Adv. Catal.* **26,** 351–392 (1977). Symbols used in this volume are listed below.

Butane	the unbranched C_4 alkane
Butanes	both butane and isobutane
Pentane	the unbranched C_5 alkane
Pentanes	includes the unbranched C_5 alkane, and its isomers isopentane and neopentane
Ethene	ethylene
Propene	propylene
Butenes	1- or 2-butene. The term demands a chain length of four carbons, hence excludes the isomeric methylpropene (isobutylene) with chain length of three carbons
Butylenes	includes both straight chain and isobutylene
Hexenes	unbranched, acyclic C_6H_{12}
Propyl, butyl	the radicals $CH_3CH_2CH_2-$, $CH_3CH_2CH_2CH_2-$
Butylbenzene	applies to $C_6H_5CH_2CH_2CH_2CH_3$; the prefix *n*-butylbenzene should be avoided
GHSV	gaseous hourly space velocity, e.g., volume of gas/volume of catalyst/hour
LHSV	liquid hourly space velocity, e.g., volume of liquid/volume of catalyst/hour
WHSV	weight hourly space velocity, e.g., weight of substrate/volume of catalyst/hour
Temperature	expressed in °C, unless otherwise indicated

1

Acid-Catalyzed Reactions

I. INTRODUCTION

Acid-catalyzed reactions are by far the most important reactions involved in the rearrangement and conversion of hydrocarbons. These reactions are responsible for isomerization of alkanes and cycloalkanes, polymerization of alkenes, catalytic cracking, alkylation of isobutane to high-octane hydrocarbons, reforming of naphthas, synthesis of alkylaromatic hydrocarbons, etc. The relevant acids include three types: (a) protonic acids, of which the commercially most important are silicophosphoric acid, sulfuric acid, and hydrogen fluoride; (b) Lewis acids such as aluminum chloride and boron trifluoride; (c) acidic oxides represented by silica–alumina (cracking catalyst), zeolites, and "acidic" alumina, which contains small amounts of halides.

It is generally agreed that acid-catalyzed hydrocarbon conversion reactions proceed by way of highly reactive, positively charged intermediates that are referred to variously as carbonium ions, cations, or carbenium ions. Since these intermediates appear prominently in the pages that follow, it will be helpful to consider their nomenclature, and the tertiary butyl case will be illustrative.

1. The *tert*-butyl group may be neutral, $(CH_3)_3C\cdot$, or it may carry a positive charge, $(CH_3)_3C^+$, or a negative charge, $(CH_3)_3C^-$. These are named, respectively, *tert*-butyl radical, *tert*-butyl carbonium ion, and *tert*-butyl carbanion. The three situations are depicted by the separate words (not suffixes) radical, carbonium ion, and carbanion. Thus, $(CH_3)_3C^+$ is *tert*-butyl carbonium ion (not *tert*-butylcarbonium ion); it should not be named either trimethyl carbonium ion or trimethylcarbonium ion.

2. *tert*-Butyl cation is also good usage.

3. The rationale underlying "carbenium" is simple. If a proton is added to carbene (methylene) the carbenium cation results, $H_2C: + H^+ \rightarrow CH_3^+$, which is comparable to the formation of ammonium ion by adding a

proton to ammonia. Thus, carbenium cation turns out to be nothing but methyl cation. Substitution of the three H's by three methyls results in the name trimethylcarbenium cation. Note that "carbenium" in such terms must be a suffix, not a separate word.

Since it offers no advantages and may be troublesome to some, "carbenium" is not used in this chapter. It should be kept in mind that an anion must be associated with a cation. In the case of $(CH_3)_3C^+$, if the anion were HSO_4^-, the full name would be *tert*-butyl hydrogen sulfate.

A comment on the formation, reaction, stability, and geometry of carbonium ions will serve to simplify later discussions of the mechanism of acid-catalyzed reactions.

A. Carbonium Ions

Carbonium ions are produced most commonly in four ways.

1. Addition of the proton of an acid to an olefin:

$$HX + \,\diagup C = C \diagdown \longrightarrow \left[H - \overset{|}{\underset{|}{C}} - \overset{+}{C} \diagdown \right] X^-$$

The reaction is an acid–base reaction, and therefore the concentration of carbonium ion depends on the strength of the acid.

2. Removal of a halide ion from an aliphatic halide by a Lewis acid, as in the action of aluminum chloride on an alkyl halide:

$$R - Cl + AlCl_3 \longrightarrow R^+ + AlCl_4^-$$

3. Oxidation of tertiary alkanes. Oxidation of isobutane by sulfuric acid is an example:

$$\underset{\underset{CH_3}{|}}{\overset{\overset{CH_3}{|}}{CH_3CH}} + 4\,H_2SO_4 \longrightarrow \underset{\underset{CH_3}{|}}{\overset{\overset{CH_3}{|}}{CH_3\overset{+}{C}}} + 2\,H_3\overset{+}{O} + SO_2 + 3\,HSO_4^-$$

4. Addition of acids to alcohols. The initial step involves the formation of a relatively stable oxonium salt, which decomposes under suitable conditions:

$$HX + R - \overset{..}{\underset{..}{O}} - H \rightleftharpoons \left[R - \overset{\overset{H}{|}}{\underset{..}{O}} - H \right]^+ X^- \rightleftharpoons R^+ + X^- + H_2O$$

Owing to their electronical-deficient nature, carbonium ions undergo a variety of reactions. These reactions, which form the basis of isomerization mechanisms, also explain how the side reactions accompanying many isomerizations come about.

1. Loss of a proton from an adjacent carbon atom:

$$H_3C-\overset{+}{\underset{\underset{H}{|}}{C}}-\overset{\underset{H}{|}}{C}-CH_3 \xrightarrow{-H^+} CH_3CH{=}CHCH_3 + H^+$$

2. Internal rearrangement by a hydride ion shift:

$$H_3C-\overset{\underset{H_3C}{|}}{C}-\overset{+}{\underset{H}{C}}-CH_3 \xrightarrow{\sim H:} H_3C-\overset{+}{\underset{\underset{H_3C}{|}}{C}}-\overset{\underset{H}{|}}{C}-CH_3$$

Such shifts occur rapidly and are a consequence of the difference in the stability of carbonium ions. The order of increasing stability of carbonium ions is primary < secondary < tertiary and alkyl < allyl.

3. Internal rearrangement by migration of R: to an adjacent carbon:

$$H_3C-\overset{\underset{H_3C}{|}}{C}-\overset{+}{\underset{\underset{H_3C}{|}}{C}}-CH_3 \xrightarrow{\sim CH_3:} H_3C-\overset{+}{C}-\overset{\underset{CH_3}{|}}{\underset{|}{C}}-CH_3$$

4. Removal of a hydride from another molecule.

(a) In the presence of 98% sulfuric acid, the intermolecular hydride removal occurs only between a tertiary cation and molecules with hydrogen bonded to tertiary carbons:

$$(CH_3)_3C^+ + (CH_3)_2CHCH_2CH_3 \longrightarrow (CH_3)_3CH + (CH_3)_2\overset{+}{C}CH_2CH_3$$

(b) Aluminum chloride/hydrogen chloride also catalyzes intermolecular transfer of hydride ion from a secondary carbon atom to a secondary or tertiary cation:

$$(CH_3)_3C^+ + CH_3CH_2CH_2CH_3 \rightleftharpoons (CH_3)_3CH + CH_3CH_2\overset{+}{C}HCH_3$$

(c) The removal of hydride from a primary carbon atom to a cation occurs intermolecularly only in the presence of very strong acids (superacids), such as FSO_3H/SbF_5:

$$(CH_3)_2CH^+ + (CH_3)_4C \longrightarrow (CH_3)_2CH_2 + (CH_3)_3CCH_2^+$$

(d) Owing to the fact that an allylic cation is more stable than the corresponding alkyl cation, 96% sulfuric acid is able to catalyze the transfer of a secondary allylic hydrogen to a cation:

$$(CH_3)_3C^+ + RCH{=}CHCH_2R^1 \longrightarrow (CH_3)_3CH + RCH{=}CH\overset{+}{C}HR^1$$

$$R = alkyl$$

5. Addition of a carbonium ion to an olefin or to an arene:

$$(a) \quad (CH_3)_3C^+ + CH_2{=}C(CH_3)_2 \longrightarrow (CH_3)_3CCH_2\overset{+}{C}(CH_3)_2$$

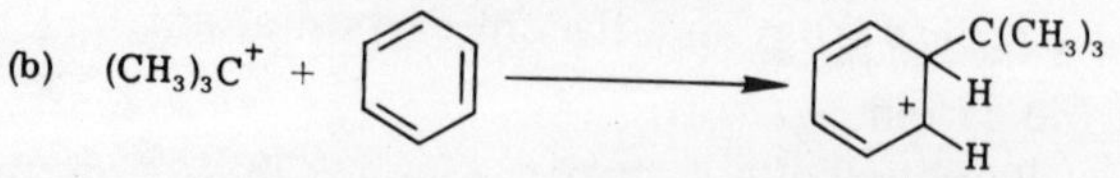

(b) $(CH_3)_3C^+$ +

6. β-Scission of a carbonium ion to another carbonium ion and an olefin:

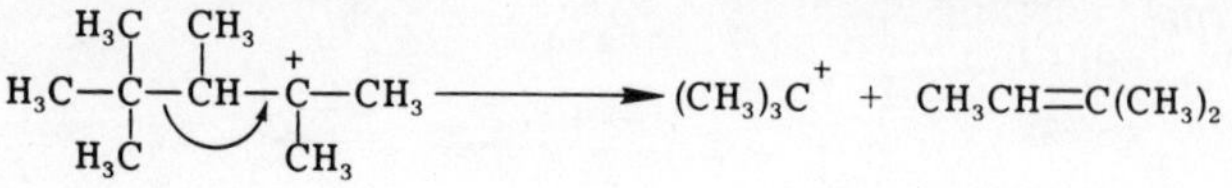

This reaction occurs most readily with ions capable of generating a tertiary cation on scission. It is to be noted that this is the reverse of reaction 5(a).

7. Rapid exchange of α-hydrogens of cations with an acid. Exchange with sulfuric acid-d_2 is an example:

$$9D_2SO_4 + (CH_3)_3C^+ \longrightarrow (CD)_3C^+ + 9DHSO_4$$

The carbonium ion is planar, and the valence bond angles of the positively charged carbon are 120°. This fact has proved extremely useful in

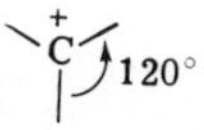

the study of carbonium ion mechanisms. If an asymmetric carbon is converted to a carbonium ion, the asymmetry is lost because of the planar character of the ion. Thus, racemization of an optically active hydrocarbon during a reaction proceeding by a carbonium ion mechanism indicates that the asymmetric carbon bore a positive charge sometime during the reaction. Also, the rate of racemization often gives valuable information concerning the mechanism of reaction.

The anions associated with cations R^+ are not shown above and usually are omitted throughout the text. The omission is for the sake of conve-

TABLE 1.1

Heat of Formation of Alkyl Cations[a]

Alkyl cation	ΔH_{298} (kcal/mol)
CH_3^+	261
$CH_3CH_2^+$	219
$CH_3CH_2CH_2^+$	208
$(CH_3)_2CH^+$	192
$CH_3CH_2CH_2CH_2^+$	201
$(CH_3)_2CHCH_2^+$	199
$CH_3CH_2\overset{+}{C}HCH_3$	183
$(CH_3)_3C^+$	167

[a] From Lossing and Semeluk (1970).

nience and does not imply that the nature of the anion has no effect on the reactivity of the cation.

The relative stabilities of alkyl cations as determined from their heat of formation are given in Table 1.1. It can be seen that *tert*-butyl cation is more stable by 16 kcal than the *sec*-butyl cation, and the latter is more stable by 16–18 kcal than the two primary butyl cations.

B. Acid Strength

The majority of the acid-catalyzed reactions of hydrocarbons are carried out in the presence of relatively strong inorganic acids. The acid strength of these acids is defined as their proton-donating capacities, quantitatively expressed as Hammet's H_0 acidity function,

$$H_0 = -\log \frac{a_{H^+}f_B}{f_{BH^+}}$$

where a_{H^+} is the total activity of the proton in the solution and f_B/f_{BH^+} is the ratio of the activity coefficients of a neutral base B and its conjugate acid BH^+. Also,

$$H_0 = pK_{BH^+} - \log \frac{C_{BH^+}}{C_B}$$

where C_{BH^+}/C_B is the directly observable concentration ratio of the indicator in its two differently colored forms, and K_{BH^+} is the thermodynamic ionization constant of its conjugate acid in terms of molar concentration.

Values of $-H_0$ for anhydrous acids are given in Table 1.2. The acids listed in the table with $-H_0 < 5$ can be classified as weak acids, with $-H_0 = 8.2-12.8$ as strong acids, and with $-H_0$ above 16.8 as superacids.

TABLE 1.2

Values of $-H_0$ for Selected Acids[a]

Acid	$-H_0$	Acid	$-H_0$
HSO_3F	12.8	BF_3/HF	16.8
BF_3/H_2O	11.4	TaF_5 (10 mol %)/HF	18.85[b]
H_2SO_4 (96%)	8.98	SbF_5 (10 mol %)/HF	24.33[b]
H_2SO_4 (98%)	9.36	SbF_5/HSO_3F	18.65
H_2SO_4 (100%)	11.0	$SiO_2-Al_2O_3$	<8.2
HF	10.2		
H_3PO_4	5		
HCO_2H	2.2		

[a] From Gillespie and Peel (1971) and Gillespie (1963).
[b] Siskin (1978).

Weak acids catalyze the conversion of olefins in which only the double bond is attacked, resulting in double-bond isomerization and also, at higher temperatures, in skeletal rearrangement of olefins and oligomerization of the more reactive olefins. The stronger acids, besides attacking the double bonds of alkenes, can also cause a hydrogen transfer reaction, leading to conjunct polymerization of olefins, skeletal isomerization of saturated hydrocarbons, and alkylation of saturated hydrocarbons. Superacids are able to attack the most stable of alkanes, such as methane and ethane, and convert them to higher molecular weight hydrocarbons (Section VII).

The acid strength values given in Table 1.2 are derived from ionization equilibria of a particular class of indicators and are usually measured at room temperature. However, the catalytic conversion of hydrocarbons is not always limited to experimental conditions of lower temperature, and in the case of the more temperature-resistant catalysts, such as $SiO_2-Al_2O_3$ or silicophosphoric acid, temperatures of the order of 300°–450°C can be used. Under these conditions these relatively weak acids can catalyze certain reactions that may occur at much lower temperatures with acids having $-H_0$ equal to 9.0.

II. ISOMERIZATION

A. Isomerization of Olefins

1. Introduction

Olefins are isomerized by contact with a wide variety of acid catalysts: strong organic acids, such as mono-, di-, or trichloroacetic and benzenesulfonic acids; mineral acids, such as hydrofluoric, perchloric, sulfuric, phosphoric, and silicophosphoric acids; acid salts of strong acids, such as potassium bisulfate; metal halides, such as ferric chloride and stannic chloride; and acid-acting oxides, such as silica–alumina and certain kinds of aluminas. The use of strong acids, such as 96% sulfuric acid, hydrogen fluoride, or aluminum chloride/hydrogen chloride, is not desirable because the isomerization is accompanied to a great extent by conjunct polymerization.

Three types of olefin isomerizations can be differentiated: (a) cis–trans isomerization; (b) double-bond migration; and (c) skeletal rearrangement, which includes change in chain branching and in cyclic olefin ring expansion and contraction.

2. Racemization and Double-Bond Migration

Racemization and double-bond migration are the simplest isomerizations that can take place in the presence of an acid. Both reactions proceed through carbonium ion intermediates. All the reactions characteristic of carbonium ions can occur during isomerization. However, by the use of a catalyst of mild acid strength and with control of the contact time and temperature, the desired olefin isomerization is usually achieved with a minimum of side reaction. In general, the contact time and temperature required are inversely related to the acid strength of the catalyst used.

The mechanism of double-bond migration in the presence of a protonic acid can be illustrated by the isomerization of 1-butene to *cis-* and *trans-*2-butene. In step 1 the proton adds to the terminal carbon to form the more

$$H^+ + CH_3CH_2CH=CH_2 \overset{1}{\rightleftharpoons} CH_3CH_2\overset{+}{C}HCH_3 \overset{2}{\rightleftharpoons} H^+ + CH_3CH=CHCH_3$$

stable secondary carbonium ion. In expelling the proton, the carbonium ion reverts to starting material, 1-butene, or forms *cis-* and *trans-*2-butene. The proton from the carbonium ion can return to the catalyst or can be transferred to another olefin molecule and thus create a chain reaction.

The reversible double-bond shift at a branch in the chain, such as in methylbutenes **1** and **2**, proceeds more readily than that of their isomer **3**. In the isomerization of **3** a *sec*-alkyl cation is involved, whereas in isomerization of **1** and **2** only the more stable *tert*-alkyl cation participates.

3. Skeletal Isomerization

The ease of skeletal isomerization depends on which carbonium ion intermediates are formed. Skeletal rearrangement requires higher temperatures or stronger acids than does double-bond isomerization. Silicophosphoric acid, aluminas (acidic), and silicoaluminas are useful catalysts for these reactions. If a primary carbonium ion is involved, the isomerization will be slow. For example, in the reversible isomerization of 1-butene to isobutylene, primary carbonium ion formation follows proton addition. Because primary carbonium ion formation is slow, more drastic condi-

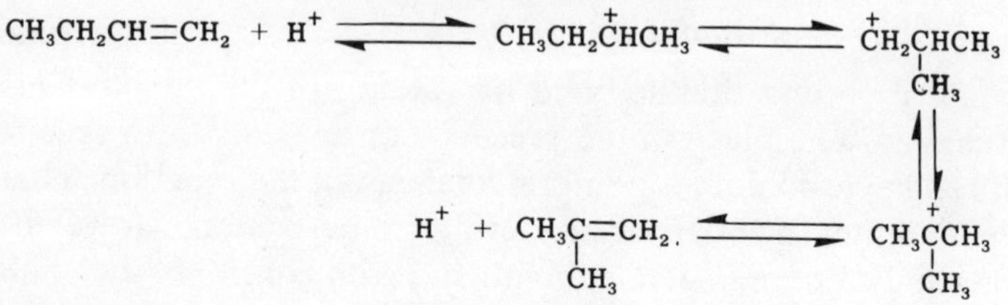

$$CH_3CH_2CH{=}CH_2 \ + \ H^+ \ \rightleftharpoons \ CH_3CH_2\overset{+}{C}HCH_3 \ \rightleftharpoons \ \overset{+}{C}H_2CHCH_3$$

$$H^+ \ + \ \underset{\underset{CH_3}{|}}{CH_3C}{=}CH_2 \ \rightleftharpoons \ \underset{\underset{CH_3}{|}}{CH_3\overset{+}{C}CH_3}$$

tions are required for skeletal isomerization, and under such conditions side reactions, such as polymerization, become competitive.

In the skeletal isomerization of *tert*-butylethylene to 2,3-dimethyl-1- and 2,3-dimethyl-2-butene, the favorable change of a secondary to a tertiary carbonium ion takes place. Such an isomerization occurs under relatively mild conditions, and, therefore, little side reaction occurs.

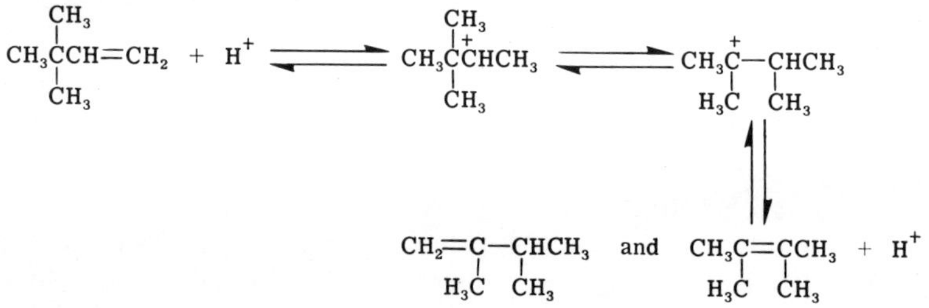

$$\underset{\underset{CH_3}{|}}{\overset{\overset{CH_3}{|}}{CH_3C}}CH{=}CH_2 \ + \ H^+ \ \rightleftharpoons \ \underset{\underset{CH_3}{|}}{\overset{\overset{CH_3}{|}}{CH_3C}}\overset{+}{C}HCH_3 \ \rightleftharpoons \ \underset{\underset{H_3C}{|} \ \underset{CH_3}{|}}{CH_3\overset{+}{C}{-}CHCH_3}$$

$$\underset{\underset{H_3C}{|} \ \underset{CH_3}{|}}{CH_2{=}C{-}CHCH_3} \quad \text{and} \quad \underset{\underset{H_3C}{|} \ \underset{CH_3}{|}}{CH_3C{=}CCH_3} \ + \ H^+$$

Cycloalkenes undergo skeletal isomerizations with a change in the ring size. The simplest example is the reversible rearrangement of methylcyclopentene.

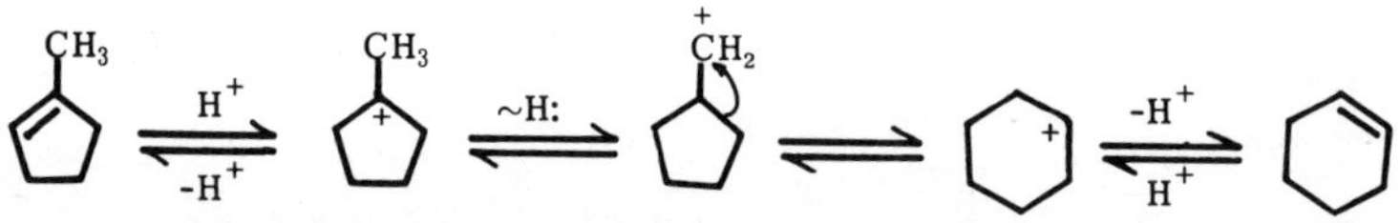

Another type of skeletal isomerization of cyclic olefins involves ring-opening reactions, which occur under relatively mild conditions. An example is α-pinene (4) reacting in the presence of 4-chloro-1-naphthalene-

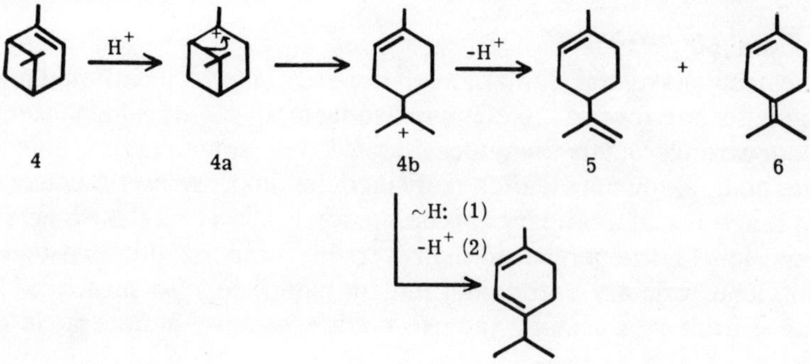

4　　　　　4a　　　　　4b　　　　　5　　　　　6

~H: (1)
-H$^+$ (2)

7

sulfonic acid to give dipentene (**5**), terpinolene (**6**), and α-terpinene (**7**). The key step in the isomerization is the formation of ion **4b** from **4a** by β-scission.

a. Alumina Catalyst. Rearrangement of the carbon skeleton of olefins takes place over aluminas. These aluminas are usually prepared by methods that make them substantially free from the presence of alkali metal ions. Aluminas obtained from the hydrolysis of aluminum isopropoxide and calcined at 600°–800°C are effective catalysts for the skeletal isomerization of alkenes and cycloalkenes. Since skeletal isomerization is a carbonium-type reaction, it can be concluded that alumina must contain intrinsic acid sites to cause the rearrangement of olefins.

The extent of rearrangement of olefins over aluminas is related to the strength of their acid sites. It was thus possible to determine the relative strength of the sites from the rate and depth of rearrangement of cyclohexene and 3,3-dimethylbutene (A) when passed over alumina. The isomerization of cyclohexene to methylcyclopentenes involving the rearrangement of cyclohexene to a less stable primary cation can be expected to proceed with greater difficulty than that of A to 2,3-dimethylbutenes (B), in which only secondary and tertiary cations participate. The B formed can in turn form 2-methylpentenes (C) by steps that involve an unstable primary cation. Therefore, the latter rearrangement should proceed more slowly than the formation of B from A or require stronger acid sites.

For similar reasons the conversion of C to an unbranched hexene should also proceed more slowly. In order to characterize the acid sites of alumina it is important, therefore, to record not only the total conversion of A but also the depth of isomerization.

From the study of the kinetics of isomerization of A over prepared alumina (Table 1.3), certain important conclusions can be derived: (a) Compound B appears to be the only primary isomerization product. (b) The fact that C was nearly equilibrated before a noticeable amount of 3-methylpentenes (D) was formed shows that hydrogen migration to produce double-bond isomers is faster than migration of methyl groups. (c) The

TABLE 1.3

Hexene Isomers Obtained from the Isomerization of 3,3-Dimethylbutene over Alumina at 350°C[a]

Compound	HLSV[b]			Equilibrium (%)
	6.0	2.0	0.5	
3,3-Dimethylbutene (A)	21.0	2.7	1.9	0.5
2,3-Dimethylbutenes (B)	73.0	39.3	23.0	11.5
2-Methylpentenes (C)	6.0	37.2	42.6	40.5
3-Methylpentenes (D)	—	19.8	31.0	33.5
n-Hexenes	—	1.0	1.5	14.1

[a] From Haag and Pines (1960).
[b] Hourly liquid space velocity, or the volume of liquid feed per volume of catalyst per hour.

observation that D is produced from C indicates that the rearrangement of cations in 3,3-dimethylbutene proceeds only by a 1,2 shift of methyl groups.

The rate of isomerization of cyclohexene to methylcyclopentenes depends to a very great extent on the temperature at which the alumina is calcined. The optimal calcination temperature of alumina prepared from the hydrolysis of aluminum isopropoxide lies between 600° and 750°C. Since the rate of isomerization depends on the strength of the acid sites of the alumina, it can be concluded that the strong catalytic sites in the aluminas are developed at these temperatures through generation on heating of not fully coordinated aluminum atoms:

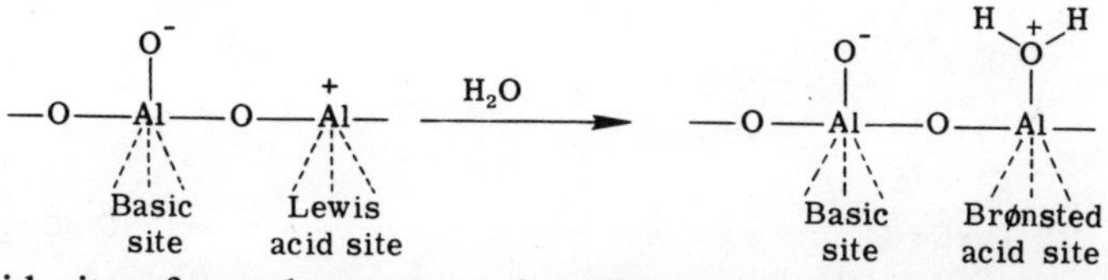

The acid sites formed are not of uniform strength since they are influenced by the inductive effect of the neighboring groups and by the bulk of the alumina. Traces of water can also convert the Lewis acid sites to Brønsted acid sites. The number of acid sites was found to be 2×10^{13} per square centimeter, and the number of acid sites strong enough to isomerize cyclohexene was about 10% of the total sites. For that reason only a small amount of alkali ions is necessary to deactivate the cyclohexene-isomerizing properties of aluminas. Furthermore, methods of depositing alkali metal cations also have a large effect on the properties of aluminas.

The striking differences in catalytic behavior between alumina impregnated with sodium hydroxide (Al_2O_3-I) and alumina obtained from sodium aluminate solutions (Al_2O_3-A) are summarized in Table 1.4, where the two catalysts with similar sodium contents are compared. For reference, a pure alumina catalyst (Al_2O_3-P) is also included.

Catalyst A is of low acid strength, as indicated by the lack of isomerization of cyclohexene and high selectivity of isomerization of compound A to compound B; in comparison, the sites in catalyst I are much stronger, and they also convert compound A to methylpentenes. The impregnation method does not always lead to a similar selective poisoning of the strong acid centers encountered in pure alumina (Al_2O_3-P). When a sodium hydroxide solution of low concentration penetrates the pores of the alumina pellets, the solute is nonselectively absorbed, causing a rapid depletion of the solution before the total volume is filled. Thus, the data obtained from the isomerization of compound B indicate that the internal surface of Al_2O_3-I contains acid sites in approximately the original acid strength distribution, whereas in Al_2O_3-A all the relatively strong acid sites were neutralized.

The strength of the acid sites of aluminas can be greatly enhanced by treating the aluminas with small amounts of hydrochloric or hydrofluoric acid. Such acid-treated alumina was found to be a very suitable support for reforming (Section VIII).

Deep-seated skeletal isomerization accompanied by a hydrogen disproportionation reaction occurs when a mixture of alkylcyclohexene and hydrogen chloride is passed over alumina at 300°C. Isopropylcyclohexene is converted to arenes, cycloalkenes, and cycloalkanes. The arenes consist

TABLE 1.4

Comparison between Impregnated and Aluminate Catalysts[a]

	Alumina[b]		
Property	P	I	A
Sodium content (%)	0	0.1	0.08
Surface area (m²/gm)	287	280	254
Conversion (%)			
Of cyclohexene	43	32	0
Of 3,3-dimethylbutene	97	87	76
Isomerization of 3,3-dimethylbutene, selectivity (%)[c]	40	65	97

[a] From Pines and Haag (1960).

[b] Key: P, Al_2O_3 prepared from aluminum isopropoxide; I, impregnated with sodium hydroxide; A, prepared from potassium aluminate.

[c] Conversion to 2,3-dimethylbutenes.

mainly of trimethylbenzenes and ethyltoluene. The skeletal structure of the alkylcyclohexenes and alkylcyclohexanes produced in the reaction is similar to that of aromatic hydrocarbons (Pines *et al.*, 1952). Hydrogen transfer accompanying skeletal isomerization is usually associated with strong acid sites of the catalysts.

   b. Silica–Alumina. Silica–alumina has strong intrinsic acid sites (see Section VI) and causes a type of hydrogen disproportionation and skeletal isomerization of alkylcyclohexenes that is similar to that catalyzed by Al_2O_3/HCl.

B. Isomerization of Saturated Hydrocarbons

1. Introduction

   Acid-catalyzed isomerization of saturated hydrocarbons was first used by the petroleum industry in the 1930's, and the reaction received theoretical examination at about the same time. Knowledge gained from such research not only led to the interpretation of the mechanism of isomerization but also pointed to the fundamental nature of the reaction and its participation in other reactions involving saturated hydrocarbons.

   The capacity to cause isomerization is characteristic of acid catalysts. The ease of isomerization depends on the strength of the acids used in the reaction. The most effective Friedel–Crafts catalysts were found to be aluminum bromide, aluminum chloride, and boron fluoride in conjunction with their corresponding hydrogen halides. Sulfuric acid and $SiO_2–Al_2O_3$ are other catalysts that have been widely studied. Their effectiveness as catalysts is, however, limited to certain branched-type hydrocarbons (Section II,B,3 and 4).

   Dual-function catalysts composed of a hydrogenation–dehydrogenation component deposited on an acid support are very useful for the isomerization of saturated hydrocarbons. The isomerization is usually carried out under hydrogen pressure.

2. Friedel–Crafts Catalysts

   a. Alkanes *i. Butane.* The isomerization of butane to isobutane was the first alkane isomerization used commercially; the isobutane thus produced was used for alkylation in the preparation of high-octane gasoline. The equilibrium composition of butane–isobutane indicates that low temperatures favor isobutane. The isomerization equilibrium constants are expressed by the following equation:

$$R \ln K_G = \frac{2318}{T} - 4.250$$

where

$$K_G = \frac{i\text{-}C_4H_{10}}{n\text{-}C_4H_{10}} \quad \text{(vapor)}$$

Early investigations showed that the isomerization of butane to isobutane and the reverse reaction proceed in the presence of catalysts containing $AlCl_3$ or $AlBr_3$. Products containing up to 62.9% of isobutane were obtained at 100°C by contacting butane with $AlCl_3/HCl$. However, more careful studies have demonstrated that pure butane does not isomerize at temperatures below 100°C when $AlCl_3/HCl$ is the catalyst (Table 1.5) or below 25°C when $AlBr_3/HBr$ is the catalyst (Table 1.6). The presence of less than 0.1% olefins as promoters is sufficient to cause isomerization. The olefins act as acceptors of protons, and the resulting cations initiate the isomerization:

$$RCH{=}CH_2 + HX + AlX_3 \rightleftharpoons [R\overset{+}{C}HCH_3]\,AlX_4^-$$

$$R\overset{+}{C}HCH_3 + CH_3CH_2CH_2CH_3 \rightleftharpoons RCH_2CH_3 + CH_3\overset{+}{C}HCH_2CH_3$$

$$CH_3\overset{+}{C}HCH_2CH_3 \rightleftharpoons \underset{\underset{CH_3}{|}}{CH_3CHCH_2^+} \overset{\sim H:}{\rightleftharpoons} \underset{\underset{CH_3}{|}}{CH_3\overset{+}{C}CH_3}$$

$$\underset{\underset{CH_3}{|}}{CH_3\overset{+}{C}CH_3} + CH_3CH_2CH_2CH_3 \rightleftharpoons \underset{\underset{CH_3}{|}}{CH_3CHCH_3} + CH_3\overset{+}{C}HCH_2CH_3 \quad \text{etc.}$$

According to the above mechanism, the cation, being constantly regenerated during a chain reaction, needs to be present in only small amounts.

The chain-initiating ion may be produced in several ways. It may be introduced in the form of alkyl halides, or it may be formed by the addition of traces of olefins to the butane.

ii. Pentanes. The isomerization of pentane to isopentane occurs even at 20°C with aluminum halide/hydrogen halide catalyst. It is accelerated by the addition of small amounts of carbonium ion precursors. The mechanism of isomerization of pentane is similar to that of butane. The absence of neopentane among the products can be rationalized in terms of the rearrangement and hydride transfer involving primary neopentyl cation. The equilibrium of pentane–isopentane can be represented by the follow-

TABLE 1.5

Effect of Olefins on the Isomerization of Butane in the Presence of Aluminum Chloride and Hydrogen Chloride at 100°C[a]

	Experiment				
	1	2	3	4	5
Charge (mol/100 mol butane)					
$AlCl_3$	5.86	5.86	5.86	5.86	5.86
HCl	3.05	3.05	3.05	3.05	3.05
Butenes	0	0.013	0.06	0.09	0.28
Reaction time (hr)	12	12	12	12	12
Product composition (mol %)					
Butane	99.9	87.6	87.0	82.8	72.3
Isobutane	<0.1	11.8	12.6	16.9	26.8
Pentanes and heavier	0.1	0.6	0.4	0.3	0.9

[a] From Pines and Wackher (1946).

ing equation:

$$R \ln K_G = \frac{1861}{T} - 1.30$$

where

$$K_G = \frac{i\text{-}C_5H_{12}}{n\text{-}C_5H_{12}} \quad \text{(vapor)}$$

Pentane isomerization in the presence of aluminum halides, even under mild conditions, is accompanied by disproportionation, which leads to hydrocarbons of lower and higher molecular weight. This reaction, if left un-

TABLE 1.6

Effect of Olefins on the Isomerization of Butane in the Presence of Aluminum Bromide and Hydrogen Bromide at 25°C[a]

	Experiment			
	1	2	3	4
Charge (mol/100 mol butane)				
$AlBr_3$	9.3	9.3	9.3	9.3
HBr	2.3	2.3	2.3	2.3
Butenes	0	0.03	0.08	0.58
Reaction time (hr)	15	15	15	15
Isobutane formed (%)	0.2	2.1	19.3	65.9

[a] From Pines and Wackher (1946).

controlled, may become the main reaction. Side reactions, however, can be inhibited by the use of hydrogen or certain organic additives, among them aromatic hydrocarbons such as benzene.

Hydrogen gas under pressure inhibits side reactions during the isomerization of pentane. In a control test in which pentane was heated with aluminum chloride under nitrogen pressure, side reactions converted much of the pentane to butanes, hexanes, and higher boiling alkanes, and the catalyst changed to a viscous red liquid (Table 1.7). Good isomerizations were effected, however, when hydrogen chloride was introduced in addition to hydrogen; with 10% by weight of HCl, 62% conversion of pentane was achieved with a selectivity to isopentane of 98%.

The hydrogen pressure employed is of importance. In comparable tests at 100°C, the minimal effective pressure is between 9 and 14 atm. Below 9 atm the production of butanes is extensive, whereas above 14 atm there is some tendency toward retarded isomerization. The minimal pressure required is higher at higher temperatures or with more active catalyst mixtures.

The critical influence of benzene concentration is strikingly demon-

TABLE 1.7

Effect of Hydrogen on the Efficiency of Isomerization of Pentane at Elevated Temperature[a]

	Experiment				
	1	2	3	4	5
Gas, 100 atm initial pressure	N_2	N_2	H_2	H_2	H_2
Temperature (°C)	125	125	125	125	125
Charge (gm)					
n-Pentane	100	100	100	100	100
Aluminum chloride	10	10	10	10	10
Hydrogen chloride	0	9	0	2	10
Catalyst recovered (gm)	16	16	10	10	10
Product composition (mol %)					
Methane	12.9	4.1	—	0.0	0.0
Ethane	0.0	3.5	—	0.0	0.0
Propane	0.0	21.0	0.0	0.0	0.0
Isobutane	27.6	28.0	0.0	0.0	0.5
Butane	18.4	9.8	0.0	0.0	0.0
Isopentane	5.5	8.0	2.8	22.4	61.0
n-Pentane	25.4	17.2	96.6	77.3	38.0
C_6^+	10.2	8.4	0.6	0.3	0.5
Approximate conversion (%)	71	80	3.4	23.7	62.0
Approximate efficiency (%)	9	11	82	94	98

[a] From Ipatieff and Schmerling (1948).

strated in Table 1.8. As increasing proportions of benzene are employed, there is less and less disproportionation. The isomerization reaction simultaneously becomes more selective and reaches a peak at a point where the amount of side reaction becomes quite small. A further increase in benzene concentration gradually inhibits isomerization

The isomerization of *n*-pentane is accompanied by a disproportionation reaction that produces mainly butanes and hexanes. If a proton is lost by a *tert*-pentyl ion, the olefin produced can condense with neighboring cations. Isomerization and β-scission of the cations leads to disproportionation products:

$$\overset{+}{CH_3CCH_2CH_3} \rightleftharpoons H^+ + CH_3C{=}CHCH_3$$
$$\underset{CH_3}{|} \qquad\qquad \underset{CH_3}{|}$$

$$CH_3C{=}CHCH_3 + CH_3CH_2\overset{+}{C}CH_3 \rightleftharpoons CH_3\overset{+}{C}{-}CH{-}\overset{CH_3}{\underset{|}{C}}{-}CH_2CH_3$$
$$\underset{CH_3}{|} \qquad\qquad \underset{CH_3}{|} \qquad\qquad \underset{H_3C}{|}\ \underset{CH_3}{|}\ \underset{CH_3}{|}$$

$$CH_3\overset{+}{C}{-}CH{-}\overset{CH_3}{\underset{|}{C}}{-}CH_2CH_3 \rightleftharpoons CH_3\overset{CH_3}{\underset{|}{C}}{-}\overset{+}{CH}{-}\overset{CH_3}{\underset{|}{C}}{-}CH_2CH_3 \rightleftharpoons CH_3\overset{H_3C}{\underset{|}{C}}{-}CH{-}\overset{+}{C}{-}CH_2CH_3$$

$$CH_3\overset{H_3C}{\underset{|}{C}}{-}CH{-}\overset{+}{C}{-}CH_2CH_3 \rightleftharpoons CH_3\overset{+}{C} + CH{=}\overset{CH_3}{\underset{|}{C}}{-}CH_2CH_3$$
$$\underset{H_3C}{|}\ \underset{CH_3}{|}\ \underset{CH_3}{|} \qquad\qquad \underset{CH_3}{|} \qquad\ \underset{CH_3}{|}\ \underset{CH_3}{|}$$

$$CH_3\overset{CH_3}{\underset{|}{\overset{+}{C}}} + CH_3CH_2CH_2CH_2CH_3 \rightleftharpoons CH_3\overset{CH_3}{\underset{|}{CH}} + CH_3\overset{+}{C}HCH_2CH_2CH_3$$
$$\underset{CH_3}{|} \qquad\qquad\qquad\qquad \underset{CH_3}{|}$$

$$CH_3CH{=}\overset{}{C}CH_2CH_2 + H^+ \rightleftharpoons CH_3CH_2\overset{+}{C}CH_2CH_3$$
$$\underset{CH_3}{|} \qquad\qquad\qquad \underset{CH_3}{|}$$

$$CH_3CH_2\overset{+}{C}CH_2CH_3 + CH_3CH_2CH_2CH_2CH_3 \rightleftharpoons CH_3CH_2\overset{}{C}HCH_2CH_3 + CH_3\overset{+}{C}HCH_2CH_2CH_3$$
$$\underset{CH_3}{|} \qquad\qquad\qquad\qquad\qquad \underset{CH_3}{|}$$

The above mechanism predicts that hexanes and butanes are formed in equal amounts. Experimentally, butanes are formed to a much greater extent. By reentering the reaction series, hexanes would eventually be converted to a C_4 fragment. However, if butanes reentered the disproportionation reaction series, they would be regenerated.

The function of the disproportionation inhibitors is to maintain a low

TABLE 1.8

Effect of Benzene on the Efficiency of Isomerization of Pentane with Aluminum Chloride/Hydrogen Chloride at 100°C[a]

	Experiment[b]					
	1	2	3	4	5	6
Feed[c]	A	B	B	B	B	B
Benzene in feed (vol %)	0.0	0.10	0.25	0.50	2.0	10.0
Product composition, benzene-free basis (mol %)						
Butanes	8.3	4.4	2.3	0.4	0.3	0.6
Isopentane	17.5	33.6	47.5	49.3	26.5	18.1
Pentane	67.9	57.3	47.3	48.4	71.3	79.8
Hexanes and homologues	6.3	4.7	2.9	1.9	1.9	1.5
Approximate conversion (%)	27.5	34	44	43	20	11
Approximate efficiency (%)	64	78	92	98	97	95

[a] From Mavity *et al.* (1948).

[b] All experiments were carried out in a continuous-flow system at 34 atm with a liquid hourly space velocity of 0.10 and with 0.10 mol HCl per mole of charge. The liquid feed, containing added benzene, was saturated with aluminum chloride at 77°C before entering the reaction chamber at 100°C.

[c] Commercial pentane feeds having the following compositions were used: A: C_5H_{12}, 95.4, mol %; hexanes and homologues, 4.6 mol %. B: $i\text{-}C_5H_{12}$, 7.4 mol %; C_5H_{12}, 91.1 mol %; hexanes and homologues 1.5 mol %.

concentration of *tert*-pentyl cation, which may lead to side reactions. Hydrogen can react with the cation to regenerate the alkane, e.g.,

$$\overset{+}{\underset{\underset{CH_3}{|}}{CH_3CCH_2CH_3}} + H_2 \longrightarrow \underset{\underset{CH_3}{|}}{CH_3CHCH_2CH_3} + H^+$$

Benzene, on the other hand, is readily alkylated with the cations to form alkylbenzenes.

$$C_6H_6 + R^+ \longrightarrow C_6H_5R + H^+$$

iii. Hexanes. Hexanes appear to be more susceptible to side reactions than are pentanes. These reactions can likewise be controlled through the use of inhibitors, such as hydrogen and aromatic hydrocarbons.

The isomerization of hexane takes place in steps; there are significant differences in the rates of these steps when either AlCl$_3$/HCl (Fig. 1.1) or BF$_3$/HF (Fig. 1.2) is used as catalyst. In both kinetic studies the reversible isomerization of 2-methylpentane to 3-methylpentane is markedly faster than other isomerizations. Such a result is consistent with a carbonium ion chain mechanism.

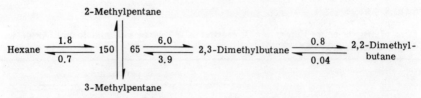

Fig. 1.1. Isomerization of hexane catalyzed by AlCl₃/HCl at 100°C. Numbers over arrows indicate rates of reaction (Evering *et al.*, 1953).

iv. Heptanes. The reaction is characterized by extensive disproportionation and sludge formation. The use of disproportionation inhibitors appears to be relatively ineffective, and side reactions make the isomerization impractical.

b. Cycloalkanes. The isomerization of cycloalkanes (cycloparaffins) is manifested not only in the migration of alkyl groups but also in ring expansion and contraction. Interconversion of five- and six-membered cycloalkanes is a quite general reaction. In such reactions the formation of isomers with ring sizes outside this range has not been found to occur to any measurable extent.

The most common types of alkyl migrations are the shifting of the position of methyl groups around the ring and the formation of methylated ring isomers from alkylcycloalkanes that possess a side chain larger than methyl. Interconversion of the cis and trans forms in polymethylated cycloalkanes proceeds with great ease.

The isomerization of cycloalkanes takes place via a carbonium ion mechanism, similar to that applied to alkanes.

i. Methylcyclopentane. and cyclohexane. The title hydrocarbons undergo a reversible isomerization. At equilibrium the concentration of methylcyclopentane in cyclohexane is 11.4% at 25°C, 20.0% at 59°C, and 33.5% at 100°C.

The isomerization of methylcyclopentane has been studied at 25°C under carefully controlled conditions using AlBr₃/HBr as catalyst. With

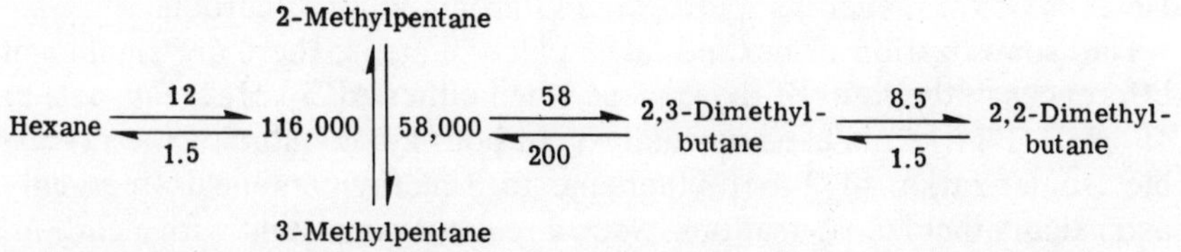

2-Methylpentane

Hexane $\xrightarrow[1.5]{12}$ 116,000 $\quad$ 58,000 $\xrightarrow[200]{58}$ 2,3-Dimethyl-butane $\xrightarrow[1.5]{8.5}$ 2,2-Dimethyl-butane

3-Methylpentane

Fig. 1.2. Isomerization of hexane catalyzed by BF₃/HF at 25°C. Numbers over arrows indicate rates of reaction (McCaulay, 1959).

TABLE 1.9

Effect of *sec*-Butyl Bromide and Benzene on the Isomerization of Methylcyclopentane[a]

Reagents (mol/100 mol methylcyclopentane)				Cyclohexane formed (mol %)
$AlBr_3$	HBr	*sec*-C_4H_9Br	Benzene	
2	1.0	0.0	0.000	0
2	0.9	0.1	0.000	51
2	0.9	0.1	0.022	22
2	0.9	0.1	0.072	5

[a] From Pines *et al* (1949).

highly purified reagents, isomerization to cyclohexane does not occur. However, the addition of 0.05 mol % of cyclohexene to methylcyclopentane resulted in a 28% conversion to cyclohexane. Cyclohexane reacted in a manner similar to methylcyclopentane. Traces of alkyl halides likewise promoted the isomerization of methylcyclopentane. Benzene, on the other hand, inhibited, or with increasing concentrations almost suppressed the isomerization of methylcyclopentane (Table 1.9).

The isomerization of methylcyclopentane is similar to that of pentane. The first step involves the transfer of a hydride from methylcyclopentane to *sec*-butyl cation (R^+) (*sec*-butyl tetrabromoaluminate) when used as an initiator. The methylcyclopentyl cation thus generated ultimately produces cyclohexane.

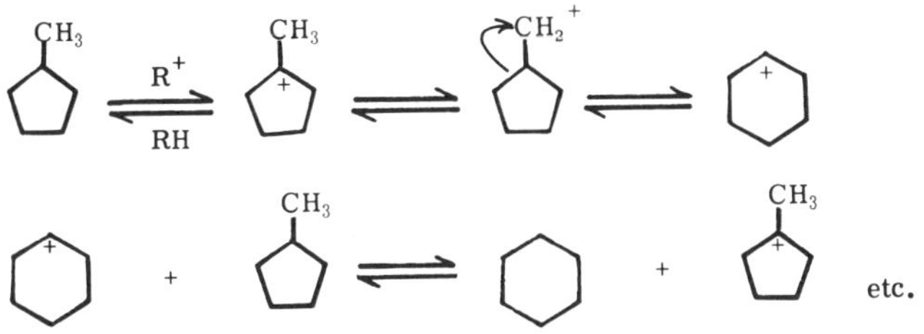

ii. Cyclopentanes and cyclohexane containing seven carbons. The equilibrium at temperatures below 100°C is over 97% on the side of methylcyclohexane. The comparative ease of isomerization of ethylcyclopentane, 1,1-dimethylcyclopentane, and 1,3-dimethylcyclopentane is summarized in Table 1.10. The rate with which these transformations occur can be related to the ease of generation of cations. Ethylcyclopentane undergoes isomerization to methylcyclohexane with great ease because only

TABLE 1.10

Comparative Ease of Isomerization of Alkylcyclopentanes at 25°C[a,b]

Alkylcyclopentane	sec-C_4H_9Br (mol%)	Duration (hr)	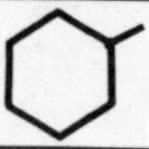 (%)
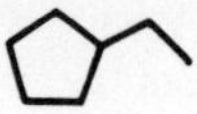	0	9	4
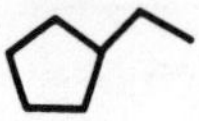	0.05	2	97
	0.05	2	39
	0.05	9	11

[a] From Pines *et al.* (1951a).
[b] Reagent: $AlBr_3$, 1.5 mol %, based on alkylcyclopentanes used.

tertiary and secondary cations are involved. In order for the isomerization to occur a chain initiator is required.

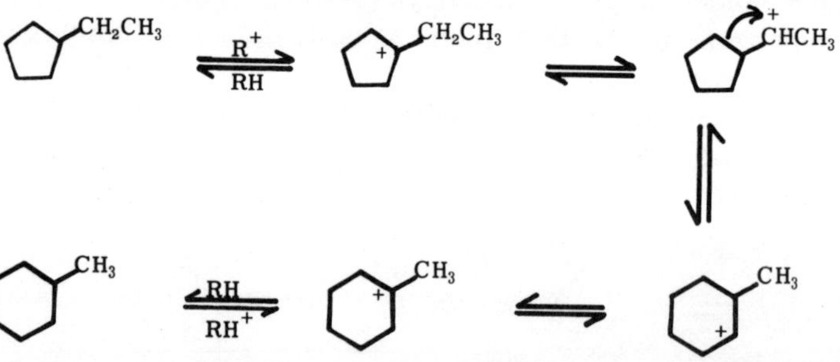

1,1-Dimethylcyclopentane reacts with great difficulty since the less stable secondary and primary cations participate in its isomerization to methylcyclohexane.

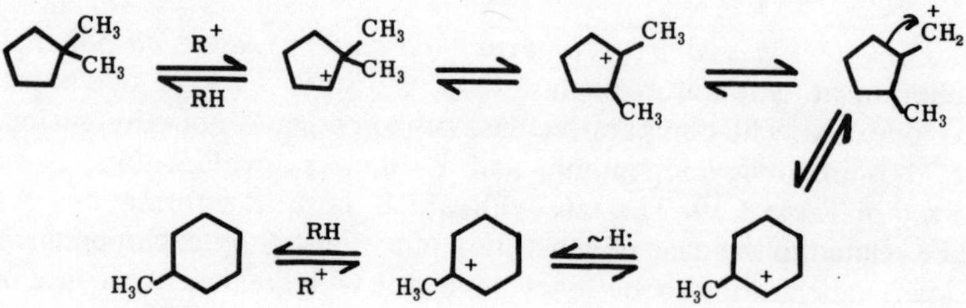

iii. Alkylcyclopentanes and alkylcyclohexanes with eight or more car-bon atoms. The isomerization of 1,1-dimethylcyclohexane, 1,2-dimethyl-cyclohexane, or ethylcyclohexane with aluminum chloride at 120°–130°C affords a mixture of dimethylcyclohexanes of the following composition regardless of which isomer was used as the starting material: 1,1- (6%); 1,2- (14%); 1,3- (58%); and 1,4-dimethylcyclohexane (20%).

The isomerization of ethylcyclohexane to dimethylcyclohexanes can be explained by the following sequence:

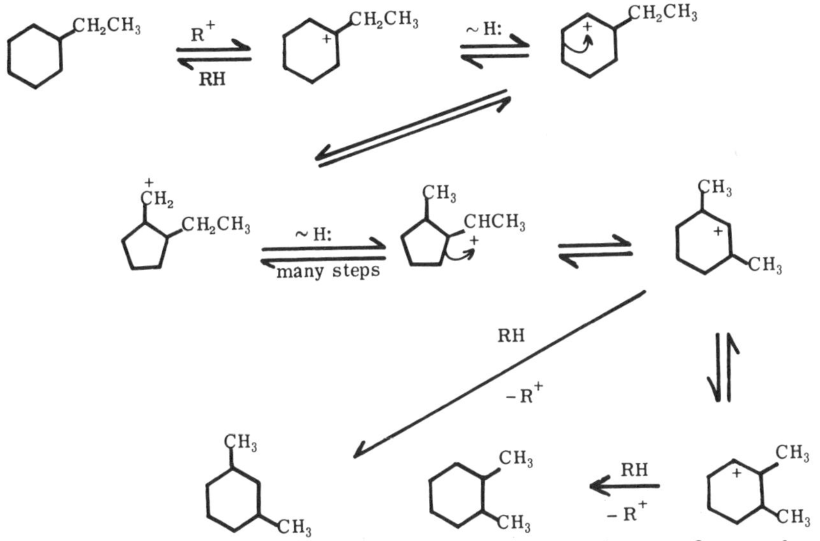

Alkylcyclopentanes and alkylcyclohexanes having three or four carbon atoms in the side chain likewise yield on isomerization polymethylcyclo-hexanes. The formation of isomeric trimethylcyclohexanes from *tert*-bu-tylcyclopentane can be explained as follows:

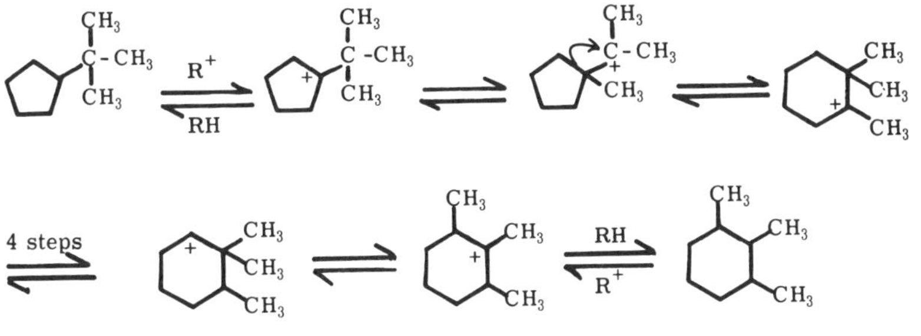

3. Sulfuric Acid Catalyst

As an isomerization catalyst sulfuric acid differs from aluminum halides in several ways. Probably the greatest difference is that sulfuric acid is capable of isomerizing only those hydrocarbons that possess a tertiary hydrogen, and only hydrocarbons possessing a tertiary hydrogen are formed. Another difference is the ease with which sulfuric acid oxidizes hydrocarbons having a tertiary hydrogen to a cation, and therefore there is no need, as with aluminum halides, to add a promoter in order to initiate the isomerization.

Tremendous differences in isomerization activity occur within a rather small range of sulfuric acid concentration. Sharp activity peaks were found at 99.8% concentration with several hydrocarbons. The effect is shown in Fig. 1.3, where the reaction rate constant for 3-methylpentane isomerization is plotted against the initial concentration of the sulfuric acid. A striking parallel is observed between the 64-fold increase in isomerization rate between 95.5 and 99.8% sulfuric acid and a similar increase in the Hammett acidity function, H_0, a measure of the capacity of an acid to donate a proton to a neutral base. The rapid decline in isomerization activity beyond the 99.8% initial concentration has been attributed to dilution of the acid with degradation products.

A suitable technique for carrying out isomerization of hydrocarbons with sulfuric acid merely involves efficient mixing of the two liquid phases for the required time at ordinary pressure. A rather narrow temperature range is applicable, most work having been confined to 0°–60°C.

The mechanism for the sulfuric acid-catalyzed isomerization of alkanes is similar to that proposed for the aluminum halide-catalyzed reactions of alkanes. The cation required for the chain initiation reaction is produced

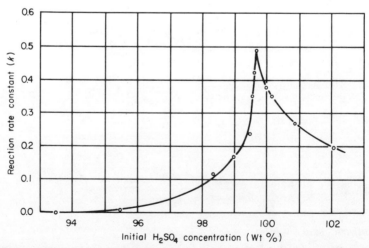

Fig. 1.3. Effect of acid concentration on reaction rate of 3-methylpentane (Roebuck and Evering, 1955).

by the oxidation of the tertiary carbon atom of the alkanes. The formation of a mixture of 2- and 3-methylpentane in the final product can be ex-

$$CH_3CHCH_2CH_3 \ + \ 4\,H_2SO_4 \ \longrightarrow \ CH_3\overset{+}{C}CH_2CH_2CH_3 \ + \ 2\,H_2O \ + \ SO_2 \ + \ 3\,HSO_4^-$$
$$\underset{CH_3}{|} \qquad\qquad\qquad\qquad \underset{CH_3}{|}$$

plained by intramolecular methyl migration and intermolecular hydride transfer between the alkane and the cations.

Very strong evidence supporting the carbonium ion mechanism has been obtained by studying deuterium–hydrogen exchange between hydrocarbons and an excess of sulfuric acid-d_2. α-Hydrogens in the cations exchange with the hydrogens of sulfuric acid. The exchange might be visualized as a rapid equilibration between cations and quasi-olefin.

$$R_2\overset{+}{C}-CH_3 \rightleftharpoons [H^+R_2C{=}CH_2] \underset{DHSO_4}{\overset{D_2SO_4}{\rightleftharpoons}} [D^+R_2C{=}CH_2] \rightleftharpoons R_2\overset{+}{C}-CH_2D \quad etc.$$

Isobutane exchanges nine of the hydrogens with deuteriosulfuric acid. The absence of C_4D_{10} supports the chain mechanism and shows that all

$$CH_3\overset{+}{C}CH_3 \ + \ D_2SO_4 \ \longrightarrow \ CD_3\overset{+}{C}CD_3 \ + \ DHSO_4$$
$$\underset{CH_3}{|} \quad (excess) \qquad\qquad \underset{CD_3}{|}$$

$$CH_3CHCH_3 \ + \ CD_3\overset{+}{C}CD_3 \ \longrightarrow \ CH_3\overset{+}{C}CH_3 \ + \ CD_3CHCD_3$$
$$\underset{CH_3}{|} \qquad \underset{CD_3}{|} \qquad\qquad \underset{CH_3}{|} \qquad \underset{CD_3}{|}$$

hydride transfers are tertiary, because a primary–tertiary exchange would eventually produce some C_4D_{10}.

For polymethylcycloalkanes, the reactions observed involve rapid cis–trans isomerization and considerably slower shifting of the position of the methyl group attached to the cycloalkane ring. For ethylcyclopentane, H_2SO_4 catalyzes ring enlargement to form methylcyclohexane.

4. Silica–Alumina Catalyst

Silica–alumina, ordinarily a cracking catalyst, can be used under mild conditions, 100°–150°C, as an isomerization catalyst for saturated hydrocarbons. The catalytic properties of silica–alumina resemble those of sulfuric acids insofar as only tertiary–tertiary hydride transfer occurs in its presence.

The nature of the acidity responsible for the catalytic activity of silica–alumina is discussed in Section VI. The catalyst could be a protonic, acid, or Lewis acid catalyst. However, once a carbonium ion is formed, the mechanism of isomerization is the same as that of isomerization catalyzed by sulfuric acid. The carbonium ion initiator can be formed through abstraction of a hydride ion from a hydrocarbon by the Lewis acid sites of the catalyst or by formation of an olefin through cracking followed by addition of a proton of the acid to the olefin.

5. Dual-Function Catalysts

Saturated hydrocarbons undergo skeletal isomerization in the presence of hydrogen under pressure when passed over specially prepared dual-function catalysts consisting of hydrogenation–dehydrogenation components deposited on acidic supports. This reaction is called re-forming and is described in Section VIII.

C. Aromatic Hydrocarbons

1. Introduction

The isomerization of alkylaromatic hydrocarbons is of three kinds: (a) positional isomerization of the alkyl groups around the ring; (b) skeletal rearrangement of the alkyl groups; and (c) intramolecular disproportionation of alkyl groups around the benzene ring, which results in a change in the number of substituents around the ring. Like saturated hydrocarbons, alkyl aromatic hydrocarbons isomerize in the presence of strong acids, such as aluminum chloride/hydrogen chloride or silica–alumina, at elevated temperatures.

2. Positional Isomerization

a. Methylbenzenes. The migration of alkyl groups in aromatic hydrocarbons can occur (a) intramolecularly, by a 1,2 shift, which results in positional isomerization, or (b) by migration to another ring, thus causing intermolecular disproportionation (alkyl transfer reaction). Both reactions usually occur together, but with methyl groups intermolecular migration can be suppressed. p-Xylene or o-xylene can be isomerized to m-xylene

TABLE 1.11

Migration Rate Constant of Alkyl Groups in Alkyltoluenes[a]

Interconversion	Methyl	Ethyl	Isopropyl
Ortho ⟶ para	0.0	2.9	19.9
Para ⟶ ortho	0.0	1.0	1.0
Ortho ⟶ meta	3.6	38.2	45.8
Meta ⟶ ortho	1.0	5.3	1.0
Para ⟶ meta	6.0	60.8	5.5
Meta ⟶ para	2.1	24.6	2.4

[a] From Allen et al. (1959, 1960).

with no direct interisomerization between o- and p-xylene. With ethyl and isopropyl groups some intermolecular migration occurs, and with *tert*-butyl groups intermolecular migration is dominant. The rate of direct ortho–para interconversion relative to ortho–meta and para–meta interconversion is a measure of the degree of inter- compared to intramolecular migration (Table 1.11).

The first step in the migration reaction is the formation of a complex between the alkylbenzenes and the catalyst. Solid complexes of methylbenzenes–HF–BF$_3$ with molar ratios of 1:1:1 were isolated. These complexes dissociated on melting, and their stabilities were in order of the basicity of the hydrocarbons:

Complex with	Color	mp (°C)
Toluene	Yellow-green	-65
m-Xylene	Yellow	-55
1,3,5-Trimethylbenzene	Yellow	-15
1,2,3,5-Tetramethylbenzene	Orange	-10

These are σ complexes because of the nature of the bond formed by the adding proton. For the HF/BF$_3$ catalyst this can be represented as follows:

Obviously, four such structures can be written differing in the carbon to which the proton bonds.

That such adducts are salts is evident from measurements of the conductivities of methylbenzenes in HF (Table 1.12). These values are compared with the values for relative basicity in BF$_3$/HF; NMR studies also support the postulated structure of σ complexes.

Increasing the number of alkyl groups bonded to the benzene ring increases the basicity of the aromatic. The more groups meta to each other, the greater the basicity of the aromatic. Thus, all the xylenes are more basic than toluene, and among the xylenes m-xylene is the most basic.

Until the basicity of aromatic hydrocarbons toward strong acids was recognized, paradoxical results were obtained in the isomerization of polymethylbenzenes. For isomerizations catalyzed by BF$_3$/HF or AlCl$_3$/HCl the distribution of isomers depends to a great extent on the concentration of BF$_3$ or AlCl$_3$. The xylenes, in the presence of a low concentration of boron fluoride in hydrogen fluoride, isomerize to a mixture

TABLE 1.12

Relative Conductivities and Basicities of Methylbenzenes[a]

Methylbenzene	Relative conductivity in HF	Relative basicity in BF_3/HF
p-Xylene	1	1
o-Xylene	1.1	2
m-Xylene	26	20
Pseudocumene (1,2,4-)	63	40
Hemimellitene (1,2,3-)	69	40
Durene (1,2,4,5-)	140	120
Prehnitene (1,2,3,4-)	400	170
Mesitylene (1,3,5-)	13,000	2,800
Isodurene (1,2,3,5-)	16,000	6,500
Pentamethylbenzene	29,000	8,700
Hexamethylbenzene	97,000	89,000

[a] From McCaulay (1964, p. 1054).

the composition of which is in agreement with thermodynamically calculated concentrations. However, at high boron fluoride ratios the m-xylene concentration approaches 100% (Table 1.13).

Similarly, the predominant tri- and tetramethylbenzene isomers produced with minor amounts of boron fluoride are 1,2,4-trimethylbenzene (66%) and 1,2,4,5-tetramethylbenzene (70%). With an excess of boron fluoride the isomerized products are 1,3,5-trimethylbenzene (100%) and 1,2,3,5-tetramethylbenzene (100%).

The difference in behavior is due to selective complex formation. With minor amounts of BF_3 most of the xylenes are uncomplexed and in a separate hydrocarbon phase:

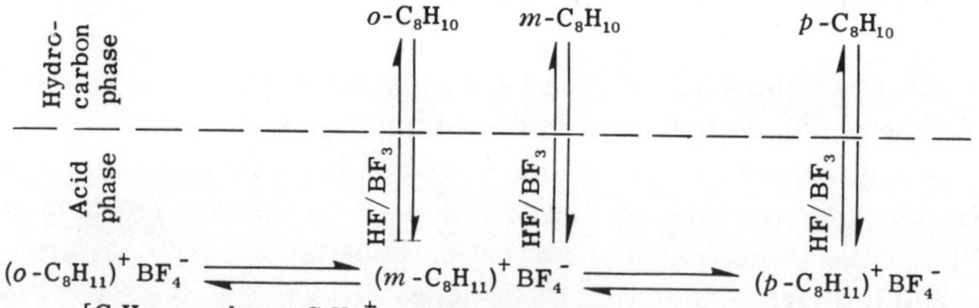

$$(o\text{-}C_8H_{11})^+ BF_4^- \rightleftharpoons (m\text{-}C_8H_{11})^+ BF_4^- \rightleftharpoons (p\text{-}C_8H_{11})^+ BF_4^-$$

$[C_8H_{10} = \text{xylene}; C_8H_{11}^+ = \text{conjugate acid}]$

Reaction in the acid layer and exchange with the hydrocarbon layer occur until the hydrocarbon layer reaches thermodynamic equilibrium. However, with excess BF_3, all the xylenes are complexed and brought into the acid phase. The complex with m-xylene is much more stable than those with the other xylenes, and consequently, at equilibrium, its concentra-

TABLE 1.13

Isomerization of Xylenes[a,b]

	Xylene				
	Meta	Meta	Ortho	Ortho	Para
BF$_3$ (mol/mol xylene)	0.09	0.65	0.10	0.72	3.0
Temperature (°C)	82	82	100	82	3
Contact time (min)	30	30	30	30	1380
Composition of xylenes (%)[c]					
o-Xylene	15	11	19	11	0
m-Xylene	64	74	61	75	100
p-Xylene	22	15	20	14	0

[a] From McCaulay and Lien (1952).
[b] Six moles HF per mole xylene.
[c] Equilibrium composition as calculated at 80°C: ortho, 18%; meta, 58%; para, 24%; at 120°C: ortho, 19%; meta, 57%; para, 24%.

tion is much higher. Removal of the catalyst by quenching with water leaves a product that is about 100% m-xylene.

The mechanism of an intramolecular 1,2 shift of p-xylene can be represented as follows:

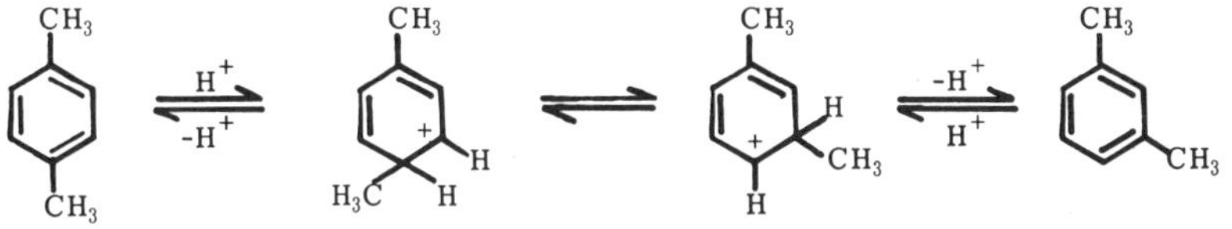

The isomerization of higher alkylbenzenes in the presence of AlCl$_3$/HCl or BF$_3$/HF is accomplished at low temperatures. Ethyl groups migrate at room temperature, and isopropyl and tert-butyl at −80°C. Another type of isomerization occurs through transalkylation (Section V,H,3).

b. Methylnaphthalenes. Dimethylnaphthalenes also undergo methyl migration. Equilibrium mixtures of the naphthalenes were obtained using as catalyst a large excess of hydrogen fluoride, with added boron fluoride concentrations (16 and 110%) based on the aromatic hydrocarbons used. The presence of BF$_3$ is not a prerequisite for the occurrence of the isomerization reaction at ambient temperature; the isomerization rate in the absence of BF$_3$, however, is very slow (Table 1.14). In accordance with the findings with xylenes (Table 1.13), the isomer distribution in the equilibrium isomerates depends on the amount of added BF$_3$.

The unique feature of the isomerization of dimethylnaphthalenes is that

TABLE 1.14

Equilibrium Concentrations of Di- and Monomethylnaphthalenes[a]

Conditions[b]	Equilibrium concentration (wt %)		
	2,6-Di $\rightleftharpoons$	1,6-Di $\rightleftharpoons$	1,5-Di
A	57	40	3
B	35	63	2
C	56	40	4
	2,7-Di $\rightleftharpoons$	1,7-Di $\rightleftharpoons$	1,8-Di
A	61	39	0
B	90	10	0
	2,3-Di $\rightleftharpoons$	1,3-Di $\rightleftharpoons$	1,4-Di
A	35	64	1
B	<1	>99	<1
	1,2-Di	Other dimethylnapthalenes	
	1-Me-$C_{10}H_7$ $\rightleftharpoons$	2-Me-$C_{10}H_7$	
A	23	77	
B	12	88	

[a] From Suld and Stuart (1964).
[b] A, benzene phase, with catalytic amount of BF_3, 70°C; B, acid phase, with equimolar amount of BF_3, 28°C; C, benzene phase, with HF alone, 70°C.

a number of discrete equilibrium sets of isomer groups exist. Although isomerization of dimethylnaphthalenes within the aromatic ring is facile, no interconversion of isomers belonging to the two different rings in naphthalene has been observed.

The 1,2 isomer does not isomerize to any of the isomers of dimethyl-naphthalenes, nor do other isomers give rise to 1,2-dimethylnaphthalene. This behavior can be accounted for by the location of the positive charge in the σ-complex intermediate:

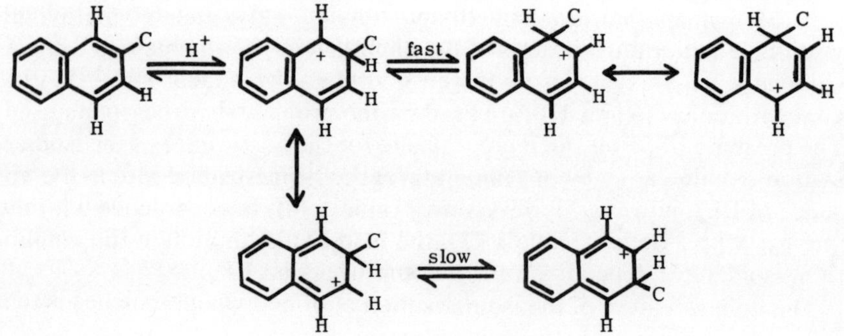

Thus, the fast $\alpha-\beta$ interconversion involves the relatively stable benzyl- and allyl-type cations, whereas the much slower $\beta-\beta$ interconversion involves an isolated secondary cation and loss of aromaticity in both rings. Similar reasoning applies to the intramolecular migration of methyl groups from one ring to the other. The consecutive methyl shifts would also involve quinoid intermediates with loss of aromaticity.

3. Intramolecular Disproportionation

a. m-Xylene. Ethylbenzene and xylenes are reversibly isomerized in the presence of SiO_2/Al_2O_3 at 400°C. Almost an equilibrium composition of ethylbenzene (11%) is produced, and its formation follows an interconversion of C_6 and C_5 ring cations.

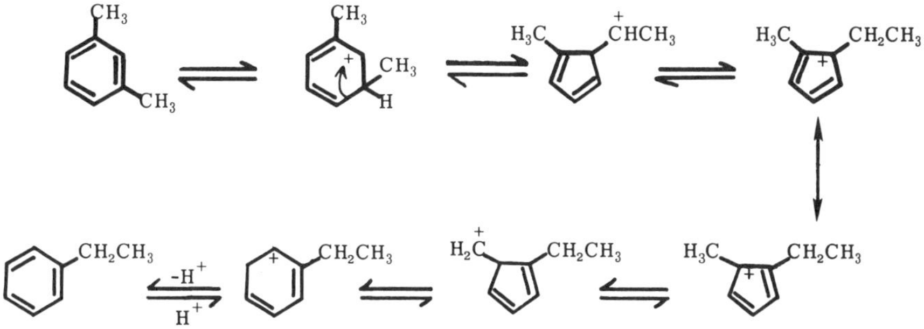

b. [1-¹⁴C]Naphthalene. Ring expansions and contractions appear to be involved when [1-¹⁴C]naphthalene in benzene solution is heated at 60°C in the presence of aluminum chloride containing traces of water. It explains scrambling of the isotope label in position 1 to positions 2 and 8a. The mechanism of this reaction involves as a primary step the protonation of naphthalene by the protonic acid formed from traces of water and $AlCl_3$ (Scheme 1.1).

4. Skeletal Isomerization of Alkylbenzenes and Related Compounds

a. Propylbenzene. Alkyl groups in alkylbenzenes undergo rearrangement when treated with $AlCl_3/HCl$. Some of the rearrangements are observed when ¹⁴C-labeled compounds are used. When propylbenzene is treated at 100°C for 6.5 hr, rearrangement of the carbon atoms occurs. The α- and β-carbon atoms interchange, but the γ-carbon atom remains unchanged. Only a very small amount of the propylbenzene rearranges to isopropylbenzene (Scheme 1.2).

The first step in the interconversion of labeled propylbenzene involves the removal of the benzylic hydrogen by a cation which is presumably present in at least trace amounts. This can then follow preferentially by

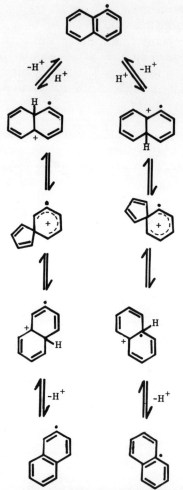

Scheme 1.1. Mechanism of isomerization of [1-¹⁴C]naphthalene (Balaban and Farcasia, 1967).

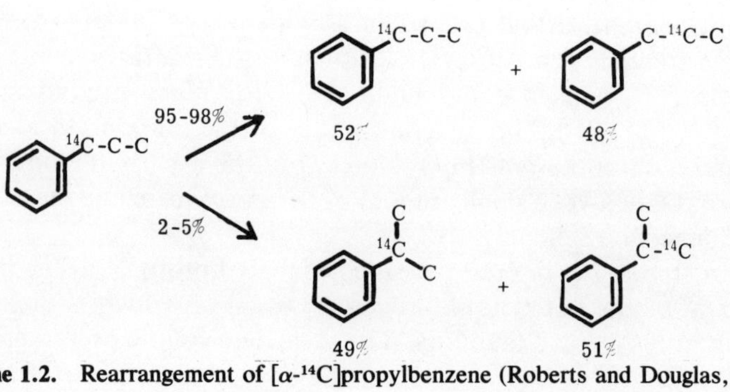

Scheme 1.2. Rearrangement of [α-¹⁴C]propylbenzene (Roberts and Douglas, 1958).

methyl migration with the participation of a phenonium ion with ultimate formation of [β-^{14}C]propylbenzene. The conversion of propyl- to isopropylbenzene is relatively slow because it involves a primary carbonium ion in the intermolecular hydride transfer step.

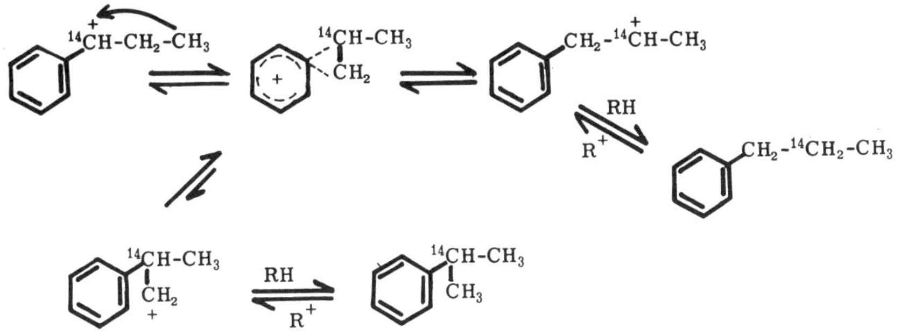

b. Butylbenzenes. *sec*-Butylbenzene and isobutylbenzene undergo interconversion in the presence of $AlCl_3$ without the formation of *tert*-butylbenzene, and this can be explained as follows:

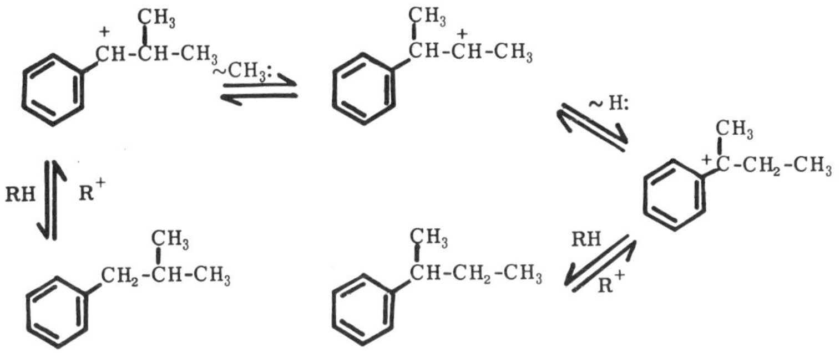

The formation of isobutylbenzene and *sec*-butylbenzene involves the formation of secondary cations, whereas that of *tert*-butylbenzene would require the formation and intermolecular transfer of a hydride to a primary cation intermediate and is therefore energetically a less favorable reaction.

c. Pentylbenzenes. *tert*-Pentylbenzene (A) undergoes interconversion at 80°C in the presence of $AlCl_3$ with B and C to produce after 24 hr an equilibrium mixture consisting of A (7%), B (30%), and C (63%).

A similar type of skeletal isomerization occurs when 2,3-dimethyl-2-phenylbutane, $C_6H_5C(CH_3)_2CH(CH_3)_2$, is submitted to the action of $AlCl_3$.

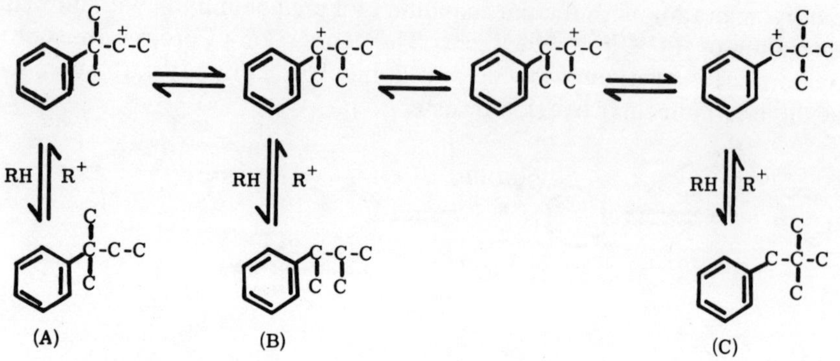

(A)   (B)   (C)

d. Octahydroanthracene. Polycyclic compounds such as octahydroanthracene and octahydrophenanthrene undergo reversible isomerization when contacted with BF_3/HF at $0°-50°C$, according to the following mechanism (Buschick, 1968):

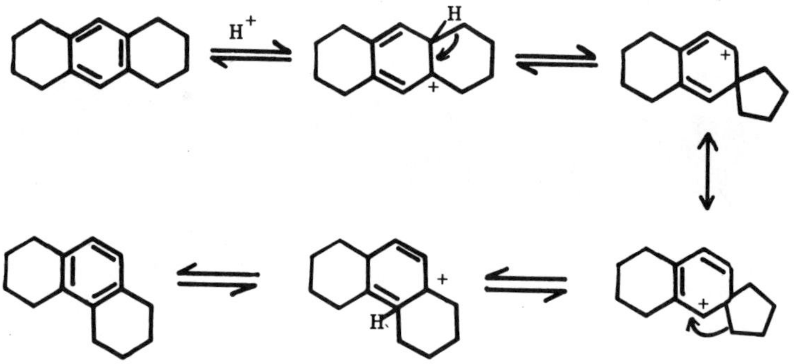

III. POLYMERIZATION OF OLEFINS

A. Introduction

The polymerization of propylene and butylenes catalyzed by acids was introduced in the 1930's as the first commercial catalytic process in the petroleum industry. Most of the fundamental studies described in the literature are concerned mainly with the polymerization of alkenes.

Alkenes may undergo two types of polymerization. The first type involves the formation of oligomers consisting of alkenes only, and it is called "true" or "pure" polymerization. The second type, named "conjunct polymerization," in which many types of reactions occur concurrently, is accompanied by hydrogen transfer, and the product consists of a complex of hydrocarbons that include alkanes, alkenes, cycloalkanes, cycloalkenes, and cycloalkadienes and in some cases also of aromatic hydrocarbons.

The "true" or "pure" polymerization may be divided into "simple" polymerization, in which only one structural monomer is involved, and "cross-polymerization" or "copolymerization," in which there is interpolymerization of two structurally different monomers.

Of the gaseous olefins, isobutylene is the most readily polymerized, and ethylene is the least reactive. The rate of polymerization is related to the ease with which a cation is formed through the interaction of the catalyst with alkenes.

Catalysts used for polymerization may consist of protonic acids such as sulfuric acid, phosphoric acid, and alkanesulfonic acid. Certain oxides, such as activated clays, silica–alumina cracking catalyst, and zeolites, are very active polymerization catalysts. Metal halide catalysts known as Friedel–Crafts catalysts, such as aluminum chloride, ferric chloride, zinc chloride, stannic chloride, and boron fluoride, in the presence of small amounts of the corresponding hydrogen halide or other proton-generating agents, have been used as polymerization catalysts.

Most of the fundamental studies relating to the elucidation of the mechanism of alkene oligomerization were made by F. C. Whitmore, who used dilute sulfuric acid as the catalyst. This catalyst was found to be efficient for the oligomerization of lower boiling branched alkenes and for the copolymerization of isobutylene with butenes. Ethylene and propene, however, do not undergo polymerization in the presence of sulfuric acid. In the presence of dilute acid, propene forms isopropyl alcohol and diisopropyl ether, and with 96% sulfuric acid the chief product is isopropyl hydrogen sulfate, which yields isopropyl alcohol on hydrolysis; ethylene forms diethyl sulfate.

The most widely used commercial catalyst for the oligomerization of gaseous alkenes is "silicophosphoric" acid, sometimes called "solid phosphoric." This catalyst is a modification of phosphoric acid, which is a liquid, and it is difficult to handle commercially due to its corrosiveness at the temperature at which the polymerization is most effective. Silicophosphoric acid is prepared by mixing 85% orthophosphoric acid with Kieselguhr (diatomaceous earth) to form a plastic composite that is calcinated at temperatures from 180° to 300°C. This material is then crushed and screened to form granular catalyst particles of about 2 to 10 mm average diameter. The final catalyst is composed, by weight, of 60% P_2O_5 and 40% Kieselguhr. For commercial use it is generally produced in the form of shaped bodies, such as cylindrical pieces, by extruding the plastic, cutting it into pieces, and calcining to give it physical strength at temperatures from 300° to 500°C.

The polymerization of alkenes in the presence of silicophosphoric acid catalyst involves passing the alkenes through a reactor containing a layer of the catalyst maintained at a chosen temperature from about 100° to about 250°C (for ethylene to 325°C) and pressures ranging from about 10 to 40 atm.

B. True Polymerization

1. Mechanism

Most of the studies of the mechanism of true polymerization of alkenes were made using dilute sulfuric acid as the catalyst and C_4 to C_7 alkenes as the monomers. The carbonium ion mechanism, with all its implications and ramifications (Section I,A), is the accepted mechanism for the acid-catalyzed polymerization of olefins.

True polymerization, depending on the monomer, catalyst, and experimental conditions, may proceed as follows: (a) without the isomerization of the original alkene monomer, as in isobutylene; (b) with isomerization of the monomer before isomerization, as shown by tetramethylethylene; or (c) with isomerization and fragmentation of the dimer produced, as encountered with triptene.

a. Isobutylene. Isobutylene reacts with 65–70% sulfuric acid at 0°C to form two dimers (**8** and **9**), **8** being dominant, and trimers **10–13**. Cation

$$CH_3-\underset{\underset{CH_3}{|}}{C}=CH_2 \;\underset{\longleftarrow}{\overset{H^+}{\rightleftharpoons}}\; CH_3-\overset{+}{\underset{\underset{CH_3}{|}}{C}}-CH_3 \;\underset{\longleftarrow}{\overset{i\text{-}C_4H_8}{\rightleftharpoons}}\; CH_3-\underset{\underset{CH_3}{|}}{\overset{\overset{CH_3}{|}}{C}}-CH_2-\overset{+}{\underset{\underset{CH_3}{|}}{C}}-CH_3$$

8a

$$\Big\updownarrow -H^+$$

$$CH_3-\underset{\underset{CH_3}{|}}{\overset{\overset{CH_3}{|}}{C}}-CH_2-\underset{\underset{CH_3}{|}}{C}=CH_2 \;+\; CH_3-\underset{\underset{CH_3}{|}}{\overset{\overset{CH_3}{|}}{C}}-CH=C-CH_3 \;\;\underset{\underset{CH_3}{|}}{}$$

                             9   **8**

8a may add to another molecule of isobutylene to produce trimer **10** or **11**, or *tert*-butyl cation may add to dimer **8** to form trimers **12** and **13**; the latter addition predominates (Scheme 1.3).

b. Tetramethylethylene (14). Monomer **14** exists in the presence of dilute acid as an equilibrium mixture composed of **14** (64%), **15** (33%), and **16** (3%).

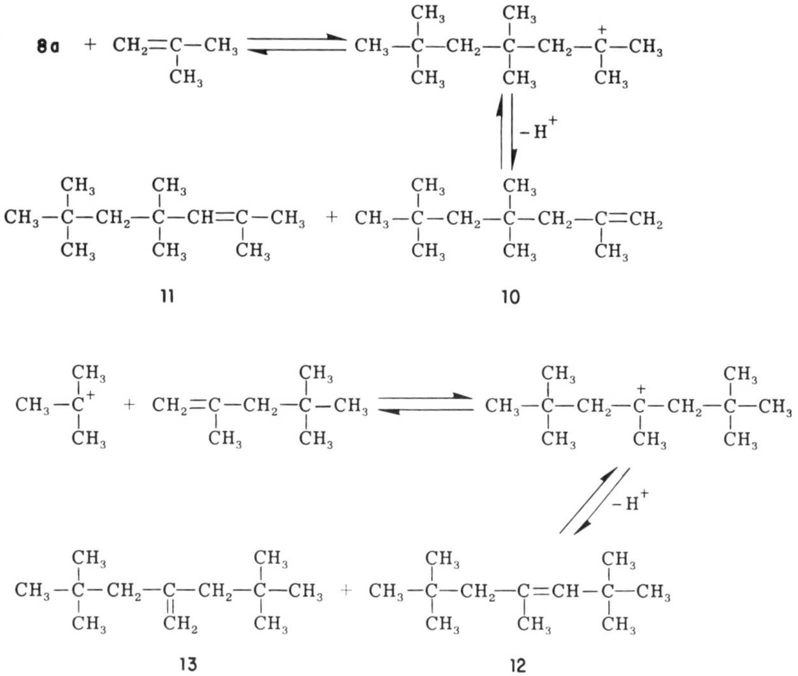

Scheme 1.3. Formation of trimers from isobutylene.

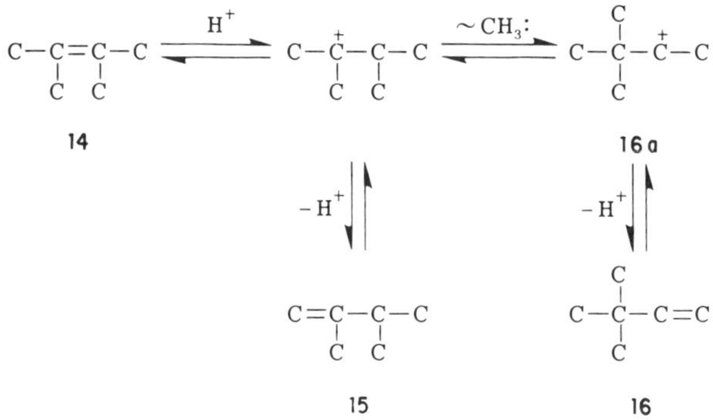

On treatment with 80% sulfuric acid tetramethylene yields three main dimers: **17** (50%), **18** (25%), and **19** (10%) (see Scheme 1.4, p. 36).

c. Triptene (20). Alkene **20**, like isobutylene, is not altered by the rearrangements that may occur in the presence of a 75% sulfuric acid catalyst. The initial polymerization of triptene, therefore, should be simple. How-

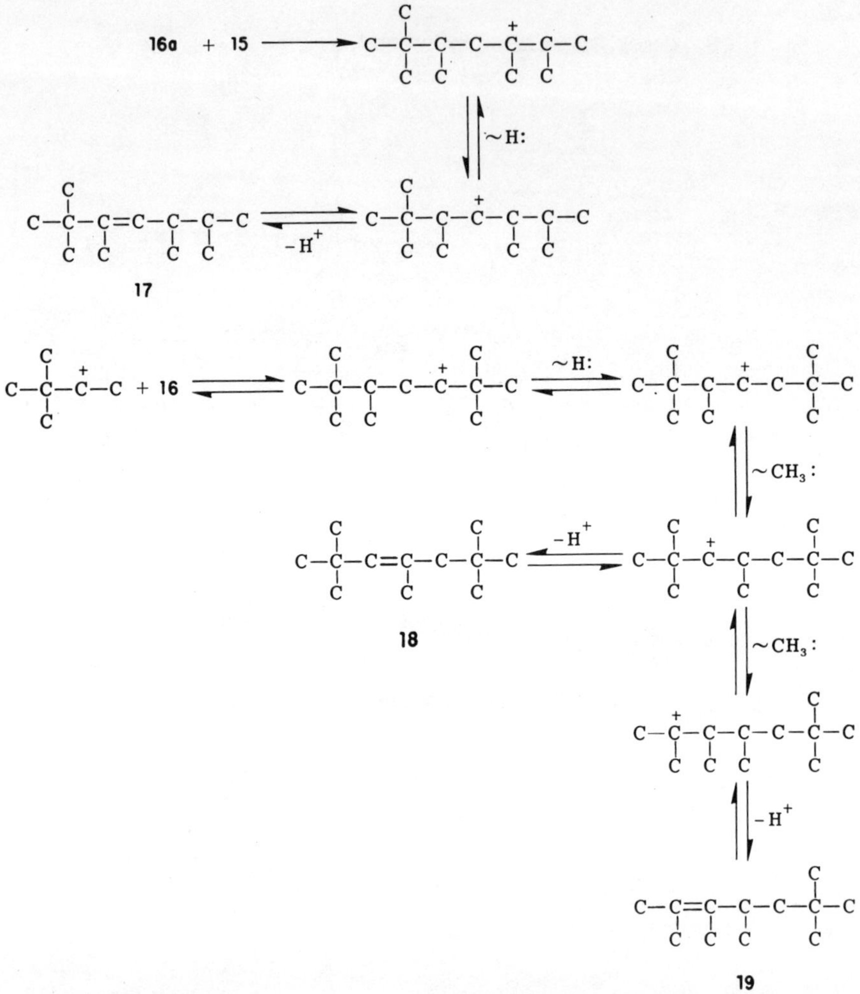

16a + 15 ⟶ [cation structure]

~H:

[structure] 17 ⇌ (−H⁺) [cation structure]

[cation] C—C⁺—C⁺—C + 16 ⇌ [structure] ⇌ (~H:) [structure]

~CH₃:

[structure] 18 ⇌ (−H⁺) [cation structure]

~CH₃:

[cation structure]

−H⁺

[structure] 19

Scheme 1.4. Formation of dimers from tetramethylethylene (Whitmore and Meunier, 1941).

ever, the complicating reactions following the polymerization, such as re-arrangements, β-scission, and repolymerization, are abundant. The products are C_6H_{12} (0.2%), C_8H_{16} (0.6%), C_9H_{18} (3%), $C_{10}H_{20}$ (12%), $C_{11}H_{22}$ (10%), $C_{12}H_{24}$ and $C_{13}H_{26}$ (9%), $C_{14}H_{28}$ (56%), and $C_{15}H_{30}$ and above (9%). The major components of the alkenes produced are compounds 21–33, the structures of which are indicated in Scheme 1.5. The reason for the diversity of the alkenes afforded by the polymerization of triptene is the ease of generation of tertiary alkyl cations through β-scission.

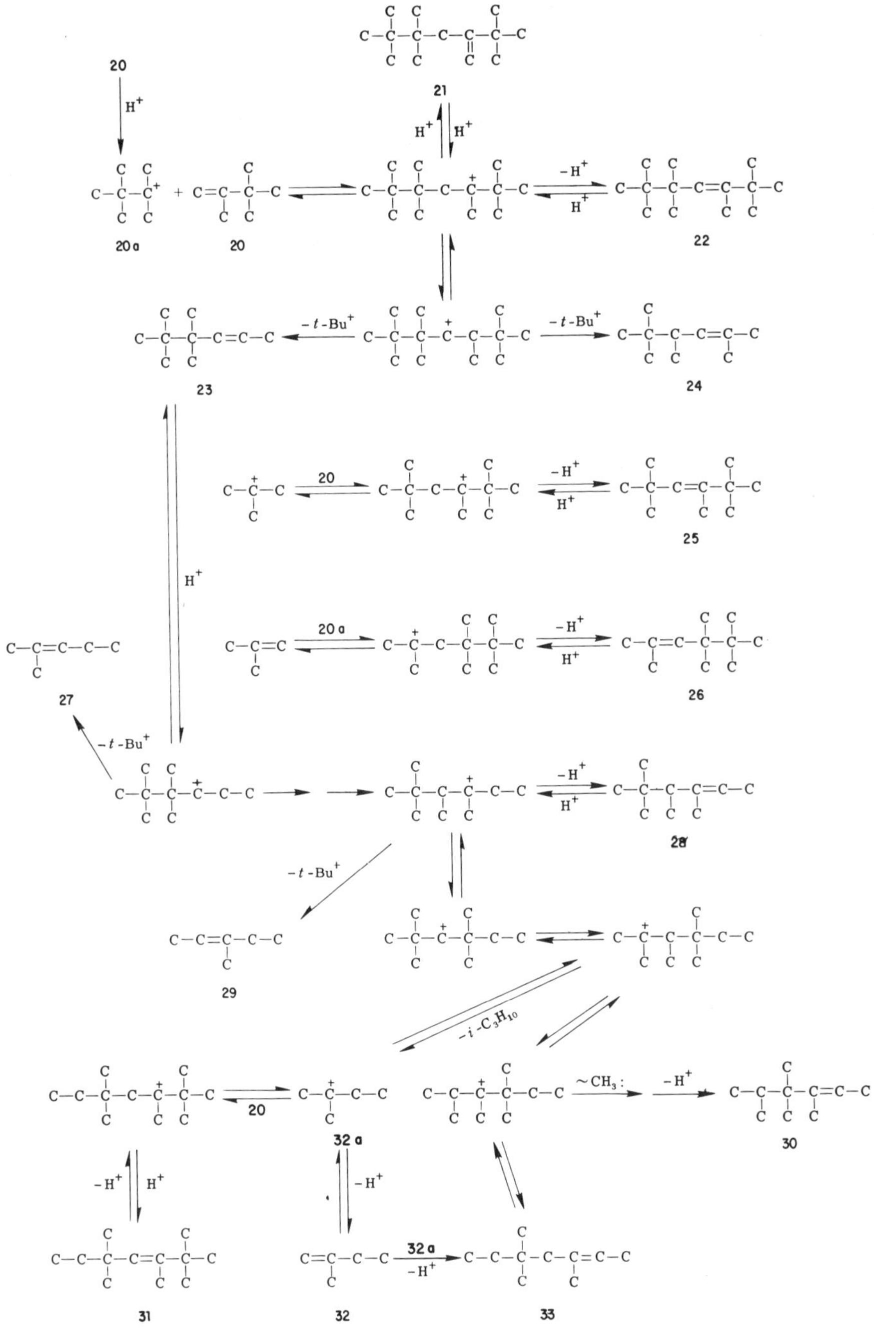

Scheme 1.5. Dimerization of triptene with sulfuric acid (from Whitmore *et al.*, 1950).

2. Copolymerization

When straight-chain alkenes such as propene and butenes are mixed with more reactive alkenes such as isobutylene and methylbutenes and then contacted with an acid catalyst, the least reactive alkene interacts with the branched alkene.

a. Isobutylene and Propene. Silicophosphoric acid catalyzes the copolymerization of isobutylene with propene at 135°C and 38 atm. An equimolar mixture of the two alkenes yielded a liquid polymer that contained 40–45% C_7H_{14} olefins. On hydrogenation the heptenes formed largely 2,3-dimethylpentane, together with a lesser amount of 2,2- and a trace amount of 2,4-dimethylpentane. A similar type of copolymer was obtained when dihydroxyfluorboric acid at 15°–20°C was used as the catalyst. In the absence of isobutylene, propene does not undergo polymerization when treated with dihydroxyfluorboric acid at 0°–40°C. The copolymerization can be represented by reactions (1) and (2).

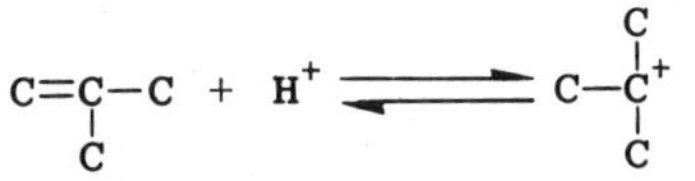

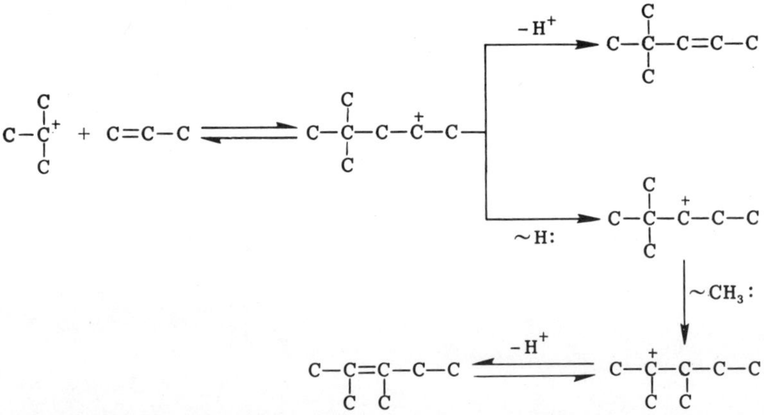

b. Isobutylene and 1-Butene. The title monomers were copolymerized in the presence of dihydroxyfluorboric acid at 20°C. The C_8 olefins obtained from the reaction were hydrogenated to the corresponding octanes; nearly half of the octanes were derived from dimers of isobutylene, and the rest were codimers of the original monomers. The hydrogenated dimers were composed of 2,2,4- and 2,3,4-trimethylpentane. The hydrogenated codimers consisted of 2,2- and 2,3-dimethylhexane and 2,2,3-trimethylpentane. The latter must have resulted from the copolymerization of isobutylene with 2-butene. The isomerization of 1-butene to 2-butene usually accompanies acid-type reactions.

C. Conjunct Polymerization

1. Introduction

When an olefin other than ethylene or propene is treated with an excess of 96–98% sulfuric acid at 0°C, it dissolves at first. On standing, two layers are formed: an upper hydrocarbon layer and a lower acid layer. Two layers are also formed when ethylene is polymerized in the presence of aluminum chloride/hydrogen chloride. Hydrogen transfer is involved in the formation of "sludge hydrocarbons" in the acid layer and of saturated compounds in the hydrocarbon layer. This combination of simultaneous reactions including polymerization, isomerization, cyclization, and hydrogen transfer is called "conjunct polymerization" as distinguished from true polymerization, by which alkenes are condensed to form olefins of higher molecular weight. The sludge hydrocarbons are cyclic unsaturated hydrocarbons consisting predominantly of five-membered mono- or polycyclic rings and having an average of two to three double bonds per molecule. The acid used in the reaction is complexed with each mole of hydrocarbons.

Conjunct polymerization is catalyzed by strong acids. Much of the information relating to this reaction is derived from alkene reactions catalyzed by concentrated sulfuric acid, hydrogen fluoride, BF_3/HF, and $AlCl_3/HCl$. Phosphoric acid at temperatures above 300°C causes conjunct polymerization of ethylene and propene, but the type of products obtained differs from those obtained with the stronger acids.

2. Sulfuric Acid

When alkenes such as butylenes, 3-methyl-1-butene, and nonenes were treated with an equal volume of 96% H_2SO_4 at 0°C, the hydrocarbons dissolved at first in acid. On standing at 0°C two layers separated. The top layer amounted to 70% by weight of the original alkenes; the sulfuric acid layer showed an approximately 20% increase in volume over that of the starting acid. Approximately 32–40% of the product in the hydrocarbon layer boiled below 220°C and was completely saturated. The higher boiling hydrocarbons consisted of a mixture of saturated, unsaturated aliphatic, and cyclic hydrocarbons.

The ratio of acid to alkenes has an effect on the nature of the product formed. With isobutylene, for example, treatment with about 50% by volume of 96% H_2SO_4 affords a product in which the hydrocarbons boiling below 250°C are completely saturated. When isobutylene is treated with 10% by volume of acid, the hydrocarbon fraction boiling below 225°C contains 50–60% of olefins.

The variation in the strength of sulfuric acid has an effect on the composition of the product. Treatment of isobutylene with 91% H_2SO_4 yields a product that contains in the fraction boiling below 200°C 63% of paraffinic hydrocarbons, whereas with 77% H_2SO_4 no paraffins are produced.

3. Aluminum Chloride

Ethylene does not react when treated with pure aluminum chloride even at superatmospheric pressure. However, in the presence of a cocatalyst such as hydrogen chloride or in the presence of moisture, a rapid drop in pressure occurs with the formation of two layers. The upper layer consists of water-white liquid paraffin hydrocarbons. The lower, catalyst layer is a viscous, dark red-brown oil and consists of addition compounds of aluminum chloride and the highly unsaturated cyclic compounds corresponding to the formula $C_nH_{2n-x}\cdot AlCl_3$.

A similar type of conjunct polymerization takes place when butylenes and higher olefins are treated with $AlCl_3/HCl$ at 0°–60°C. A reaction equivalent to conjunct polymerization also occurs when certain paraffins, such as 2,2,4-trimethylpentane, are subjected to the action of $AlBr_3$ at 25°C. The product of the reaction consists of a mixture of isomers of octanes as well as of C_4 to C_7 alkanes and higher boiling compounds, including "catalyst sludge oil."

4. Constitution of Catalyst Sludge

Catalyst sludges obtained from conjunct polymerization are usually dark-colored. Such sludges contain hydrocarbon components and generally all of the catalyst. The sludges on hydrolysis with water form an exceedingly complex mixture composed of highly unsaturated cyclic hydrocarbons containing conjugated double bonds.

There appears to be no simple correlation between the molar amount of sludge hydrocarbon products and the amount of aluminum chloride, except on the basis of 1 mol of $AlCl_3$ being present per double bond of the sludge hydrocarbon.

When hydrogen fluoride is used as catalyst, the acid complex is composed of 6 mol of hydrogen fluoride per double bond, with the catalyst presumably acting as $H_5F_6^-H^+$.

The sludge hydrocarbons generally have a fairly wide boiling range, starting from 69°C at 25 torr pressure. The boiling point at 90% overhead is 236°C at 25 torr. The boiling point may change with the type of alkenes used for conjunct polymerization, the catalyst, and the temperature of reaction. Since the sludge hydrocarbons are mixtures of a very large number of complex molecules, the analytical data on the whole polymers can suggest only the structure of an "average" molecule. Formulas have been proposed for probable structures of monocyclic and bicyclic molecules. On the basis of existing evidence, it has been concluded that all of the monocyclic hydrocarbons contain five-membered rings having at least one conjugated double bond. The following probable structures of typical monocyclic molecules have been proposed (almost all of the R's represent alkyl, alkenyl, cycloalkyl, and cycloalkenyl groups or hydrogen):

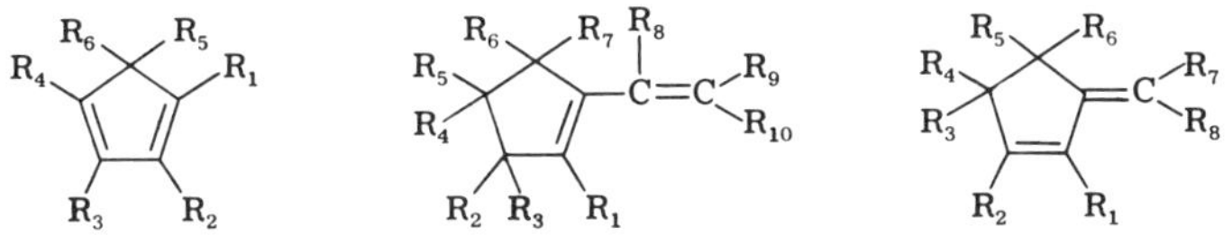

5. Mechanism

Conjunct polymerization of alkenes involves the formation of alkanes in the hydrocarbon layer at the expense of hydrogen from the olefins, the latter being converted to cyclopentenyl cations and cyclopentadienyl hydrocarbons, which form complexes with the strong acid catalysts. This reaction involves hydrogen transfer. The alkanes consist mainly of branched-chain hydrocarbons.

Owing to the complexity of the process as evidenced by the numerous types of hydrocarbons produced, there is no simple stepwise mechanism; however, it is possible to give a general, simplified picture of how the cyclopentyl cations could be produced.

The mechanism of formation of alkylcyclopentadiene in the presence of strong acids such as 96% H_2SO_4 at 0°C can be illustrated using 2,2,4-trimethyl-1-pentene as a model compound (Scheme 1.6).

The governing force in the formation of the allylic cation is the greater stability of the allylic over the alkyl cation. The cyclization reaction is facilitated by the high stability of the cyclopentenylic cation, which is allylic.

Support for this mechanism comes from observations that stable cyclopentyl cations are readily prepared from di- and trimethyl-1,3,5-hexatrienes in 96% H_2SO_4. It was also observed by NMR that the sludge hydrocarbons obtained from 2,2,4-trimethylpentenes contain a substantial amount of 1,2-dimethyl-3-alkylcyclopentenyl cations.

6. Phosphoric Acid

The polymerization of ethylene in the presence of 90% phosphoric acid at 250°–330°C and under an initial pressure of 50–65 atm produced a mix-

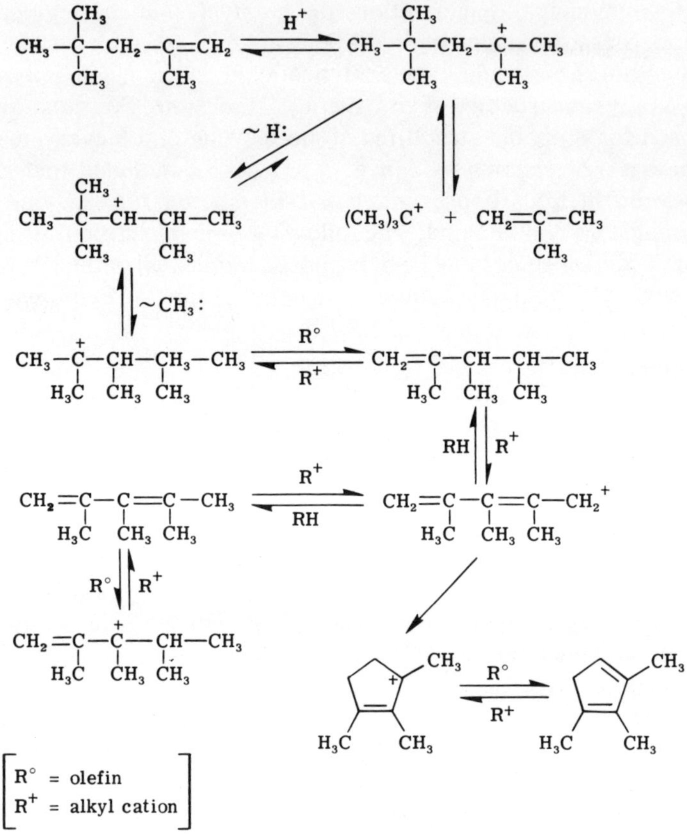

Scheme 1.6. Mechanism of conjunct polymerization (Pines, 1972).

ture of paraffinic, olefinic, naphthenic, and aromatic hydrocarbons. Depending on the reaction conditions, 31–46% of the product distilled below 110°C and 70% below 300°C. The concentration of paraffins was highest in the lowest boiling fractions; the aromatic hydrocarbons, on the other hand, appeared in the fractions boiling at 225°C and higher. The naphthenic hydrocarbons were present in the fractions boiling above 110°C. An interesting aspect of this polymerization was the formation of isobutane, the percentage of which increased with the temperature at which the polymerization was carried out. At 250°C the weight yield of isobutane produced was 2.5%, and at 330°C it amounted to 18.8% (Ipatieff and Pines, 1935).

The conjunct polymerization of ethylene in the presence of phosphoric acid can be explained by a mechanism similar to that described for the stronger acids. Isobutane could thus be produced by the dimerization of ethylene, followed by skeletal isomerization and hydride transfer between the liquid polymers produced and the *tert*-butyl cation:

$$2\,CH_2{=}CH_2 \xrightarrow{\ H^+\ } CH_3CH_2CH_2\overset{+}{C}H_2 \rightleftharpoons CH_3CH_2\overset{+}{C}HCH_3 \rightleftharpoons \overset{+}{C}H_2CHCH_3$$

$$\overset{+}{C}H_3\overset{|}{\underset{CH_3}{C}}CH_3 + RCH_2CH{=}CHR' \rightleftharpoons (CH_3)_3CH + R\overset{+}{C}HCH{=}CHR'$$

Owing to the relatively weak acidity of phosphoric acid as compared with 96% sulfuric acid and the high temperature of reaction, stable complexes between the acid and the cyclopentadienes are not formed. The cyclopentadienes can thus undergo skeletal isomerization to cyclohexadienyl cations, which by a release of a proton afford aromatic hydrocarbons. Such skeletal isomerization is described in Section VIII,C,4,b.

Propene likewise undergoes conjunct polymerization in the presence of 90% H_3PO_4 at 300°C under experimental conditions described for ethylene. The product consists of alkanes, alkenes, cycloalkanes, cycloalkenes, and aromatic hydrocarbons.

D. Dimerization of Styrenes

Of the alkenylbenzenes the oligomerization of styrenes has been the most widely studied.

1. Styrene

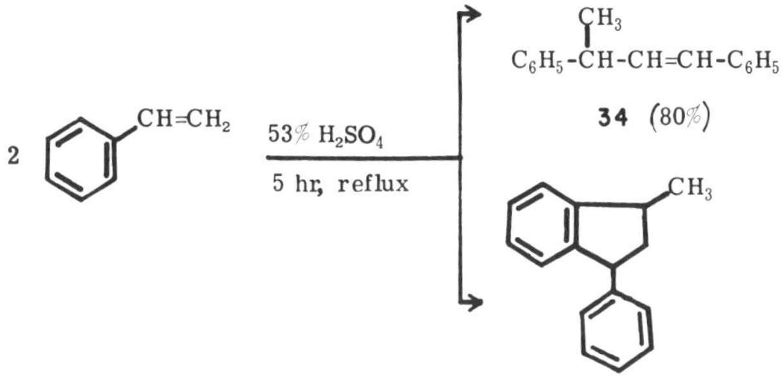

Styrene undergoes facile polymerization in the presence of acids to produce high-molecular-weight polymers. However, under controlled conditions and using dilute sulfuric acid as catalyst, it is possible to dimerize styrene to form an unsaturated dimer (**34**) as the main product and a cyclic dimer (**35**) as a minor product. A similar dimerization carried out under reflux for 40 hr and in the presence of 65% H_2SO_4 produces dimer **34** in 13% and dimer **35** in 87% yield.

2. α-Methylstyrene and β-Methylstyrene

These homologues of styrenes form unsaturated dimers **36** and **37**, respectively, when heated under reflux with 43% H_2SO_4. α-Methylstyrene dimerizes mainly to a saturated compound (**38**) when the concentration of sulfuric acid is increased to 62%.

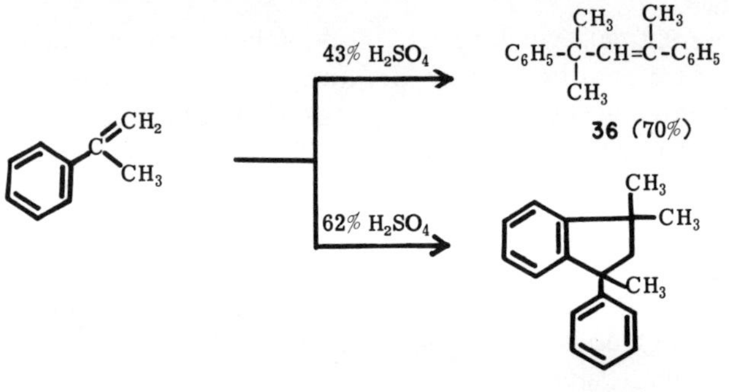

$$C_6H_5-\overset{\overset{\displaystyle CH_3}{|}}{\underset{\underset{\displaystyle CH_3}{|}}{C}}-CH=\overset{\overset{\displaystyle CH_3}{|}}{C}-C_6H_5$$

36 (70%)

38 (82%)

$$C_6H_5-CH=CH-CH_3 \xrightarrow{\text{43\% } H_2SO_4} C_6H_5-\overset{\overset{\displaystyle C_2H_5}{|}}{CH}-\overset{\overset{\displaystyle CH_3}{|}}{C}=CH-C_6H_5$$

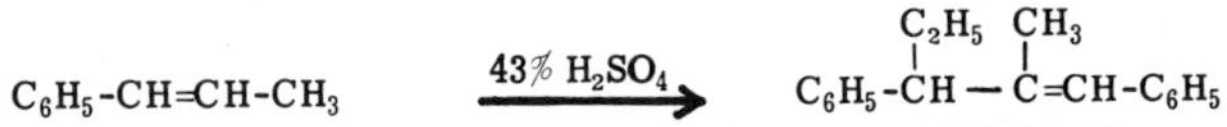

37 (70%)

3. α-Ethylstyrene

α-Ethylstyrene undergoes dimerization when treated with a catalytic amount of $SnCl_4$ in nitrobenzene/carbon tetrachloride solvent at 0°C. After 1 hr a 78% yield of unsaturated dimer **39** is produced; when the contact time is extended to 24 hr, a cyclic dimer, 1-methyl-1,3-diethyl-3-phenylindan (**40**), is formed in 46% yield. The dimerization reaction is represented as follows (Overberger *et al.*, 1958):

The conversion of **39a** to **40a**, which can be considered to be cycloalkylation of a benzenoid ring, usually requires more drastic conditions than a simple dimerization reaction, because it involves the transitional destruction of aromaticity.

4. o-Methyl-α-methylstyrene and p-Methyl-α-methylstrene

The two title monomers form saturated dimers when treated with sulfuric acid in toluene solvent; the para isomer produces dimer **41**, whereas the meta isomer forms dimers **42** and **43** in almost 1:2 ratios.

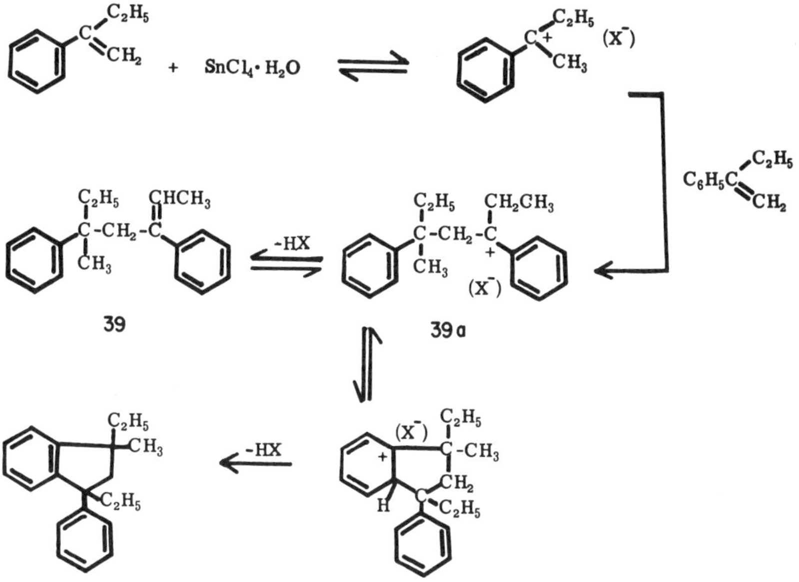

39   39 a

40   49 a

$[X^- = SnCl_4OH]$

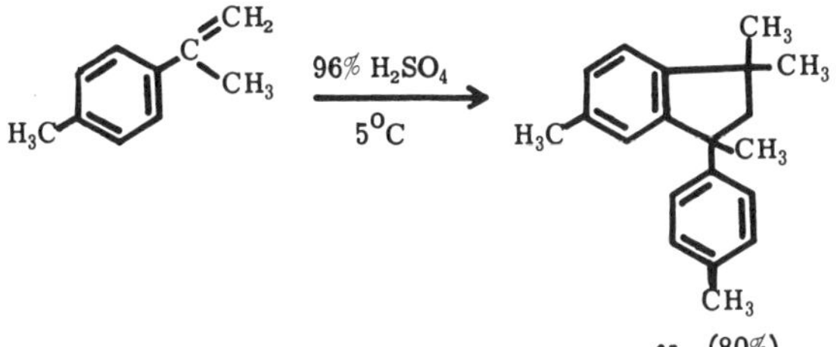

41 (80%)

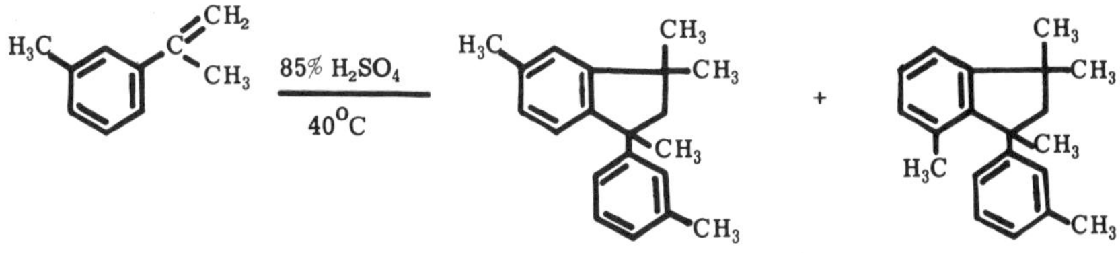

42   (74%)   43

E. Cycloalkenes

The study of the acid-catalyzed oligomerization of cycloolefins is limited. One of the most interesting reactions is the dimerization of cyclopentene to octahydronaphthalene in the presence of P_2O_5 and traces of water. After 10 days of reflux, a 45% yield of dimer was obtained, the remainder being high-molecular-weight polymers. The course of the reaction is interpreted as follows (Criegee and Riebel, 1953):

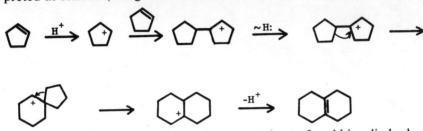

This type of conversion of a cyclopentene ring to fused bicyclic hydrocarbons is probably indirectly involved in the production of naphthalene hydrocarbons in the petrochemical industry.

F. Addition Polymerization (Macropolymerization)

1. Introduction

Acid-catalyzed addition polymerization, also called macropolymerization, of olefins may lead to the formation of hydrocarbons of high molecular weight. Although the chemistry of macropolymerization is outside the scope of this book, it was nevertheless deemed advisable to present a brief discussion of it since this might lead to a better understanding of the chemistry of acid-catalyzed polymerization reactions.

Alkenes undergo conjunct polymerization when treated with strong acids at a temperature of 0°C or higher. Conjunct polymerization is accompanied by an intermolecular hydrogen transfer reaction that creates saturated hydrocarbons and unsaturated cyclic hydrocarbons containing two or more conjugated double bonds per molecule. These unsaturated hydrocarbons form complexes with the acids, as described in Section III,C.

Macropolymerization is likewise catalyzed by strong acids. The competing conjunct polymerization reaction can be avoided by a suitable selection of temperature, strength of acids, and olefins.

Acids most suitable as catalysts for macropolymerization are Friedel–Crafts type of metal halides, such as BF_3, $AlCl_3$, $SbCl_3$, and $BiCl_3$, but such Lewis acids require trace amounts of protonic acids to generate activity. Hydrogen halides are efficient cocatalysts for this reaction, although traces of water that could form hydrogen halide *in situ* also can be used as cocatalyst. Macropolymerization is carried out at low temperatures: $-70°$ to $-20°C$.

The olefins most suitable for undergoing macropolymerizations are those that form stable cations on protonation. The more stable the cations, the lower can be the temperature of reaction and the less likely the occurrence of such secondary reactions as hydrogen or alkyl rearrangements. The order of stabilities of cations is tertiary > secondary > primary. The ease of protonation of the lower alkenes is $(CH_3)_2C{=}CH_2 \simeq (CH_3)_2C{=}CHCH_3 > CH_3CH{=}CH_2 > CH_2{=}CH_2$. The protonation of conjugated dienes such as isoprene or styrenes (phenylethylenes) is facilitated by the formation of resonance-stabilized allylic and benzylic cations, e.g., $(CH_3)_2\overset{+}{C}CH{=}CH_2 \leftrightarrow (CH_3)_2C{=}CH\overset{+}{C}H_2$, $C_6H_5\overset{+}{C}HCH_3$. The following steps are involved in the polymerization:

Initiation:

$$\begin{array}{c} -CR \\ \parallel \\ -C \\ \mid \end{array} + HX \longrightarrow \begin{array}{c} \overset{+}{-}CR \\ \mid \\ -CH \\ \mid \end{array} X^-$$

(A)

Propagation:

$$A + \overset{\mid}{\underset{\mid}{C}}{=}CR \longrightarrow \begin{array}{c} R \\ \mid \\ -C-\overset{\mid}{\underset{\mid}{C}}-\overset{+}{C}R \\ \mid \\ -CH \\ \mid \end{array} X^- \xrightarrow{n\,(C{=}CR)} \begin{array}{c} R \quad\quad R \quad H \\ \mid \quad\quad \mid \quad \mid \\ -C-(C-C)_n-C-\overset{+}{C}R \\ \mid \quad\quad \mid \quad \mid \quad \mid \\ -CH \\ \mid \end{array} X^-$$

(B)

Termination:

$$B \xrightarrow{-H} \begin{array}{c} R \quad\quad R \\ \mid \quad\quad \mid \\ -C-(C-C)_n-C{=}CR + HX \\ \mid \quad\quad \mid \\ CH \\ \mid \end{array}$$

The molecular weight of the polymer depends on the strength of the acid. The stronger the acid is, the less nucleophilic is its anion X^- and the less likely it is to remove the proton from the growing polymer cation. The use of $AlCl_3/HCl$ gives rise to high-molecular-weight polymers, and 80% sulfuric acid produces smaller oligomers. Likewise, styrene gives high-molecular-weight polymers with 96% sulfuric acid, whereas with weak acid, such as 50% H_2SO_4, it forms only a dimer.

The molecular weight also depends on temperature. The propagation reaction requires the least activation energy. At low temperature it predominates over the termination reaction, but at higher temperatures the relative rates of termination increase and polymers of lower molecular weight result.

A more complicated termination step may occur at a higher temperature wherein the H of an alkene, namely, an allylic hydride from the monomer, is transferred to the growing polymer cation:

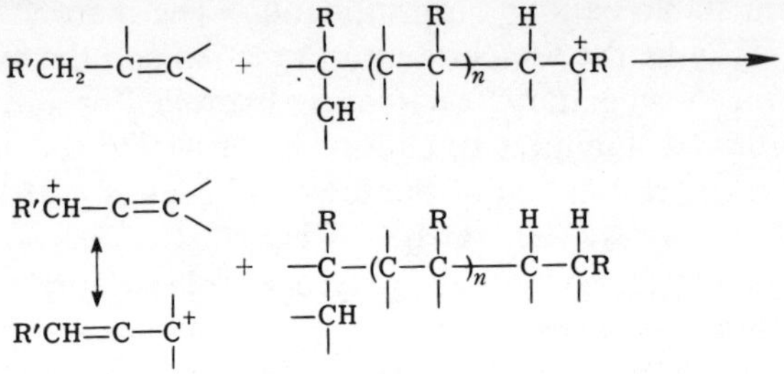

The allylic cation may complex with the catalyst and hence deactivate the catalytic action of the acid as explained in Section III,C on conjunct polymerization.

2. Isobutylene

One of the most striking examples of cationic macropolymerization of alkenes is the low-temperature reaction of isobutylene, which was discovered in 1940 by R. H. Thomas *et al.* at the Esso Laboratories. This affords a polyisobutylene with a molecular weight of about 200,000. The copolymerization of isobutylene with 2–3% of isoprene at −75° to −100°C by means of small amounts of Friedel–Crafts catalysts produces the commercially important "butyl rubber" that contains enough unsaturation from isoprene to permit vulcanization. Butyl rubber, due to its low porosity, was once used in fabricating inner tubes of tires.

The molecular weight of the product obtained from the polymerization of isobutylene depends on the temperature of reaction. At the boiling point of isobutylene, −6°C, the polymerization of isobutylene in the presence of BF_3 results in an oily product, after a long induction period. The molecular weight of polyisobutylene increases from 10,000 to 200,000 when the temperature is decreased from −10° to −90°C.

Diluents are employed in the polymerization to moderate the violence of the reaction and thereby create better temperature control. A mixture of isobutylene and isoprene in liquid ethane can be copolymerized in the presence of small amounts of $AlCl_3$ in ethyl chloride. Boiling off part of the liquid ethane is one way of controlling the heat of reaction. The smaller the quantity of catalyst used, the higher is the molecular weight of the polyisobutylene formed, but a critical concentration of catalyst, approximately 0.03 mol %, has to be exceeded for the production of the polymer. The catalyst efficiency is highest at the greatest dilution.

The overall extent of macropolymerization with a decrease in temperature can be attributed to the relatively high activation energy required for the expulsion of a proton as compared with the relatively low activation energy of the propagation reaction, in which only the more stable tertiary cations participate.

The growing isobutylene polymer cation adds readily to isoprene, forming a resonance-stabilized allylic cation.

$$
\begin{array}{ccc}
& \text{CH}_3 & \text{CH}_3 \\
& | & | \\
\text{CH}_3\text{C}-(\text{M})_{\overline{n}}-\text{CH}_2\text{C}^+ & + & \text{CH}_2=\text{CHC}=\text{CH}_2 \\
| & & \\
\text{CH}_3 & \text{CH}_3 &
\end{array}
\longrightarrow
\begin{array}{c}
\text{CH}_3 \qquad\qquad \text{CH}_3 \\
| \qquad\qquad\qquad | \\
\text{CH}_3\text{C}-(\text{M})_{n+1}-\text{CH}_2\text{CH}=\overset{+}{\text{C}}\text{CH}_2 \\
| \\
\text{CH}_3
\end{array}
$$

$$
\text{M} = (-\text{CH}_2\underset{\underset{\text{CH}_3}{|}}{\overset{\overset{\text{CH}_3}{|}}{\text{C}}}-)
\qquad\qquad
\begin{array}{c}
\text{CH}_3 \qquad\qquad \text{CH}_3 \\
| \qquad\qquad\qquad | \\
\text{CH}_3\text{C}-(\text{M})_{n+1}-\text{CH}_2\overset{+}{\text{CHC}}=\text{CH}_2 \\
| \\
\text{CH}_3
\end{array}
$$

3. Unbranched Alkenes

Unbranched alkenes, unlike isobutylene, produce polymeric material of moderate molecular weight, up to ca. 20,000, when allowed to react at $-10°$ to $-20°$C in the presence of Friedel–Crafts catalysts. The relatively low molecular weight of the polymers is attributed to a more ready rearrangement, termination and hydrogen transfer reaction accompanying polymerization.

Ethylene and propene form primary and secondary cations, respectively, in the first stages of polymerization. Since these cations are less stable than the tertiary cation produced from isobutylene, they require higher temperatures to undergo polymerization. These polymers are used as additives in compounding lubricants.

4. Viscous Polymers

Strong acids have been used for polymerizing alkenes to oils of medium molecular weight with little concurrent conjunct polymerization. Commercial processes for polymerizing the isobutylene in a butane–butene stream (about 50% butanes, 25% isobutylene, and 25% butenes) produce viscous polymers (molecular weight 2000–4000), with nearly 100% unsaturation. This polymerization yields very small amounts of acid-soluble oils, which usually accompany conjunct polymerization. If $AlCl_3/HCl$ is used as catalyst at $0°-30°$C, this overlaps the conditions favoring conjunct polymerization. The secondary reactions are avoided, however, with short contact time, about 4 sec, employing 1000 gm of hydrocarbon per gram of catalyst hourly (D. A. McCaulay, private communication).

IV. ALKYLATION OF SATURATED HYDROCARBONS

A. Introduction

Since the discovery in 1932 by V. N. Ipatieff and H. Pines in the laboratories of Universal Oil Products Company that isoparaffins can be made to react with olefins at ordinary temperatures, the alkylation of saturated hydrocarbons has attracted considerable interest, from both a theoretical and practical point of view. The reaction of isobutane with olefinic gases offers an economical means for the production of high-octane motor fuel. The estimated daily production of the alkylate in 1978 in the United States alone was of the order of about 45 million gallons.

Two types of acid-acting catalysts are effective for alkylation: (a) the Friedel–Crafts type, which includes $AlCl_3$, $AlBr_3$, $ZrCl_4$, and BF_3 in conjunction with small amounts of the corresponding hydrogen halides; and (b) protonic acids, of which sulfuric acid and liquid hydrogen fluoride are the principal acids used.

For the protonic acids alkylation proceeds readily with saturated hydrocarbons containing a tertiary carbon–hydrogen bond. Normal alkanes, such as butane, do not undergo alkylation, whereas isobutane, isopentane, and isohexane react with alkenes to form alkylates. Similarly, cyclopentane is inert in the alkylation reaction, whereas methylcyclopentane reacts readily with alkenes. All of the alkenes from ethylene through butylenes were studied extensively in the alkylation of isobutane, and the products obtained from these reactions form the main basis for the interpretation of the mechanism. Ethylene was observed to be the most sluggish of these olefins; its reaction with isobutane proceeds in the presence of Friedel–Crafts catalysts, such as $AlCl_3/HCl$ or BF_3/HF, but not in the presence of protonic acids.

Sulfuric acid of 96–100% concentration and substantially anhydrous hydrogen fluoride are the catalysts used industrially for the alkylation of isobutane with propene and butylenes. The acid strength drops during the course of alkylation owing to the formation of sludge hydrocarbons similar to those encountered in conjunct polymerization (Section III,C). With sulfuric acid, the acid strength can drop to 90% without serious effect on the product obtained by alkylation. If the acid strength is too low, polymerization of the alkenes becomes the main reaction. For hydrogen fluoride, complexation with unsaturated hydrocarbons has a smaller effect. Catalysts containing 80% HF are still active for alkylation.

Depending on the alkenes and the catalyst, the temperature at which the alkylation is carried out may range between 0° and 35°C. With sulfuric acid, temperatures above 30°C are to be avoided because of the oxidation effect of the acid. The alkylation with propene requires higher temperatures and greater acid strength than the alkylation with butylenes.

Polymerization and conjunct polymerization are the fastest reactions under alkylation conditions. These reactions are minimized in commercial operation by providing high isobutane to olefin concentrations at the reaction zone. First, by isobutane recycle, the reactor feed is kept at a nominal isobutane to olefin ratio of about 4:1. Then, the feed is added to a large, continuously stirred reactor containing hydrocarbon–acid mixture at a slow enough rate (about 0.2 gm feed per gram of catalyst each hour) to keep the internal isoparaffin to olefin ratio in the neighborhood of 1000:1. At this high ratio the C_8^+ cation from the addition of *tert*-butyl cation to the butenes feed preferentially abstracts hydride from an isobutane rather than adding to or taking hydrogen from an olefin molecule, present in only extremely low concentration.

B. Alkylation of Alkanes

The majority of the studies of alkylation reactions were made with isobutane as the alkane. Isolated experiments were made with other alkanes containing a tertiary carbon atom attached to hydrogen, such as isopentane and 2-methylpentane, both of which react with alkenes to form alkylates. However, most of the knowledge used to interpret the mechanism of reaction was derived from the study of isobutane, and for that reason the reaction of alkenes with isobutane is discussed in more detail.

1. Ethylene

Isobutane reacts with ethylene under pressure in the presence of either $AlCl_3/HCl$ or BF_3/HF at 25°–35°C or 0°–5°C, respectively. The product consists of 45% of hexanes composed of 2,3-dimethylbutane (70–90%), 2,2-dimethylbutane (3%), and 2-methylpentane.

The best results have been obtained with a catalyst composed of liquid aluminum chloride sludge hydrocarbon similar to that derived from conjunct polymerization. The activity of the sludge is presumably due to that portion of aluminum chloride which is dissolved but not bound in the form of a stable complex. Alkylation of isobutane with ethylene at 60°C in the presence of the sludge and hydrogen chloride as promoter afforded an alkylate containing 81% by volume of C_6H_{14} and about 12% of C_8H_{18} hydrocarbons. Nine-tenths of the hexane fraction was composed of 2,3-dimethylbutane, whereas the octane fraction was essentially 2,2,4-trimethylpentane.

2. Propene and Butylenes

The alkylation of isobutane with the title olefins in the presence of $AlCl_3$/HCl at room temperature is accompanied by much side reaction, which can be diminished by carrying out the alkylation at lower temperatures. Thus, at $-30°C$ the alkylate from isobutane and propene contained 42% C_7H_{16} and 20% $C_{10}H_{22}$ alkanes. The heptane fraction consisted chiefly of 2,3-dimethylpentane admixed with some 2,4-dimethylpentane.

Alkylations with propene and butylenes in the presence of 100% sulfuric acid were made in a batch-type system using a high-speed mechanical mixer. The experimental conditions and results are given in Table 1.15. Similar experiments using a hydrogen fluoride catalyst are summarized in Table 1.16.

TABLE 1.15

Composition of Products Obtained from the Alkylation of Isobutane in the Presence of Sulfuric Acid[a,b]

Alkylate	Alkene used		
	Propene	2-Butene	Isobutylene
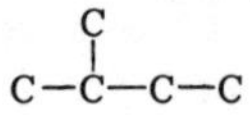 C—C(C)—C—C (2-methylbutane skeleton, with branch C)	—	—	7–9
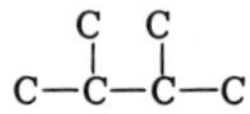 C—C(C)—C(C)—C	—	4–6	8–10
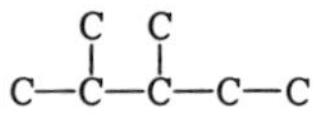 C—C(C)—C(C)—C—C	62–66	—	—
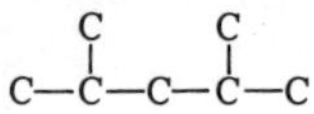 C—C(C)—C—C(C)—C	8–12	—	—
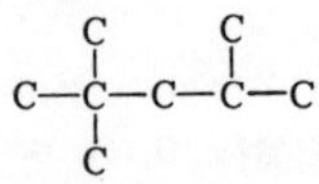 C—C(C)—C—C(C)(C)—C	5–9	34–38	24–28
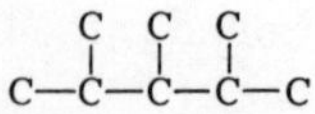 C—C(C)—C(C)—C(C)—C or C—C(C)—C(C)(C)—C—C 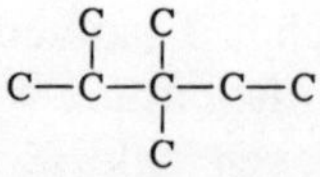	6–10[c]	51–55	30–34

[a] From McAllister et al. (1941).

[b] Sulfuric acid, 100% concentration; mole ratio of isobutane to alkene, about 5; volume ratio of acid to hydrocarbons, 0.7; temperature, 10°C; temperature for propene, 30°C; contact time, 10 min. Values are given as weight percent of alkylates.

[c] Probably only 2,3,4-trimethylpentane (see text).

TABLE 1.16

Composition of Products Obtained from the Alkylation of Isobutane with Alkenes in the Presence of Hydrogen Fluoride[a,b]

Alkylate	Alkene used			
	Propene	1-Butene	2-Butene	Isobutylene
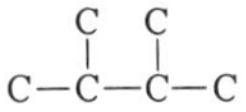 C—C(C)—C(C)—C	1.3	0.6	2.5	1.5
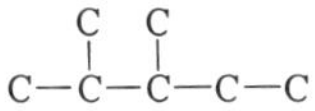 C—C(C)—C(C)—C—C	43.4	1.7	1.4	2.7
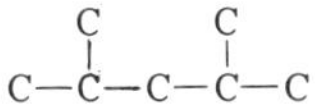 C—C(C)—C—C(C)—C	7.3	1.3	2.4	2.3
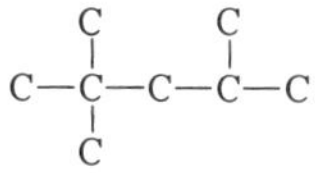 C—C(C)(C)—C—C(C)—C	3.7	29.5	37.9	49.0
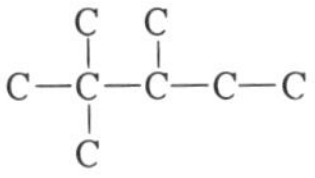 C—C(C)(C)—C(C)—C—C	0	0.9	2.4	1.5
 C—C(C)—C(C)—C(C)—C	0.9	14.1	19.4	9.4
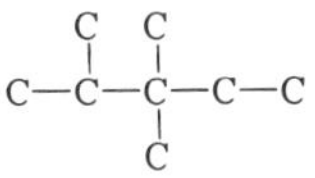 C—C(C)—C(C)(C)—C—C	0.4	8.2	10.1	6.8
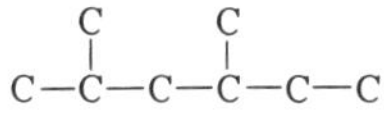 C—C(C)—C—C(C)—C—C	0.2	4.9	2.6	3.3
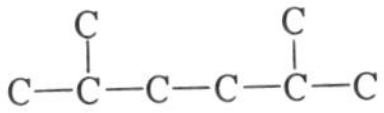 C—C(C)—C—C—C(C)—C	0.3	1.9	2.8	2.9
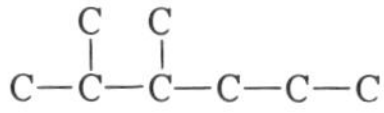 C—C(C)—C(C)—C—C—C	0.4	25.2	3.4	2.4
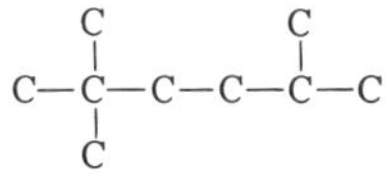 C—C(C)(C)—C—C—C(C)—C	1.5	6.0	7.2	9.7
Uncalculated[c]	40.2	4.7	5.6	5.2

[a] From Kennedy (1958, p. 8).

[b] Isobutane/alkene ratio: 5.1 ± 0.5; temperature: 10°C; contact time (minutes): C_3H_6, 11.8; 1-C_4H_8, 2.9; 2-C_4H_8, 3.8; i-C_4H_8, 5.2. Values given as volume percent of depentanized alkylate.

[c] Unindified alkanes boiling above 125°C.

The multiplicity of products obtained from the alkylation is striking. Examination of these tables shows that isomerization accompanying alkylation has been extensive. The fact that the product of the reaction of isobutane with propylene forms not only C_7H_{16} but also C_8H_{18} alkanes indicates that the reaction is complicated and is accompanied by "secondary" reactions. Similarly, the alkylation of isobutane with butylenes produces, besides octane hydrocarbons, also C_6H_{14} and C_7H_{16} alkanes.

3. Mechanism

The puzzling information presented above was explained by L. Schmerling by means of a carbonium ion mechanism. The reaction of isobutane with propene will be used as the starting point to describe the mechanism. Two types of reactions must be considered: (a) a "normal" reaction leading to the formation of C_7H_{16} hydrocarbons and (b) an "abnormal" reaction, which causes the production of C_8H_{18} and C_6H_{14} alkanes.

In the "normal" reaction the starting point is the conversion of propene to isopropyl cation:

$$CH_3CH{=}CH_2 + HX \rightleftharpoons (CH_3)_2\overset{+}{C}H\ X^-$$

The anion X^- (as $AlCl_4^-$, HSO_4^-, $H_{n-1}F_n^-$, etc.) will usually be omitted from equations for the sake of brevity. The isopropyl cation, which is secondary, reacts with isobutane to bring about an intermolecular carbonium ion transfer. This generates the more stable tertiary carbonium ion:

$$(CH_3)_2\overset{+}{C}H + (CH_3)_3CH \longrightarrow (CH_3)_2CH_2 + (CH_3)_3C^+$$

Under some circumstances $(CH_3)_3C^+$ may lose H^+ to form isobutylene, but under alkylation conditions it adds to the double bond of propene to form a C_7 cation:

$$CH_3{-}\underset{\underset{\displaystyle CH_3}{|}}{\overset{\overset{\displaystyle CH_3}{|}}{C}}{}^+ + CH_2{=}CH{-}CH_3 \rightleftharpoons CH_3{-}\underset{\underset{\displaystyle CH_3}{|}}{\overset{\overset{\displaystyle CH_3}{|}}{C}}{-}CH_2{-}\overset{+}{C}H{-}CH_3$$

44

This cation, being secondary, undergoes an intramolecular cationic shift that involves a vicinal hydride. This shift paves the way to a more stable tertiary cation by another cationic shift that involves wandering of a vicinal methyl group:

$$\mathbf{44} \rightleftharpoons CH_3{-}\underset{\underset{\displaystyle CH_3}{|}}{\overset{\overset{\displaystyle CH_3}{|}}{C}}{-}\overset{+}{C}H{-}CH_2{-}CH_3 \rightleftharpoons CH_3{-}\overset{+}{\underset{\underset{\displaystyle H_3C}{|}}{C}}{-}\underset{\underset{\displaystyle CH_3}{|}}{CH}{-}CH_2{-}CH_3$$

45

Intermolecular reaction of cation **45** with isobutane gives rise to 2,3-dimethylpentane, the major product, plus *tert*-butyl cation, which reacts with propene as outlined above:

$$\textbf{45} + \underset{\underset{\displaystyle CH_3}{|}}{\overset{\overset{\displaystyle CH_3}{|}}{CH_3-CH}} \rightleftharpoons \underset{\underset{\displaystyle CH_3 \quad CH_3}{|\quad\;\;|}}{CH_3-CH-CH-CH_2-CH_3} + \underset{\underset{\displaystyle CH_3}{|}}{\overset{\overset{\displaystyle CH_3}{|}}{CH_3-\overset{+}{C}}}$$

Some of the above C_7 cation may isomerize by related cationic shifts before transferring its charge in the reaction with isobutane. This accounts for the small yield of 2,4-dimethylpentane in the products.

$$\textbf{45} \rightleftharpoons \underset{\underset{\displaystyle CH_3 \quad CH_3}{|\quad\;\;|}}{CH_3-CH-\overset{+}{C}-CH_2-CH_3} \rightleftharpoons \underset{\underset{\displaystyle CH_3 \quad CH_3}{|\quad\;\;|}}{CH_3-CH-CH-\overset{+}{CH}-CH_3}$$

$$\Updownarrow$$

$$\underset{\underset{\displaystyle CH_3 \qquad CH_3}{|\qquad\quad|}}{CH_3-CH-CH_2-\overset{+}{C}-CH_3} \rightleftharpoons \underset{\underset{\displaystyle CH_3 \qquad CH_3}{|\qquad\;\;|}}{CH_3-CH-\overset{+}{CH}-CH-CH_3}$$

$$Me_3C^+ \Updownarrow Me_3CH$$

$$\underset{\underset{\displaystyle CH_3 \qquad CH_3}{|\qquad\quad|}}{CH_3-CH-CH_2-CH-CH_3}$$

In view of the above-described mechanism of alkylation, it should now be clear why butane or pentane does not react with alkenes in the presence of H_2SO_4. For alkylation to take place it is imperative that intermolecular carbonium ion exchange occur between the cation adduct and the alkane. As the above steps show, a secondary carbonium ion may undergo vicinal shift intramolecularly to create another secondary carbonium ion on the way toward formation of a tertiary carbonium ion, but no such carbonium ion exchange occurs intermolecularly between either a tertiary or a secondary carbonium and an unbranched alkane.

If $AlCl_3/HCl$ is the acid catalyst, unbranched alkanes (as pentane) may undergo partial alkylation, but the reaction is accompanied by conjunct polymerization because a hydrogen attached to a secondary carbon atom is less readily abstracted than one attached to a tertiary carbon, and therefore the rate of conjunct polymerization is faster than that of alkylation.

The formation of C_8H_{18} alkanes, which may be considered to be products of "abnormal" reaction, can be explained by the usual carbonium ion mechanism. The *tert*-butyl cation produced in the alkylation of isobutane with propene can lose a proton, and the resulting isobutylene then may react with the *tert*-butyl cation (Scheme 1.7).

The dimethylhexanes formed from the alkylation of isobutane with butylenes are the products of interaction of isobutane with 1-butene. The

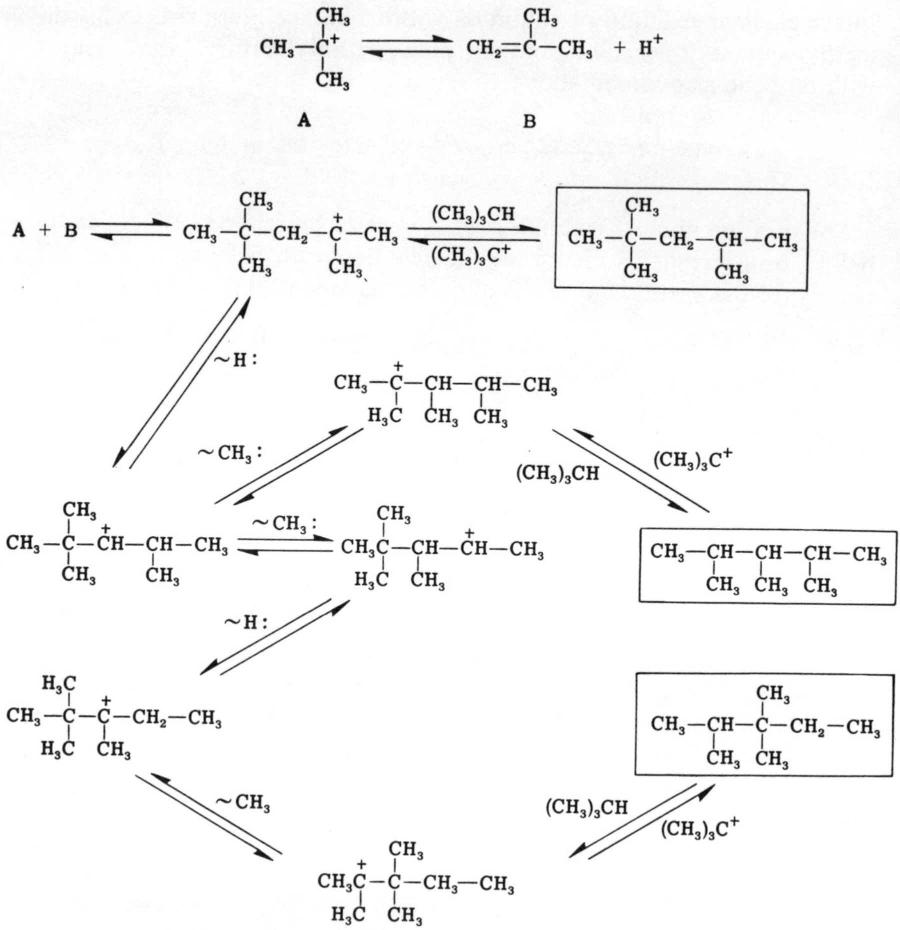

$$CH_3-\overset{\underset{|}{CH_3}}{\underset{|}{\underset{CH_3}{C^+}}} \rightleftharpoons CH_2=\overset{\underset{|}{CH_3}}{C}-CH_3 \ + \ H^+$$

A   B

$$A + B \rightleftharpoons CH_3-\overset{CH_3}{\underset{CH_3}{C}}-CH_2-\overset{+}{\underset{CH_3}{C}}-CH_3 \underset{(CH_3)_3C^+}{\overset{(CH_3)_3CH}{\rightleftharpoons}} \boxed{CH_3-\overset{CH_3}{\underset{CH_3}{C}}-CH_2-\overset{}{\underset{CH_3}{CH}}-CH_3}$$

~H:

$$CH_3-\overset{+}{\underset{H_3C}{C}}-\overset{}{\underset{CH_3}{CH}}-\overset{}{\underset{CH_3}{CH}}-CH_3$$

~CH3:

$$(CH_3)_3CH \qquad (CH_3)_3C^+$$

$$CH_3-\overset{CH_3}{\underset{CH_3}{\overset{+}{C}}}-CH-CH-CH_3 \overset{~CH_3:}{\rightleftharpoons} CH_3C-\overset{}{\underset{H_3C}{CH}}-\overset{+}{\underset{CH_3}{CH}}-CH_3 \qquad \boxed{CH_3-CH-CH-CH-CH_3 \ (CH_3\ CH_3\ CH_3)}$$

~H:

$$\overset{H_3C}{\underset{H_3C}{C}}CH_3-\overset{}{C}-\overset{+}{\underset{CH_3}{C}}-CH_2-CH_3$$

$$\boxed{CH_3-CH-\overset{CH_3}{\underset{CH_3}{C}}-CH_2-CH_3}$$

~CH3

$$(CH_3)_3CH \qquad (CH_3)_3C^+$$

$$CH_3\overset{+}{C}-\overset{CH_3}{\underset{H_3C}{\overset{}{C}}}-\overset{}{\underset{CH_3}{C}}-CH_3-CH_3$$

Scheme 1.7. Mechanism of "abnormal" alkylation (Schmerling, 1964).

fact that the yield of dimethylpentenes from an HF-catalyzed reaction is much greater with 1-butene than with 2-butene is an indication that the rate of addition of *tert*-butyl cation to 1-butene is greater than the rate of isomerization of 1-butene to 2-butene.

The disproportionation reaction occuring during the alkylation of isobutane with butylenes and leading to the formation of isopentane, 3,3-dimethylbutane, and small amounts of dimethylpentanes is similar to the disproportionation reactions encountered in the skeletal isomerization of pentanes and also in the polymerization of alkenes. The reaction involves the addition of cations to alkenes, skeletal isomerization of the cation adducts, and β-scission, as described in Section II,B,2,a,ii.

A more detailed study of the alkylation reaction using isobutane with ^{14}C-labeled butylenes revealed that the initial product of reaction is derived from conjunct polymerization of butylenes. At a steady state, i.e., after sludge hydrocarbon complexes are formed, the predominant reaction is direct alkylation. The radioactive carbon distribution also showed that the formation of fragments smaller than octanes proceeds via the β-scission of C_{12}^+ intermediates. It was thus concluded that the sludge hydrocarbon complexes help to dissolve the isobutane, thus assisting in the intermolecular carbonium ion exchange (Hofmann and Schriesheim, 1962).

C. Alkylation of Cycloalkanes

The alkylation of cycloalkanes containing tertiary carbon atoms proceeds under conditions similar to those used for the alkylation of isobutane. The products of reaction of methylcyclopentane and propene at $-42°C$ in the presence of $AlBr_3/HBr$ consist chiefly of 1-methyl-2-ethylcyclohexane and a smaller amount of $C_{12}H_{22}$ hydrocarbons, presumably a mixture of dimethyldecahydronaphthalenes. The formation of these products can be explained by a mechanism proposed for the alkylation of isobutane with propene (Pines and Ipatieff, 1948).

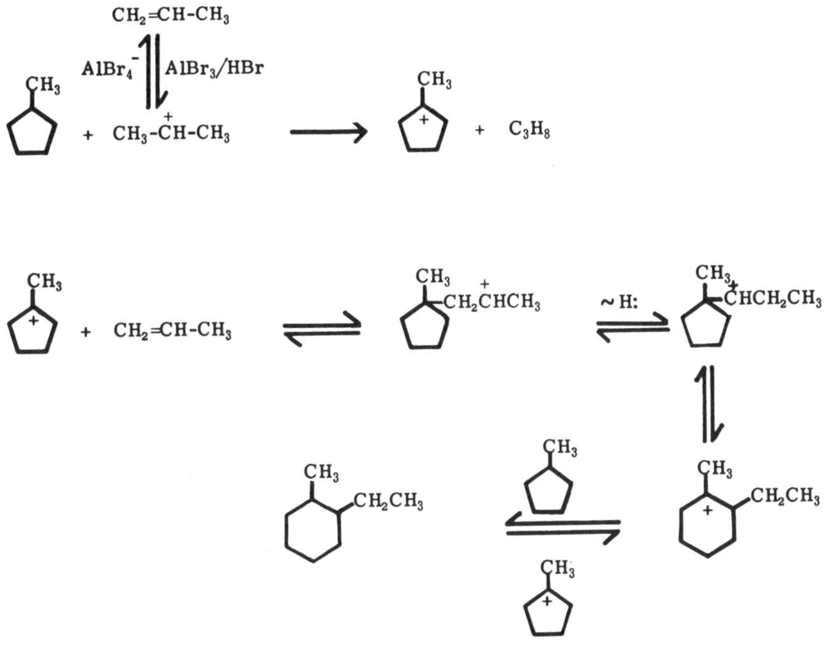

The formation of $C_{12}H_{22}$ may be explained as occurring by the following steps:

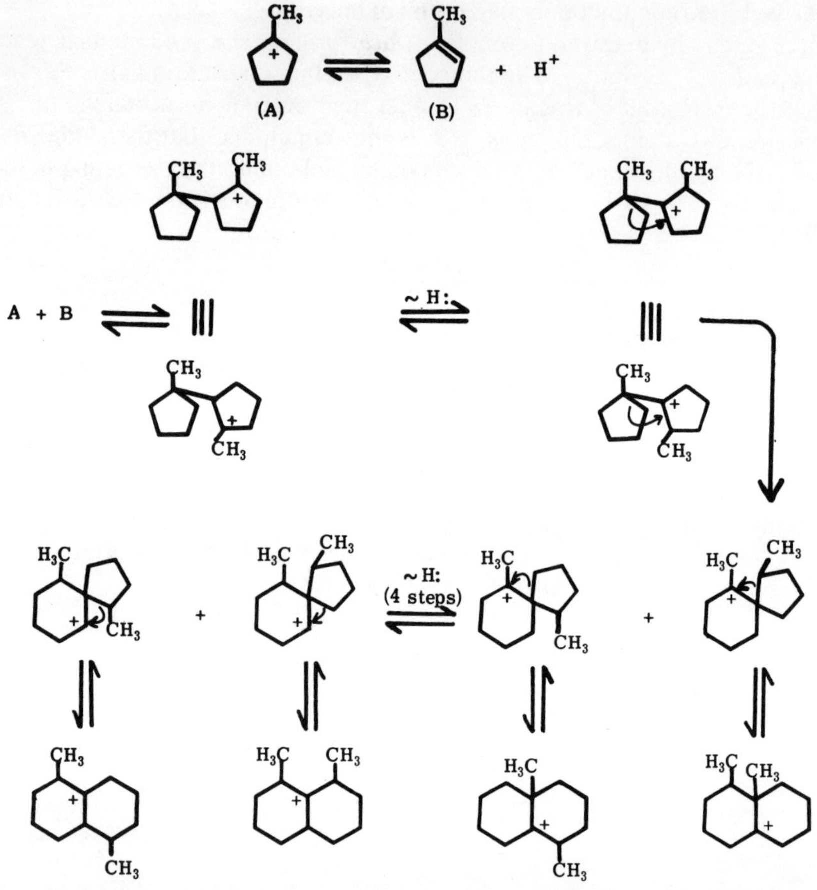

The reaction of the bicyclic cations with the excess of methylcyclopentane yields the corresponding dimethyldecahydronaphthalenes and methylcyclopentyl cation. The mechanism of formation of the bicyclic compounds is similar to that of the dimerization of cyclopentene (Section III,E).

Alkylation of methylcyclopentane with 1- and 2-butene in the presence of 100% sulfuric acid at ~ 15°C afforded chiefly 1,3-dimethyl-5-ethylcyclohexane and 1,3-dimethyl-4-ethylcyclohexane. Methylcyclohexane likewise reacts with alkenes to give presumably alkylated cyclohexanes. The structures of the resulting compounds were, however, not established.

V. ALKYLATION OF AROMATIC HYDROCARBONS

A. Introduction

The alkylation of aromatic hydrocarbons with olefins is of great industrial importance. Uses include styrene production from ethylbenzene, synthesis of acetone and phenol from isopropylbenzene (cumene), and manufacture of sodium alkylbenzenesulfonates for detergents from an alkylbenzene that is made from a long-chain alkene. In related reactions the alkylation of phenols is used for preparing large tonnages of butyl-, octyl-, nonyl, and dodecylphenols.

The introduction of alkyl and related groups, such as cycloalkyl, alkenyl, and aralkyl, onto an aromatic ring may vary considerably. The nature of the reactants may be quite varied, as may that of the catalysts and the experimental conditions. The catalysts most frequently used are protonic acids and metal halides of the Friedel–Crafts type.

The alkylation of benzene and toluene has been most extensively studied. The alkylation of polyalkylated benzene presents an interesting mechanistic problem inasmuch as steric factors have to be considered. The alkylation of polycyclic arenes has been less extensively investigated. The alkylation of phenols, owing to their strong nucleophilic character, requires mild experimental conditions and catalysts. The alkylation of thiophene, furan, and thiophenol likewise requires mild conditions. Alkylation reactions can be carried out in liquid phase or with gaseous olefins in vapor phase, in either a flow system or batchwise.

B. Catalysts

The catalysts used for the ring alkylation of aromatic hydrocarbons consist of three categories of acids: (a) protonic acids, (b) Friedel–Crafts catalysts, and (c) oxide catalysts.

1. Protonic Acids

The activity of protonic acid catalysts decreases in the following order: $HF > H_2SO_4 > H_3PO_4 > C_2H_5SO_3H$. The choice of the catalyst depends not only on its activity, but on various other considerations.

Phosphoric acid or its modification, silicophosphoric or so-called solid phosphoric acid (Section III,A), is used commercially for the reaction of propene with benzene to form isopropylbenzene. Silicophosphoric acid catalyzes the vapor-phase ethylation of benzene to form ethylbenzene. In contrast, reaction of higher alkenes with this catalyst is not recommended because of side reactions, such as skeletal isomerization, which accompany alkylation.

Sulfuric acid does not catalyze the ethylation of benzene, and it is not satisfactory for the reaction of propene with benzene to form isopropylbenzene. Sulfuric acid is, however, an effective catalyst for the alkylation of benzene with higher alkenes. Because of the sulfonating and oxidizing properties of sulfuric acid, alkylations in the presence of this catalyst are carried out at temperatures below 25°C as compared with 60°–350°C in the alkylation reactions catalyzed by silicophosphoric acid.

Hydrogen fluoride is probably the most efficient catalyst for the alkylation of butenes and higher alkenes with benzene. The control of temperature is less critical than with sulfuric acid, and the catalyst is readily recoverable.

2. Friedel–Crafts Catalysts

The general sequence of catalytic activity of Friedel–Crafts catalysts can be considered to be as follows: $AlBr_3 > AlCl_3 > GaBr_3 > GaCl_3 > FeCl_3$, $SbCl_5 > ZrCl_4 > BF_3 > ZnCl_2 > BiCl_3$. Completely anhydrous metal halides are inactive as catalysts for the alkylation of aromatic hydrocarbons. They require a cocatalyst. The addition of HCl or HBr, alkyl halide, or small amounts of alcohol or water activates the metal halides. The function of the hydrogen halide is to react with the alkenes to produce alkyl halides, which in the presence of the metal halides can generate the activated alkyl complex.

3. Oxide Catalysts

Of the oxide catalysts only silica–alumina cracking catalyst and zeolites should be considered. Silica–alumina catalysts have been used for the alkylation of benzene with ethylene and propene. A number of crystalline aluminum silicates (zeolites) have been used for the alkylation of benzene (Section VI,C).

C. Mechanism

The alkylation of arenes, exemplified by the reaction of benzene with olefins, proceeds by a carbonium ion mechanism. The first step in the alkylation is the interaction of the alkene with the catalyst. The cation on the polarized complex is produced by protonation of the double bond.

$$\overset{\diagdown}{\underset{\diagup}{C}}=\overset{\diagup}{\underset{\diagdown}{C}} + HX \rightleftharpoons \overset{\diagdown}{\underset{\diagup}{C}}H-\overset{+}{\underset{\diagdown}{C}}\overset{\diagup}{} \quad X^-$$

The formation of the intermediates occurs when protonic acids are used as catalysts or when Friedel–Crafts catalysts containing a cocatalyst or a "promoter" are used. Metal halides without the promoter, such as hydrogen halide or traces of water, do not catalyze the alkylation of benzene. The reaction of alkenes with $AlCl_3$ and small amounts of HCl as cocatalyst can be represented as follows:

$$\diagdown C = C \diagup + HCl \xrightarrow{AlCl_3} \diagdown CH - \overset{+}{C} \diagup AlCl_4^-$$

If one starts with alkyl halide and $AlCl_3$, a comparable product results:

$$RX + AlCl_3 \longrightarrow R^+ + AlCl_3X^-$$

The prior formation of hypothetical acids such as $HAlCl_4$, $HAlBr_4$, or HBF_4 does not occur. In a generalized Friedel–Crafts carbonium ion scheme the $R^+MX_4^-$ may be considered to be the catalyst:

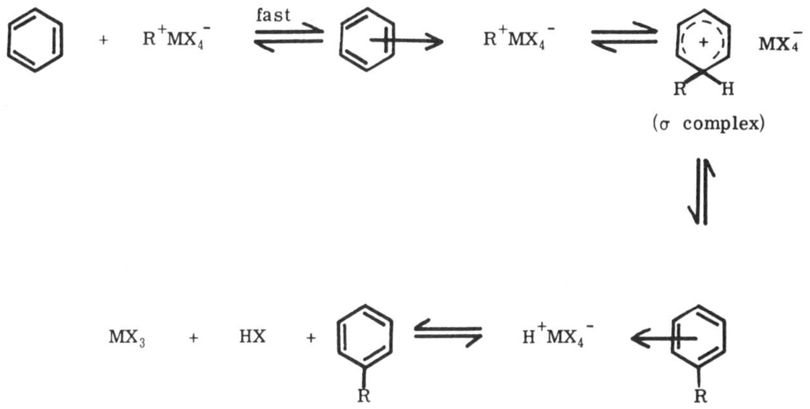

(σ complex)

The MX_3 then reacts again to form more $R^+MX_4^-$.

Solid 1:1:1 σ complexes are prepared by reacting mesitylene with ethyl fluoride and boron fluoride at temperatures of from $-75°$ to $-15°C$. These complexes decompose at $0°C$ to form the corresponding ethyl aromatics.

When R^+ is a developed carbonium ion, as from secondary alkyl or cycloalkyl cations, "side reactions" usually associated with cations, such as intra- or intermolecular carbonium ion shift and skeletal isomerization, may accompany the alkylation of aromatics.

D. Orientation

The problem of orientation of alkyl groups in di- and polyalkylated benzene has been a subject of intensive study. The initial distribution of the isomers is influenced by the strength and amount of catalyst and also by the temperature and duration of reaction. The alkyl groups on the benze-

TABLE 1.17

Orientation of Alkyl Groups in the Alkylation of Toluene[a]

	Percentage of					
Alkylation	Para	Meta	Ortho	Two para/meta	$\Delta S\ddagger$	Relative rate
Methylation	26	14	60	3.7	−20.0	1.0
Ethylation	34	18	48	3.8	−21.5	13.7
Isopropylation	36.5	21.5	42	3.4	−19.3	20,000
t-Butylation	93	7	0	26.6	—	—

[a] From Allen and Yats (1961).

noid ring, the reactivity of the substituents, and the steric requirements also influence the orientation of the alkyl groups.

The initial distribution of isomers from a monoalkyltoluene was established by the study of Friedel–Crafts methylation, ethylation, and isopropylation of toluene using a variety of catalysts and alkyl derivatives (Table 1.17). The initial isomer distribution, the ratio of para to meta substitution, is constant, which demonstrates that methylation, ethylation, and isopropylation have about the same positional selectivity. Since equal selectivity implies equal reactivity, the enormous increase in the absolute rates was interpreted as resulting from the quantity of polarized alkyl derivatives involved in each reaction.

The high ratio of para isomer to meta isomer composition for tert-butyltoluene indicates a high selectivity and therefore a low reactivity. The rapid rate of tert-butylation of toluene must thus be due to the high rate of formation of polarized alkyl species. The absence of the ortho tert-butyltoluene can be ascribed to steric effect. The constancy of entropy of activation in the methylation, ethylation, and isopropylation of toluene is an indication that no significant change in mechanism is involved.

The primary product of alkylation of methylbenzenes may undergo positional isomerization. The latter depends on the temperature of alkylation, the strength of the acids, and the duration of the reaction. The following distribution of product obtained from the alkylation of o-xylene in the presence of BF_3/H_3PO_4 catalyst demonstrates the effect of temperature on orientation:

T (°C)	ortho (% yield)	meta (% yield)
20	40	60
60	20	80
80	4	96

The distribution of alkylates from the reaction of *m*-xylene with propene is influenced by the strength of the acids used as catalysts; the ratio of 1,3,5 isomer (46) to 1,2,4 isomer (47) decreases with decreasing strength of the acidity. When $AlCl_3/HCl$ was used, the monoalkylate contained 99% of isomer 46. However, with the addition of an increasing

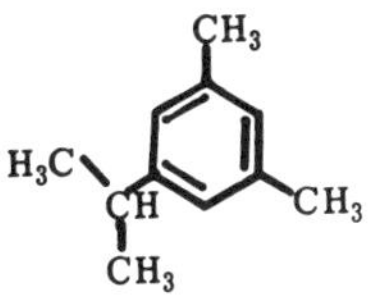

46 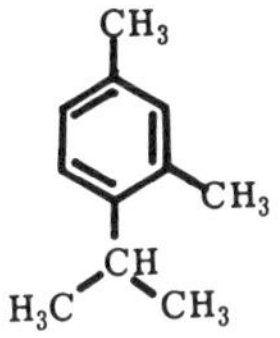   47

amount of water to $AlCl_3$, up to 2 wt %, the ratio of 47 to 46 steadily increased. With BF_3/H_3PO_4, the alkylate contained 65–75% of isomer 47.

E. Steric Effect

The amount of ortho substitution obtained during the alkylation of alkylbenzene is greatly dependent on the steric requirement of the alkyl substituent on the benzene ring and on the incoming alkyl group. In the isopropylation of toluene, isopropylbenzene, and *tert*-butylbenzene, the ratio of meta to para partial rate factors is approximately the same for each arene. However, the partial rate factors for isopropylation in the ortho position of toluene is 2.4, that of isopropylbenzene is 0.37, and that of *tert*-butylbenzene is zero.

The steric requirements of entering primary and secondary alkyl groups are modest compared to those of a tertiary alkyl group. Nevertheless, attempts to introduce three isopropyl groups adjacent to each other have not been successful.

F. Relative Reactivity

The relative reactivity of benzene and its homologues has been studied, often with contradictory results. It is generally assumed that the alkylation of monoalkylbenzenes is faster than that of benzene. It was reported that the rate ratios of alkylation of toluene vs benzene with alkyl halides is greater than 1.8. The relative rates of propylation of toluene vs benzene with propene at 40°C was about 2.1 in the presence of either BF$_3$/ether or AlCl$_3$/nitromethane. Likewise, the relative rates of ethylation of ethylbenzene vs benzene was 2.1.

G. Reaction of Benzene with Olefins

1. Ethylene

From a practical viewpoint AlCl$_3$/HCl is the most useful catalyst for the ethylation of benzene. Although the reaction proceeds at atmospheric pressure and temperatures of 55°–75°C, its rate greatly increases with the application of pressures of 10–20 atm. At the beginning the rate of reaction is slow until an active catalyst is produced by complexing of AlCl$_3$ with the polyethylated benzene formed during the reaction. The use of a large excess of benzene permits almost complete conversion of benzene to monoethylbenzene. The transalkylation activity of the catalyst also makes possible the conversion of polyethylbenzene to monoethylbenzene through the reaction with benzene.

Silica–alumina cracking catalyst causes the ethylation of benzene at 150°–275°C and 20–50 atm, but the catalyst loses its activity due to depositions of carbonaceous material and requires frequent regeneration.

Silicophosphoric acid catalyzes the formation of ethylbenzene. At 275°C, 50 atm, liquid hourly space velocity (LHSV) of 2, and a benzene to ethylene ratio of 10:1, 90% of ethylene was converted. The ratio of mono- to di- to polyethylbenzene formed was 38:1:0.3. The diethylbenzene fraction contained 30% of sec-butylbenzene, a product of dimerization followed by the alkylation of benzene:

$$C_2H_5^+ \xrightarrow{\ C_2H_4\ } C_4H_9^+ \xrightarrow{\ C_6H_6\ } C_4H_9-C_6H_5 + H^+$$

Zeolite ZSM-5 (Chapter 8, Section II,B) was found to be an effective catalyst for the ethylation of benzene. Ethylation occurs satisfactorily at 425°C and 14–28 atm using a high molar ratio of benzene to ethylene and WHSV of 300–400. In a pilot plant operation, cycles of 2 to 4 weeks were achieved before the catalyst was regenerated. The overall yield of ethylbenzene was 99.6% (Lewis and Dwyer, 1977).

2. Propene

The alkylation of benzene with propene to form isopropylbenzene (cumene) occurs over the same catalyst used for ethylation reactions, but under milder conditions. Silicophosphoric acid is the preferred commercial catalyst for the production of cumene. In a continuous-flow operation the preferred conditions are 250°–260°C, 20–25 atm, LHSV 3.5, and benzene to propene molar ratio of 3:1. About 94% of the propene is converted to form mono- with some diisopropylbenzene. For good catalyst life, controlled amounts of water are added to the feed.

3. Butenes

1-Butene and 2-butene are similar in behavior to propene; however, under conditions of high temperature skeletal isomerization may occur, with the resulting formation of *tert*-butylbenzene. For that reason catalysts more active than silicophosphoric acid, such as hydrogen fluoride, 96% sulfuric acid, or $AlCl_3/HCl$ should be used. With these catalysts temperatures of 0°–25°C are functional.

4. Unbranched C_5 and Higher Olefins: Double-Bond Migration

With pentenes and higher olefins alkylation is accompanied by an apparent double-bond migration. When 1-pentene is reacted with benzene in the presence of 96% sulfuric acid, a 65% yield of pentylbenzenes is obtained consisting of 2 parts of 2-phenylpentane (**48**) and 1 part of 3-phenylpentane (**49**).

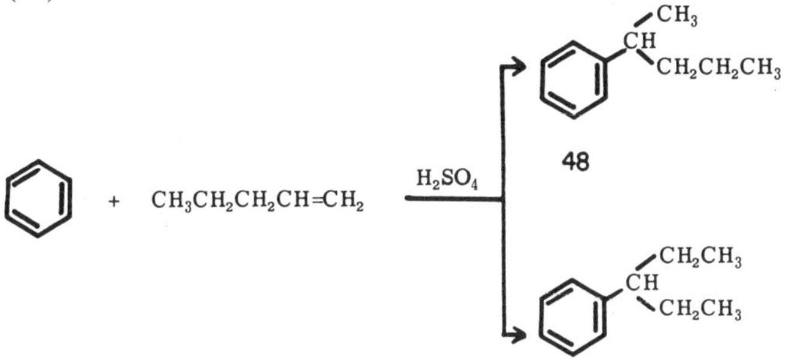

The alkylation of benzene with 1-dodecene in the presence of HF, $AlCl_3/HCl$, and 96% H_2SO_4 forms a mixture of secondary dodecylbenzenes (Table 1.18). Isomerization occurs before alkylation since on prolonged treatment of pure dodecylbenzenes with HF and H_2SO_4 the hydrocarbons are unchanged. It appears that the cation generated from the protonation of the 1-dodecene with the acid undergoes rapid carbonium ion shifts before reaction with benzene:

$$C_7H_{15}CH_2CH_2CH_2CH{=}CH_2 \xrightarrow{H^+} C_7H_{15}CH_2CH_2CH_2\overset{+}{C}HCH_3 \rightleftharpoons$$
$$C_7H_{15}CH_2CH_2\overset{+}{C}HCH_2CH_3 \rightleftharpoons C_7H_{15}CH_2\overset{+}{C}HCH_2CH_2CH_3 \rightleftharpoons C_7H_{15}\overset{+}{C}HC_4H_9 \quad \text{etc.}$$

The absence of the 1-phenyl isomer is attributed to the instability of a primary cation.

TABLE 1.18

Alkylation of Benzene with 1-Dodecene[a]

Conditions			
Catalyst	HF	AlCl$_3$/HCl	H$_2$SO$_4$
Moles of catalyst	50	0.1	5 ml (98%)
1-Dodecene (mol)	5	2	0.15
Benzene (mol)	50	10	1.5
Temp (°C)	16 ± 3	30–53	0–10
Products (%)			
Monoalkylate	92	68	72
Dialkylate	5	21	9
Ratio of isomers			
1-Phenyldodecane	0	0	0
2-Phenyldodecane	20	32	41
3-Phenyldodecane	17	22	20
4-Phenyldodecane	16	16	13
5-Phenyldodecane	23	15	13
6-Phenyldodecane	24	15	13

[a] From Olson (1960).

5. Branched Olefins: Double-Bond Migration

3-Methyl-1-butene, unlike 1- or 2-pentene, forms only one monoadduct on reaction with benzene in the presence of HF or H$_2$SO$_4$ as catalyst:

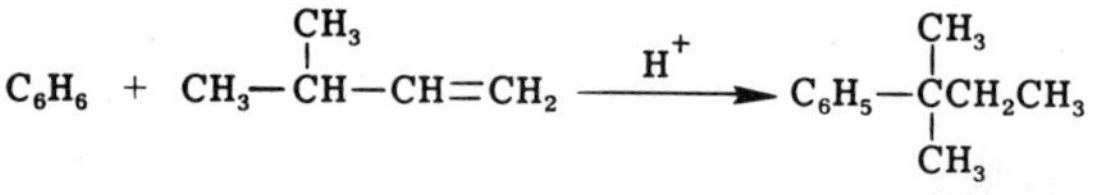

$$C_6H_6 \ + \ CH_3-\underset{\underset{CH_3}{|}}{CH}-CH=CH_2 \ \xrightarrow{\ H^+\ } \ C_6H_5-\underset{\underset{CH_3}{|}}{\overset{\overset{CH_3}{|}}{C}}CH_2CH_3$$

With AlCl$_3$ as catalyst the reaction is more complicated because *tert*-pentylbenzene once formed may undergo skeletal isomerization to afford 2-phenyl-3-methylbutane.

4-Methylcyclohexene in the presence of either HF or 96% H$_2$SO$_4$ similarly produces only 1-methylcyclohexylbenzene.

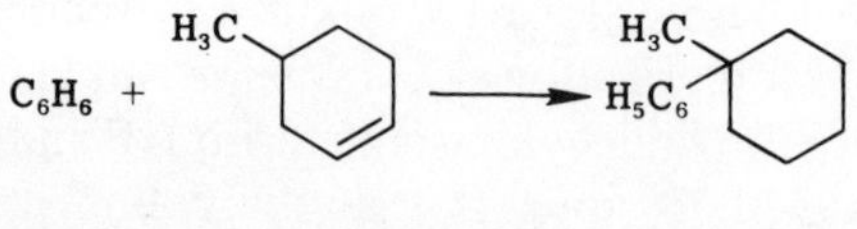

6. Olefins Containing a Quaternary Carbon Atom: Depolyalkylation

Olefins derived from di- or trimerization of isobutylene on reaction with benzene in the presence of alkylating catalysts, such as 96% sulfuric acid, form only *tert*-butyl- and di-*tert*-butylbenzenes, the latter being predominantly para. A similar reaction occurs when dimers of methylbutenes are used for alkylation. This type of alkylation, which is accompanied by scission of the alkenes, is called depolyalkylation. The scission occurs before alkylation.

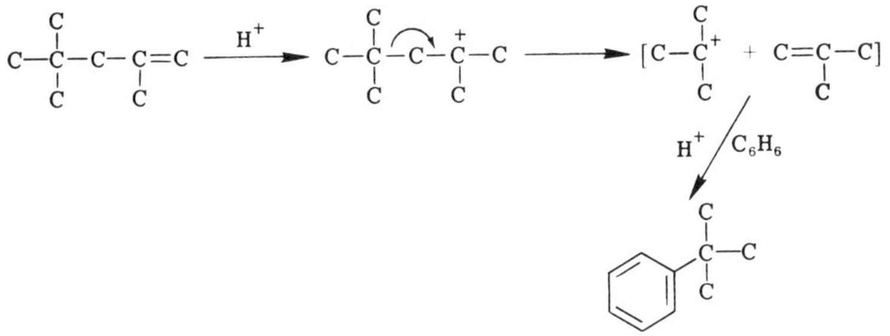

Only alkenes containing a quaternary carbon atom are capable of generating through β-scission the relatively stable tertiary cation. For that reason dodecylbenzenes and higher alkylbenzenes are manufactured from tetramers and pentamers of propene, but not from trimers of isobutylene or from copolymers of isobutylene with other alkenes. Dodecylbenzenes and their homologues are used for making detergents.

Although benzene with diisobutylene forms only depolyalkylation products, the more reactive aromatic nuclei, such as naphthalene, phenols, and thiophene, yield *tert*-octyl derivatives. *tert*-Octyltoluene can also be obtained from toluene and diisobutylene with HF at $-40°C$ and with $AlCl_3/HCl$ at $-65°C$; at higher temperatures depolyalkylation occurs.

7. Branched-Chain Olefins: Skeletal Isomerization

Skeletal isomerization may accompany the alkylation of benzene when the reaction is carried out in the presence of strong acids, such as $AlCl_3/HCl$, $ZrCl_4/HCl$, or BF_3/HF. With weaker acids, such as phosphoric acid, skeletal isomerization may occur, but at higher temperature, and the isomerization appears to occur before alkylation. In the presence of strong acids the alkylbenzenes once formed undergo skeletal isomerization, according to the mechanism discussed in Section II,C,4.

The results of the alkylation of benzene with methylbutenes illustrate the effect of experimental conditions on the composition of the monoalkylbenzenes formed (Table 1.19). The data demonstrate that *tert*-pentylbenzene must be the primary product formed, which then, in contact with the catalyst, undergoes rearrangement to 2-methyl-3-phenylbutane.

H. Carbonium Ion Transfer Accompanying Alkylation

Under certain experimental conditions carbonium ion transfer, generally referred to in the literature as hydride transfer reaction, may occur

TABLE 1.19

Alkylation of Benzene in the Presence of $AlCl_3$/HCl: Effect of Temperature and Time[a]

Alkene	Temp (°C)	Time (min)	Composition of C_6H_5R: R = —C(C)(C)—C—C	Composition of C_6H_5R: R = —C(C)—C(C)—C
C—C(C)=C—C	−40	52	100	0
	0	30	75	25
	21	34	55	45
C=C(C)—C—C	−40	52	100	0
	0	41	55	45
	21	36	0	100
C—C(C)—C=C	0	1.5	90	10
	0	300	35	65

[a] From Friedman and Morritz (1956).

when aromatic hydrocarbons are contacted with olefins in the presence of acid catalysts. Some of the transfer reactions that occur most frequently are described in this section.

1. Reaction of Benzene with Alkanes: Dehydroalkylation

When benzene is contacted with a mixture of isoparaffins and branched-chain olefins in the presence of catalysts, carbonium ion transfer may occur whereby the isoparaffins may donate hydrogen to the ole-

TABLE 1.20

Effect of Olefin Type on Dehydroalkylation of Benzene[a,b]

	Olefin: 4-Methyl-2-pentene	Olefin: 4-Methyl-1-pentene	Olefin: 1-Hexene
Charge (molar), isobutane/olefin/benzene	40/2/1	50/2/1	30/2/1
Yield of *p*-di-*t*-butylbenzene (mol % on benzene charged)	91	70	42
Yield of C_6H_{14}[c] (mol % on olefin charged)	88	67	19

[a] From Kelley and Lee (1955).
[b] Catalyst: 90% alkylation acid (H_2SO_4).
[c] Predominantly methylpentanes.

fins and react with benzene to form alkylbenzenes. The olefins are thus converted to saturated hydrocarbons. This reaction is catalyzed by strong acids such as HF, 96% H_2SO_4, or BF_3/HF at 0°–30°C. The reaction can be represented by the following equations:

$$C_nH_{2n} + H^+ \longrightarrow (C_nH_{2n+1})^+$$

$$(C_nH_{2n+1})^+ + (CH_3)_3CH \rightleftharpoons C_nH_{2n+2} + (CH_3)_3C^+$$

$$(CH_3)_3C^+ + C_6H_6 \longrightarrow C_6H_5C(CH_3)_3 + H^+$$

For C_4 and higher unbranched alkenes, the carbonium ion transfer is accompanied by simultaneous skeletal isomerization so that branched-chain paraffins are produced. Low yields were observed for this reaction when 1-, 2-, or 3-hexene was used as the olefin and isobutane as the hydrogen donor; for branched C_6 olefins the yields were higher (Table 1.20).

2. Dehydrocyclialkylation

p-Dialkylbenzenes do not undergo alkylation by branched-chain olefins because of steric interference with the ortho position. However, *p*-dialkylbenzenes of which at least one alkyl group contains an α tertiary hydrogen, such as isopropyl, *sec*-butyl, or cyclopentyl, rapidly transfers its benzylic hydrogen to a tertiary cation produced from the interaction of the acid catalyst with the olefin. Subsequent steps in the reaction result in ring closure with the ultimate formation of indans. A similar type of reaction was found with *m*-dialkylbenzenes.

a. *p*-Cymene and 4-Methylcyclohexene. The title hydrocarbons in the presence of either HF or H_2SO_4 at 0°–10°C afforded 73% of 1,3,3,6-tetramethyl-1-*p*-tolylindan (**51**) as the major product and 20% of **52**, in addition to methylcyclohexane. The reaction was interpreted as follows (Ipatieff *et al.*, 1948):

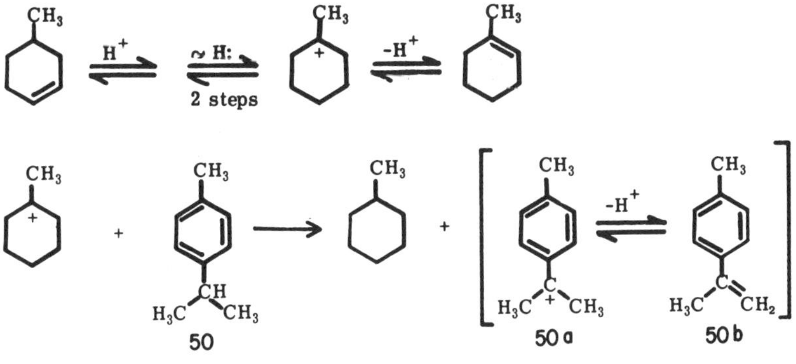

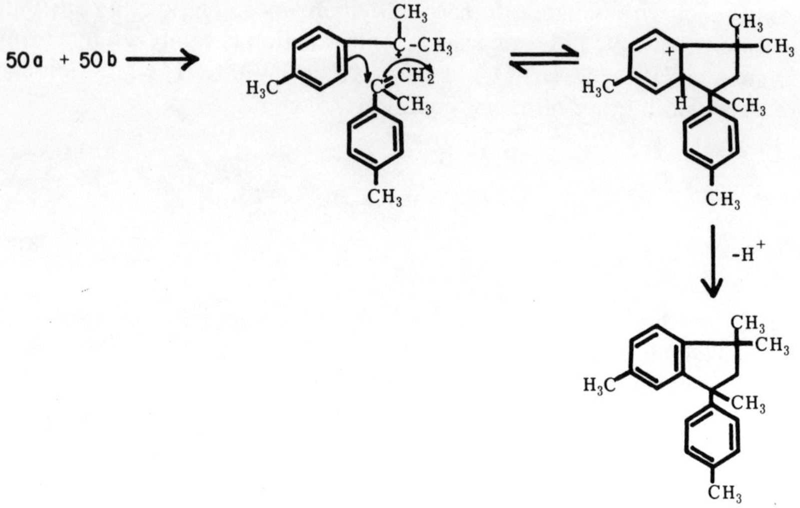

50 a + 50 b ⟶

51

Compound **52** is formed as follows:

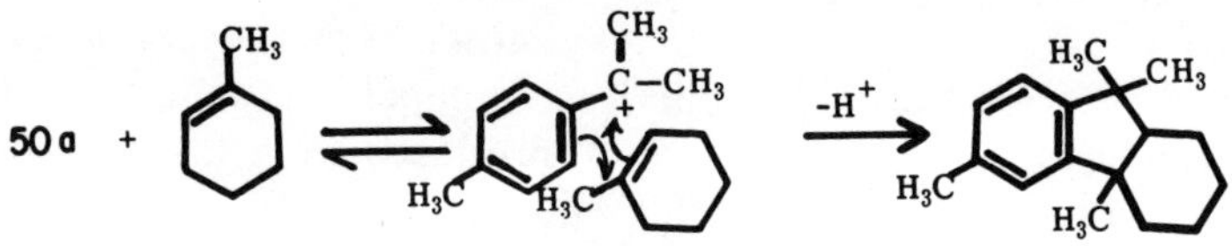

52

b. *p*-Cymene and Isobutylene. The title hydrocarbons in a molar ratio of 1:0.8 on reaction in the presence of HF form compounds **51** (16%), **53** (71.7%), and **53a** (6.6%). Compound **53** is the product of interaction of **50a** with isobutylene.

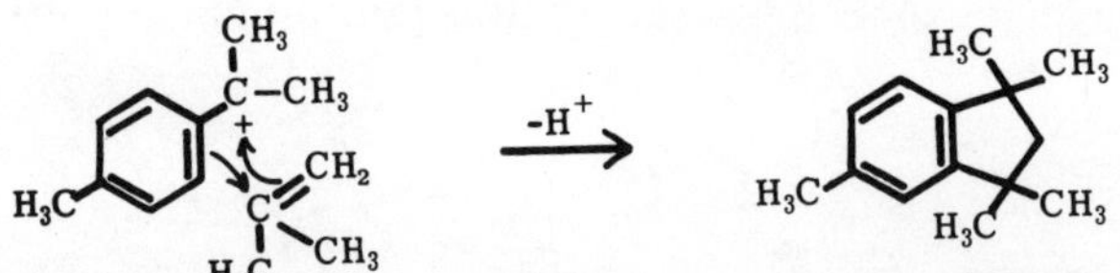

53

Compound **53a** was interpreted as being formed by the addition of **50a** to a dimer of isobutylene.

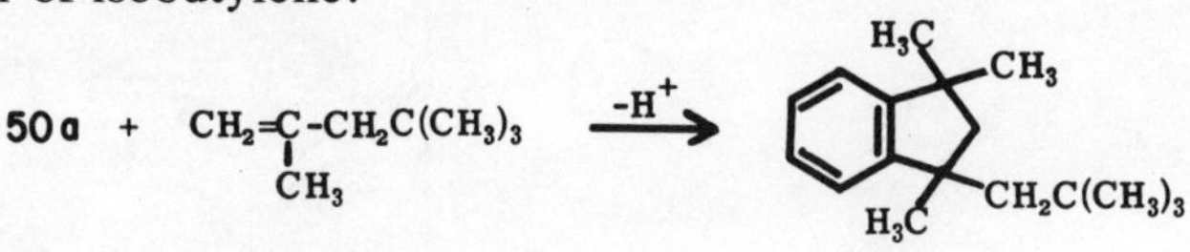

$$50a + CH_2=\overset{\underset{\displaystyle CH_3}{|}}{C}-CH_2C(CH_3)_3 \xrightarrow{-H^+}$$

53 a

c. *p-sec*-Butyltoluene and Isobutylene. The following dehydrocyclialkylated hydrocarbons were formed when 0.05 *M* butyltoluene and 0.077 *M* isobutylene were treated with 3 gm of HF at 0°C: **54** (71.7%), **55** (9.1%), **56** (7.8%), and **57** (11.3%) (molar composition of indians) (Elgavi and Pines, 1978).

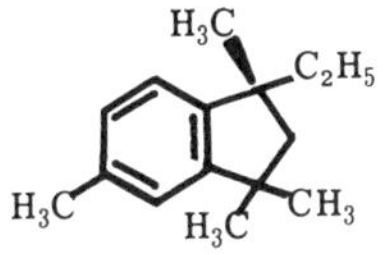

54

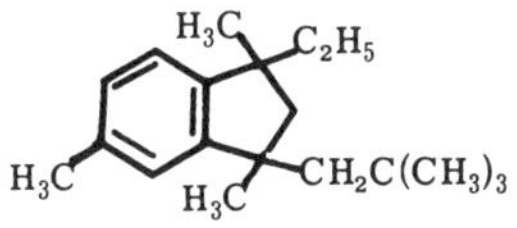

55

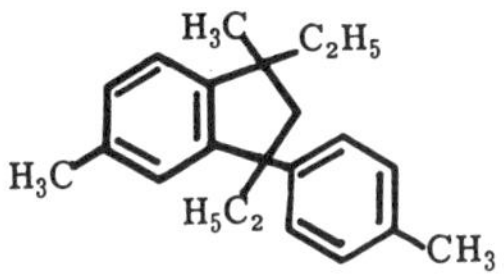

56

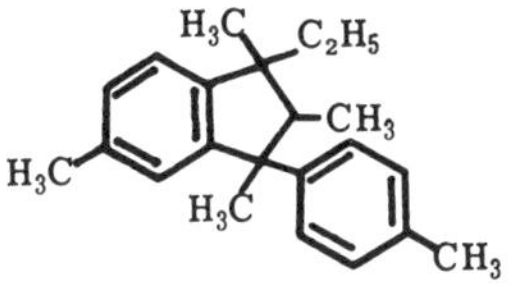

57

d. *p*-Diisopropylbenzene and Isobutylene. The title aromatic hydrocarbon with an excess of isobutylene in the presence of HF forms 1,1,3,3,5,5,-7,7-octamethylhydrindacene (**58**) in 33% yield.

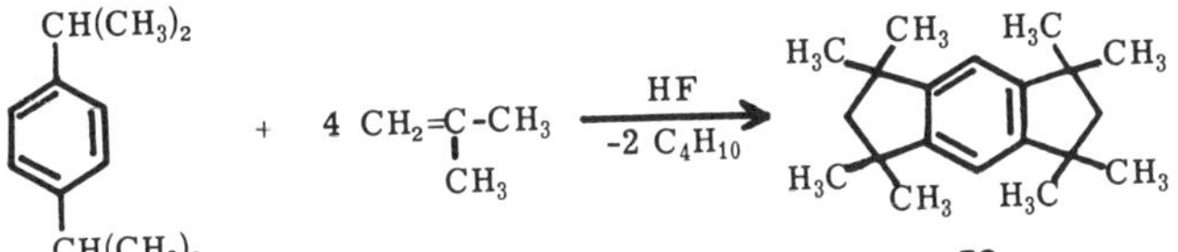

58

3. *Diarylalkanes from p-Dialkylbenzene: Transalkylation*

p-Alkyltoluenes, in which the alkyl group contains a secondary benzylic carbon atom, form diarylalkanes on reaction with 4-methylcyclohexene in the presence of either HF or 96% H_2SO_4. On protonation the latter forms, through an intramolecular hydrogen shift, a tertiary cyclohexyl cation, as shown in Section V,H,1,a. The reaction proceeds as follows:

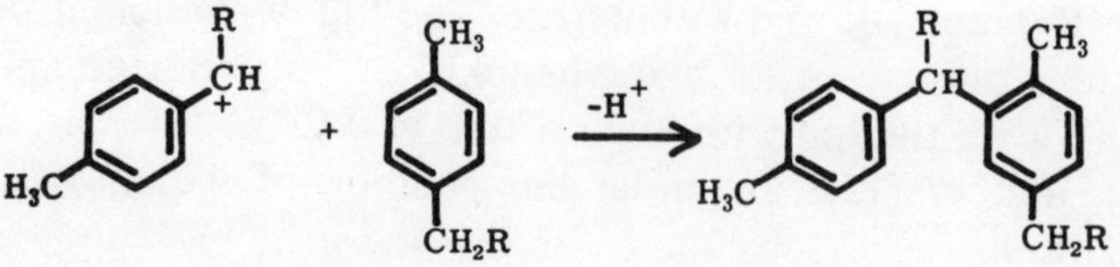

$[R = CH_3, C_2H_5, CH(CH_3)_2]$

The aralkylation occurs readily because the entering group, being a secondary cation, is not hindered by the methyl group in the ortho position.

The introduction of methyl groups ortho to the alkyl group has an inhibiting effect on the reaction, and in the case of 1,3-dimethyl-4-ethylbenzene the reaction was accompanied by transalkylation (Pines and Arrigo, 1958).

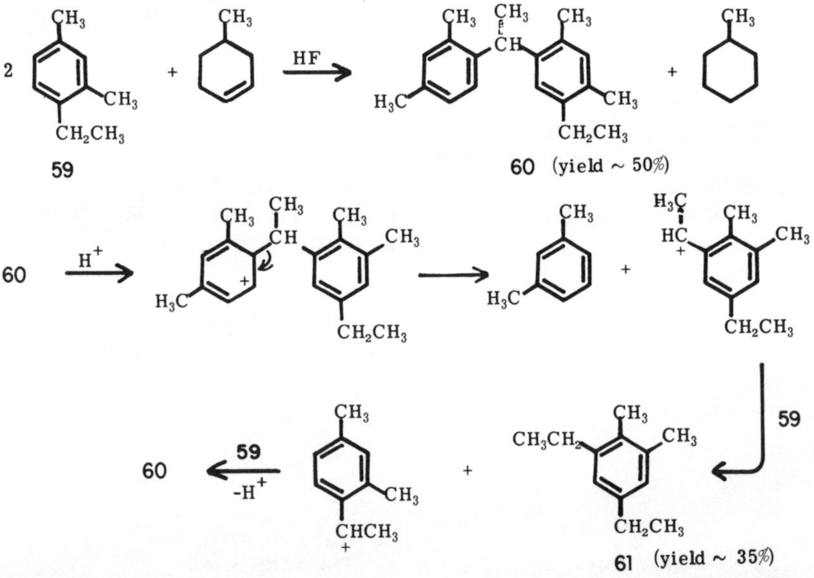

Transalkylation and Isomerization. Isomerization and transalkylation are the two principal reactions when di- and trialkylbenzenes are passed over $SiO_2-Al_2O_3$ or cation-exchanged zeolites (for zeolites see Section VI,C,2). Positional insomerization of polyalkylbenzenes may proceed by two mechanisms: (a) an intramolecular 1,2 shift (Section II,C,2) and (b) transalkylation. Below 200°C isomerization via transalkylation predominates (Scheme 1.8).

Disproportionation of toluene. Toluene undergoes disproportionation in the presence of acid catalysts such as $AlCl_3/HCl$, $SiO_2-Al_2O_3$, and zeolites. The products of reaction are benzene and an equilibrium mixture

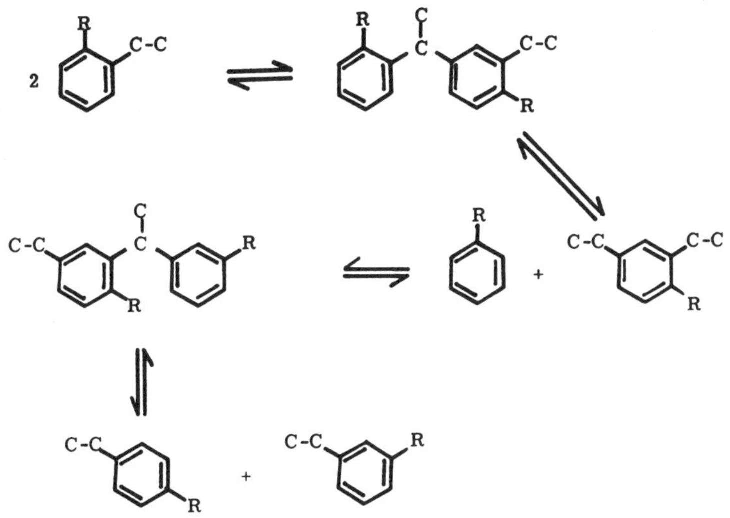

Scheme 1.8. Isomerization of alkylbenzenes via transalkylation (Csicsery, 1969).

of xylenes. However, it was found that by modifying zeolite ZSM-5 (Chapter 8, Section II,B) in such a way as to obtain smaller pore openings and to increase crystal size from <0.5 μm to 3 μm, it is possible to produce the more desirable p-xylene selectively.

The diffusivity of p-xylene in zeolite ZSM-5 is $>10^3$ times faster than

TABLE 1.21

Disproportionation of Toluene[a]

	Catalyst		
	ZSM-5	ZSM-5 (11% Mg)	Thermodynamic equilibrium (%)
Temp (°C)	500	550	
WHSV[b]	30	3.5	
Conversion (wt %)	13.2	10.9	
Product (wt %)			
Benzene	5.5	4.9	
Toluene	86.8	89.2	
Xylenes	7.5	5.9	
Composition of xylenes (%)			
o-Xylene	19	2	26
m-Xylene	46	10	51
p-Xylene	35	88	23

[a] From Chen et al. (1979).
[b] WHSV—see Terminology and Abbreviations.

that of o- and m-xylene. The diffusional character of ZSM-5 can be modified significantly by impregnating it with aqueous phosphoric acid or magnesium acetate. The introduction of 8.5% phosphorus or about 11% magnesium into the crystals reduces the pore openings and channel dimensions of the ZSM-5 crystals, favoring the formation of the less bulky p-xylene by permitting it and small molecules to diffuse out of the catalyst more rapidly (Table 1.21).

p-Xylene, about 2.5×10^9 kg of which are produced annually in the United States alone, is oxidized to terephthalic acid, a major component of polyester fibers (Chapter 5, Section V,C,1).

I. Reaction of Benzenoid Hydrocarbons with Diolefins

1. Introduction

Most of the reactions of aromatic nuclei with diolefins have been studied with conjugated dienes, especially with butadiene and isoprene. The reaction can lead to alkenyl-substituted arenes, diarylalkanes, or cyclized products. A mild acid catalyst and a low temperature favor the formation of alkenylarenes. With increasing acid strength and higher temperatures, diarylation and cyclization are favored.

2. Benzene and 1,3-Butadiene

The product of reaction of the title hydrocarbons in the presence of 5 wt % alkanesulfonic acid is 1-phenyl-2-butene (**62**). At higher temperature diarylbutanes are also formed.

When benzene and butadiene in a molar ratio of 4.5:1 are passed over silicophosphoric acid at 27 atm and 215°C, a 59% yield of the phenylbutene is produced. With stronger acids such as $AlCl_3$ at 50°C or concentrated H_2SO_4 at 0°C, a mixture of diphenylbutanes is formed (Table 1.22). The reaction is thought to follow the steps shown on the next page.

TABLE 1.22

Isomer Ratio of Diphenylbutanes[a]

Catalyst	Temp (°C)	Isomer (%)		
		1,1 (**63**)	1,2 (**64**)	1,3 (**65**)
$AlCl_3$	20	25	29	46
$AlCl_3$	50	23	30	47
H_2SO_4	0–5	50	31	19

[a] From Inukai (1966).

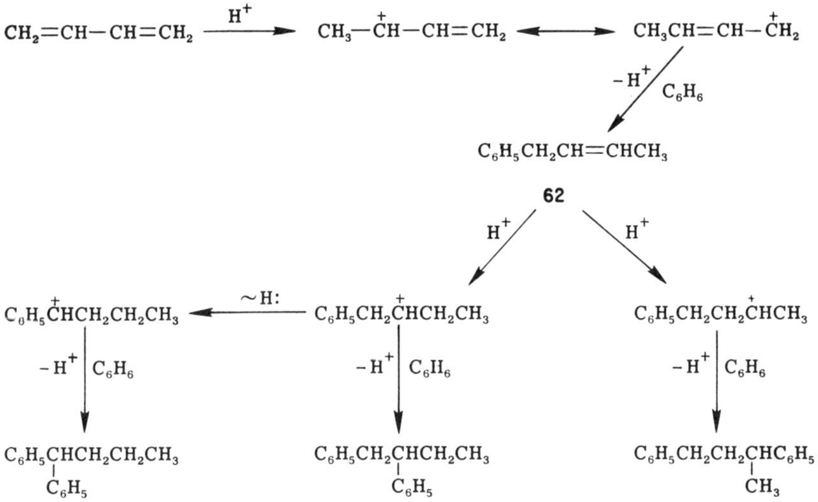

$$CH_2=CH-CH=CH_2 \xrightarrow{H^+} CH_3-\overset{+}{C}H-CH=CH_2 \longleftrightarrow CH_3CH=CH-\overset{+}{C}H_2$$

$$\xrightarrow[\quad C_6H_6 \quad]{-H^+}$$

$$C_6H_5CH_2CH=CHCH_3$$

62

$$H^+ \swarrow \qquad \searrow H^+$$

$$C_6H_5\overset{+}{C}HCH_2CH_2CH_3 \xleftarrow{\sim H:} C_6H_5CH_2\overset{+}{C}HCH_2CH_3 \qquad C_6H_5CH_2CH_2\overset{+}{C}HCH_3$$

$$\xdownarrow[C_6H_6]{-H^+} \qquad \xdownarrow[C_6H_6]{-H^+} \qquad \xdownarrow[C_6H_6]{-H^+}$$

$$\begin{array}{ccc} C_6H_5CHCH_2CH_2CH_3 & C_6H_5CH_2CHCH_2CH_3 & C_6H_5CH_2CH_2CHC_6H_5 \\ \quad | & \quad | & \quad | \\ C_6H_5 & C_6H_5 & CH_3 \\ \textbf{63} & \textbf{64} & \textbf{65} \end{array}$$

3. *Benzenoid Hydrocarbons and Isoprene*

Isoprene and benzenoid hydrocarbons having an unoccupied vicinal position combine in the presence of sulfuric acid to give indans. The cyclialkylation reaction involves several steps and, as applied to *p*-xylene, can be represented as follows:

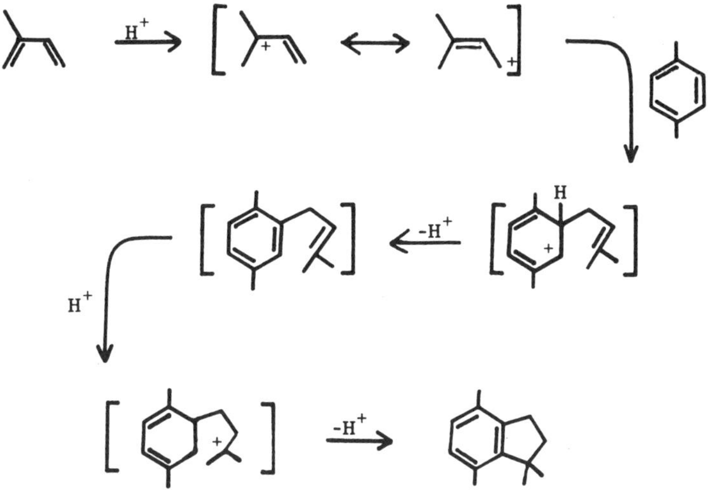

66

The primary and allylic cations, being more reactive and less sterically hindered, add preferentially to the aromatic ring.

Indans such as 1,1,4,7-tetramethylindan (66) may react with another molecule of isoprene to form hydroindacenes 67a and 67b.

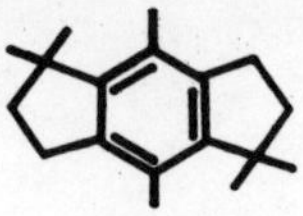

67a 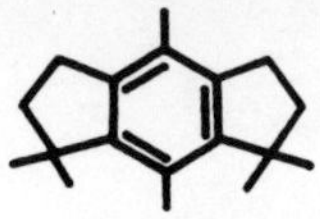   67b

The optimal sulfuric acid concentrations and reaction temperatures for cyclialkylations of xylenes are 96% below $-15°C$; 93% from $-10°$ to $5°C$; and 85% at $10°-25°C$.

a. Monoalkylbenzenes. Monoalkylbenzenes in reaction with isoprene afford predominantly those isomers that require initial attack at positions ortho or para to alkyl groups (Table 1.23).

b. Polymethylbenzenes. o-Xylene reacts with isoprene in the presence of 85% sulfuric acid to give a 47% yield of a mixture of tetramethylindans 68, 69, and 70 in respective concentrations of 16, 38, and 46%.

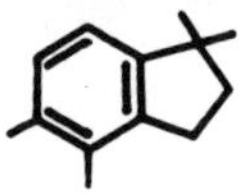

68 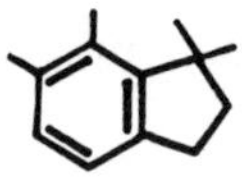   69 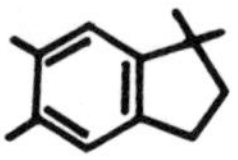   70

TABLE 1.23

Cyclialkylation Products[a] from Alkylbenzenes and Isoprene[b]

R	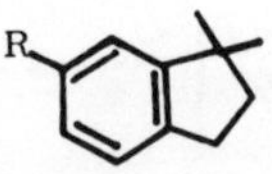 (%)	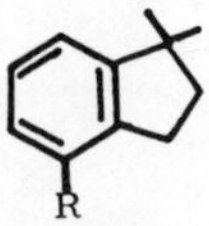 (%)
CH_3	68–71	29–32
C_2H_5	79–83	17–21
$CH(CH_3)_2$	88–92	8–12
$C(CH_3)_3$	98	2

[a] Absolute yields of indans ranged from 55 to 70%.
[b] From Eisenbraun et al. (1968).

m-Xylene, on cyclialkylation in the presence of 93% H_2SO_4 at $-5°$ to 0°C, gave a 68% yield of **71**, whereas 1,2,4-trimethylbenzene afforded a 70% yield of **72**.

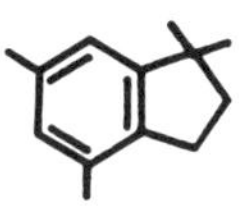

71 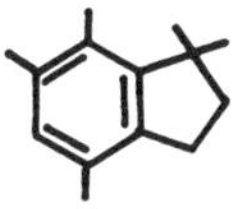   72

c. Cymenes. *m*-Cymene, under the same conditions as for *m*-xylene, gives a 76% yield of **73**, whereas *p*-cymene affords a 62% yield of a high-purity bicyclic compound corresponding to structure **74**.

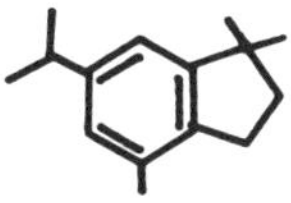

73 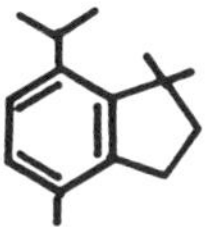   74

J. Alkylation with Cyclopropane and Cyclobutane Ring Hydrocarbons

Of the cycloalkanes only hydrocarbons containing cyclopropane or cyclobutane rings undergo reaction with benzenoid hydrocarbons to form alkylbenzenes. The alkylation reaction is catalyzed by 96% H_2SO_4, HF, or $AlCl_3$ and proceeds at 0°–25°C.

1. Alkylcyclopropanes

The products of the reaction of benzene with alkylcyclopropanes are those expected from the protonation of the cyclopropane ring to form alkyl cations (Table 1.24). The alkylation is accompanied by the usual wandering of hydrogen (or methyl), which with *n*-alkylcyclopropanes leads to the formation of mixtures of phenylalkanes, similar to those encountered in Section V,G,4. The major product from isopropylcyclopropane is 2-methyl-2-phenylpentane, but the concurrent formation of one-fifth as much of the 3-methyl-3-phenyl isomer indicates that some methyl migration takes place before alkylation.

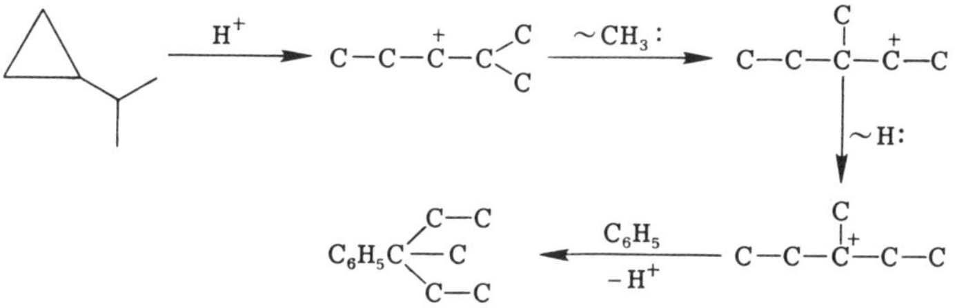

TABLE 1.24

Reaction of Alkylcyclopropanes with Benzene[a]

Alkylcyclopropane	Catalyst	Monoalkyl-benzene	Structure	Composition (%)
(cyclopropane)	HF	48	C_6H_5C⟨$^C_{C-C}$	100
(ethylcyclopropane)	HF	41	C_6H_5C⟨$^C_{C-C-C}$	67
			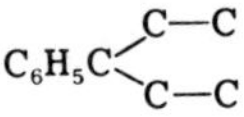 C_6H_5C⟨$^{C-C}_{C-C}$	33
	H_2SO_4	47	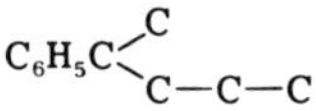 C_6H_5C⟨$^C_{C-C-C}$	67
			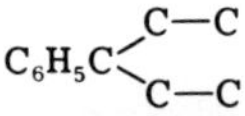 C_6H_5C⟨$^{C-C}_{C-C}$	33
	$AlCl_3$	51	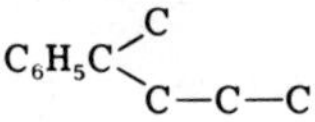 C_6H_5C⟨$^C_{C-C-C}$	67
			C_6H_5C⟨$^{C-C}_{C-C}$	33
(methylcyclopropane)	HF	43	C_6H_5C⟨$^C_{C-C-C}$ ⎫ C_6H_5C⟨$^{C-C}_{C-C}$ ⎬	10–14
			C_6H_5C⟨$^C_{C}$$-C-C$	83–87
(propylcyclopropane)	HF	85	C_6H_5C⟨$^C_{C-C-C-C}$	40
			C_6H_5C⟨$^{C-C}_{C-C-C}$	46
(isopropylcyclopropane)	HF	55	C_6H_5C⟨$^C_{C}$$-C-C-C$	80
			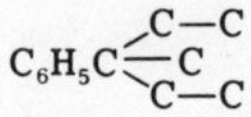 C_6H_5C⟨$^{C-C}_{C-C}$	15

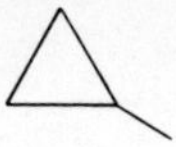

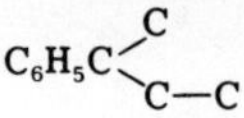

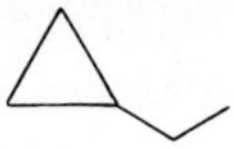

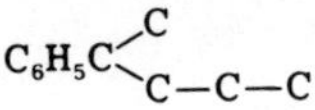

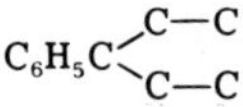

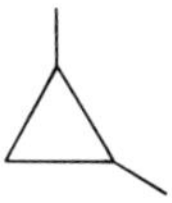

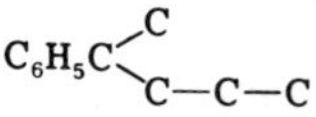

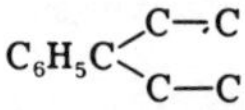

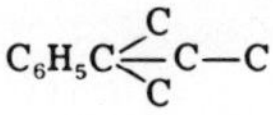

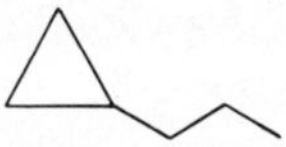

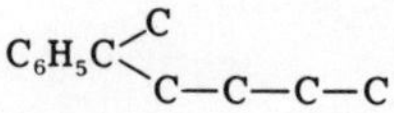

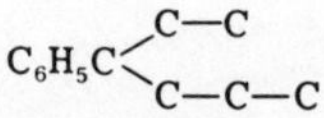

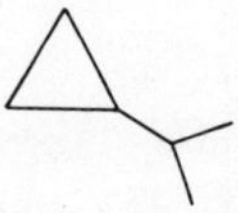

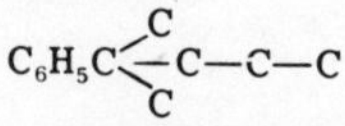

[a] From Pines *et al.* (1951b, 1953).

TABLE 1.25

Ratios of Isopropyl to *n*-Propyl Product in the Alkylation of Benzene Derivatives with Cyclopropane[a,b]

Reactant	Temp (°C)	Yield (%)		Relative yield (%)	
		Monopropyl	Polypropyl	*n*-Propyl	Isopropyl
p-Xylene	25	61	33	92	8
Toluene	25	61	22	83	17
Benzene	25	62	28	70	30
	45	67	24	59	41
Chlorobenzene	25	76	12	31	69
o-Dichlorobenzene	25	43	20	<4	>96

[a] From Deno *et al.* (1968).
[b] The cyclopropane was introduced over a period of 25 min into a fourfold excess of the aromatic and 2.5% AlCl₃ catalyst.

2. Cyclopropane

The products of the reaction of cyclopropane with benzenoid hydrocarbons in the presence of acid catalysts are a mixture of *n*-propyl- and isopropylarenes. The ratio of *n*-propyl to isopropyl products depends to a considerable extent on the relative reactivities of the arenes. As the reactivity decreases the amount of *n*-propyl decreases and the proportion of isopropyl increases (Table 1.25). It appears that, unlike alkylcyclopropanes, cyclopropane forms protonated cyclopropane species. In the presence of the less reactive aromatic hydrocarbons, the c-$C_3H_7^+$ has more opportunity to isomerize to the more stable isopropyl cation.

3. Methylcyclobutane

The reaction of benzene with methylcyclobutane in the presence of 96% H_2SO_4 at 2°–4°C afforded pentylbenzenes of which *tert*-pentylbenzene was identified.

VI. CATALYTIC CRACKING

A. Introduction

Before World War II the cracking of hydrocarbons to produce gasoline was carried out thermally, either in liquid or vapor phase, at 425°–525°C and pressures of 3–30 atm. Thermal cracking consisted of a series of decomposition and condensation reactions that took place at elevated temperatures, and the resulting product was highly olefinic and not selective.

The superiority of catalytic cracking over thermal cracking is largely dependent on reactions other than those associated with simple bond ruptures.

Catalytic cracking originally involved the use of silica–alumina catalyst, which is considered to be a solid acid, and the reactions occurring on this catalyst involve carbonium ions. Since about 1965, the cracking catalyst requirement has progressively shifted from synthetic SiO_2–Al_2O_3 to catalysts containing zeolites. The magnitude of the use of catalytic cracking is witnessed by the fact that in 1975 the United States alone processed 12 million barrels of oil per day (1 barrel = 42 gal) and of this about one-third was catalytically cracked.

B. Silica–Alumina-Catalyzed Cracking

Catalytic cracking of hydrocarbons in the presence of silica–alumina is associated with an accelerated severance of carbon–carbon bonds as compared with thermal cracking carried out under similar conditions. From two distinctly different sets of fragmentation patterns of hydrocarbons, it can be concluded that thermal cracking occurs by a free-radical or homolytic split of the C—C bond, whereas catalytic cracking takes place by a heterolytic or ionic rupture of the bond in which a carbonium ion mechanism is involved.

A comparison of reaction characteristics based on products obtained from thermal and catalytic cracking of various types of hydrocarbons is given in Table 1.26. The results show that hexadecane cracks thermally to form C_2 as the major product with a large amount of C_4 to C_{15} unbranched α-olefins, whereas the main products from catalytic cracking consist of C_3 to C_6 hydrocarbons, and the aliphatic hydrocarbons are mostly branched. From a practical viewpoint branched alkanes have a higher octane number than the straight-chain hydrocarbons.

In commercial catalytic cracking the vapors of gas oil, the fraction of petroleum boiling approximately between 200° and 500°C, are contacted with SiO_2–Al_2O_3 catalyst at about 450°–550°C, usually for a few seconds, and at a pressure of 1–2 atm. The conventional gas oil consists of a mixture of paraffins, naphthenes, and aromatics, and many reactions occur simultaneously, such as skeletal isomerization, dealkylation, cyclization, and hydrogen transfer with formation of aromatics.

A deposition of carbonaceous material is characteristic of catalytic cracking. This material, called coke, is formed within the catalyst particles and must be burned off periodically in order to restore the catalyst activity. In order to secure continuous operation and to stabilize the aver-

TABLE 1.26

Products of Thermal and Catalytic Cracking[a]

Hydrocarbon	Thermal cracking	Catalytic cracking
Hexadecane (cetane)	Major product is C_2 with large amount of C_1 and C_3; large amount of C_4 to C_{15} n-α-olefins; few branched aliphatics	Major product is C_3 to C_6; few unbranched α-olefins above C_4; aliphatics mostly branched
Alkylaromatics	Cracked within side chain	Cracked next to ring
Normal olefins	Double bond shifts slowly; little skeletal isomerization	Double bond shifts rapidly; extensive skeletal isomerization
Olefins	Hydrogen transfer is a minor reaction and is non-selective for tertiary olefins	Hydrogen transfer is important and is selective for tertiary olefins
	Crack at about same rate as corresponding paraffins	Crack at much higher rate than corresponding paraffins
Naphthenes	Crack at lower rate than paraffins	Crack at about same rate as paraffins with equivalent structural groups
Alkylaromatics (with propyl or larger substituents)	Crack at lower rate than paraffins	Crack at higher rate than paraffins

[a] From Greensfelder *et al.* (1949).

age activity of the catalyst, various engineering schemes, such as fluidized and moving-bed catalytic cracking, have been devised.

In fluidized catalytic cracking the catalyst is in the form of a fairly fine powder of 100–300 mesh size, which, during use, flows from the reactor vessel to the regenerator vessel and back continuously. For cracking in moving beds, catalysts of about 3 mm in diameter are used. The catalyst is steadily withdrawn from the bottom of the reactor and lifted by elevator buckets to the top of the regenerator. The catalyst usually spends about 2 min in the regenerator for each minute in the reactor.

1. Preparation and Characteristics of Silica–Alumina

The catalytic activity of SiO_2–Al_2O_3 is ascribed to the acidic character of the catalyst; silica alone is inactive as a cracking catalyst, and alumina alone has very minor cracking activity. The proper combination of SiO_2–Al_2O_3 is much more active than either of the components. The catalyst is prepared from hydrogels or hydrous oxides; mixtures of the anhydrous oxide with one hydrous oxide does not produce an active catalyst.

An effective method for preparing an $SiO_2-Al_2O_3$ catalyst consists of adding 6 N sulfuric acid to sodium silicate solution of an appropriate concentration to give the indicated silica content. The silica gel slurry thus formed is aged with slow agitation before addition of a 33% alum solution (by weight of alumina). The pH of the mixture is raised to 4.6 with 14% ammonia solution. Filtration and washing are then carried out until the sodium content is less than 0.005% on a dry weight basis. Drying and heat treatment of the washed gel are then carried out, with the dried material being calcined to 700°C in about 60 min.

A convenient laboratory method for the preparation of $SiO_2-Al_2O_3$ catalyst consists of mixing a desired proportion of aluminum isopropoxide and ethyl orthosilicate. The mixture is hydrolyzed by aqueous ethanol, dried, and calcined.

Good correlations have been found between total amount of acids, Brønsted and Lewis type, and the catalytic activities of $SiO_2-Al_2O_3$. The acidities are usually measured by amine titration, and the catalytic activities by decomposition of isopropylbenzene and the polymerization of propene. The amount of acid reaches a maximum when the Al_2O_3 content is between 10.3 and 25.1 mol %, probably closer to the lower level. Since the catalytic activity is related to the intrinsic acidity of $SiO_2-Al_2O_3$, it is important that alkali metals are completely removed during the preparation of the catalyst.

Structure of Silica–Alumina Catalysts. The structure of silica–alumina catalysts bears a resemblance to that of organic polymers in which the characteristic groups are repeated in two or three dimensions. Each silicon is attached to four oxygen atoms, and the oxygen atoms are located at the corners of an octahedron with silicon in the center. Aluminum can be visualized as sharing its valences with four oxygen atoms, which are spaced around it. Silica–alumina catalysts do not give an X-ray diffraction pattern that would allow one to determine their structure. It is generally believed, however, that the acid centers, whether of the Brønsted or the Lewis type, owe their existence to an isomorphous substitution of an occasional trivalent aluminum for a tetravalent silicon in the silica lattice. This would lead to the structure represented in Fig. 4 at the top of p. 83.

The aluminum atom tends to acquire a pair of electrons to fill the p shell, creating a Lewis acid in the absence of water and a Brønsted acid in the presence of a water molecule. The Hammett acidity function H_0, indicating the acid strength of the Brønsted acid site, is < -8.2, or equivalent

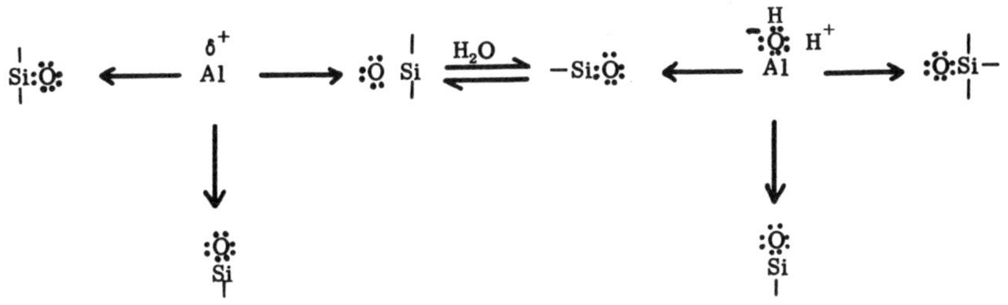

$$\text{Si:}\ddot{\text{O}}: \longleftarrow \overset{\overset{\delta^+}{|}}{\underset{|}{\text{Al}}} \longrightarrow :\ddot{\text{O}}\ \overset{|}{\underset{|}{\text{Si}}} \overset{\text{H}_2\text{O}}{\underset{\rightleftharpoons}{}} -\text{Si:}\ddot{\text{O}}: \longleftarrow \overset{\overset{\overset{\text{H}}{\overset{|}{:\ddot{\text{O}}:}}}{|}}{\underset{|}{\text{Al}}}\ \text{H}^+ \longrightarrow :\ddot{\text{O}}\text{:Si}-$$

$$\downarrow \qquad\qquad\qquad\qquad\qquad \downarrow$$

$$:\overset{\ddot{\text{O}}:}{\underset{|}{\text{Si}}} \qquad\qquad\qquad\qquad :\overset{\ddot{\text{O}}:}{\underset{|}{\text{Si}}}$$

Lewis Acid Site   Brønsted Acid Site

(Arrows indicate displacement of electrons toward the Si–O group)

Fig. 1.4.   Structure of silica–alumina catalyst.

in acidity to 90% sulfuric acid (Table 1.2). The number of acid sites as calculated from the n-butylamine titration method using methyl red as the indicator is equal to 0.55 mmol/gm. From the relatively low number of acid sites it is not surprising that only small amounts of alkali metals or basic impurities will deactivate the catalyst.

The acid sites of silica–alumina are not of equal strength, and from the calorimetric data of the heat of adsorption of butylamine it was possible to establish the existence of acid sites with strength less than that for $H_0 = -4.8$, or equivalent in acidity to about 65% H_2SO_4.

2. Hydrocarbon Conversions over Silica–Alumina

a. **Alkanes and Alkenes.** Catalytic cracking of hydrocarbons does not occur indiscriminately, but in definite ways. The products from the cracking of hexadecane show that even at a low rate of conversion there are a total of 340 mol of product for every 100 mol of hexadecane converted, consisting mainly of C_3 to C_6 hydrocarbons that include a large amount of isoparaffins. The relative rate of cracking depends on the length of the paraffinic chain: for n-C_7H_{16}, 3%; n-$C_{12}H_{26}$, 18%; and n-$C_{16}H_{34}$, 42%. For alkenes the rate of cracking is higher, and it was estimated that for hexadecenes it is 90%.

The products obtained from cracking unbranched alkanes and unbranched alkenes in the presence of SiO_2–Al_2O_3 can be explained by applying carbonium ion mechanisms (Section I,A). The formation of a polarized species from alkane can occur through the removal, by the Lewis acid sites of the catalyst, of hydrogen and its bonding electrons from one of the secondary carbon atoms of the alkane, followed by reactions of organic cations (Scheme 1.9).

The formation of all other branched-type alkenes and alkanes can be represented by a similar series of reactions.

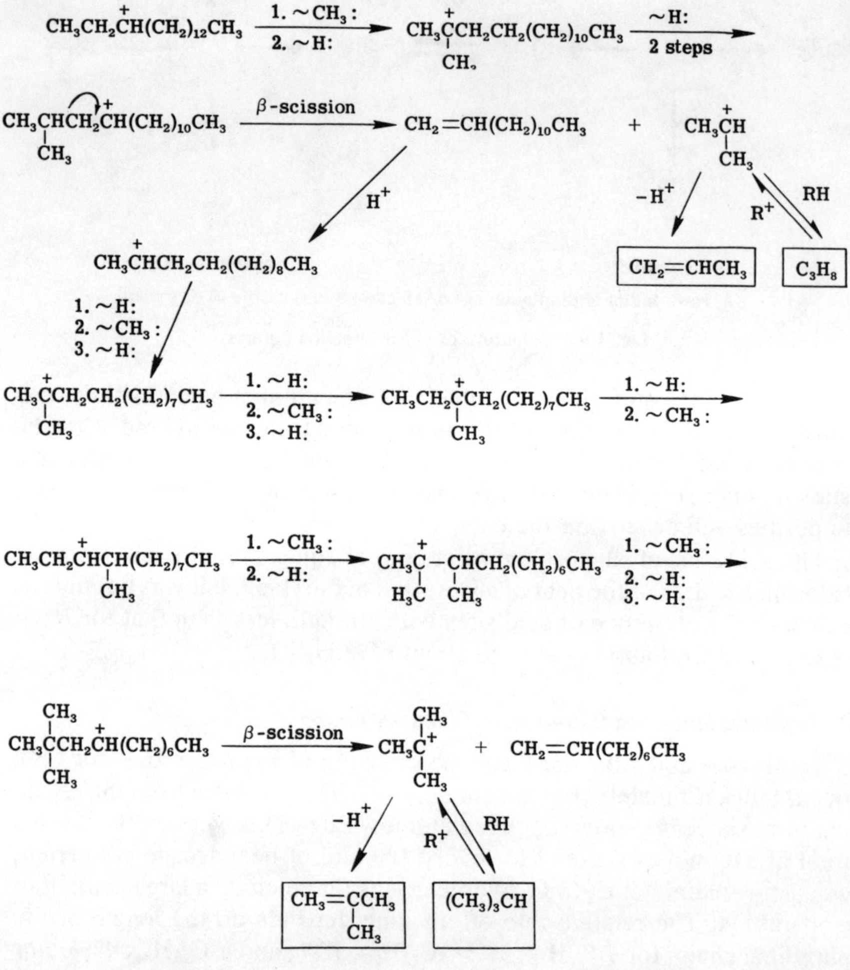

$$CH_3CH_2\overset{+}{C}H(CH_2)_{12}CH_3 \xrightarrow[\text{2. }\sim H:]{\text{1. }\sim CH_3:} CH_3\overset{+}{C}CH_2CH_2(CH_2)_{10}CH_3 \xrightarrow[\text{2 steps}]{\sim H:}$$
$$\underset{CH_3}{|}$$

$$CH_3\overset{\frown}{CH}CH_2\overset{+}{C}H(CH_2)_{10}CH_3 \xrightarrow{\beta\text{-scission}} CH_2=CH(CH_2)_{10}CH_3 \quad + \quad CH_3\overset{+}{C}H$$
$$\underset{CH_3}{|} \qquad\qquad\qquad\qquad\qquad\qquad\qquad\qquad\qquad\qquad\qquad\qquad \underset{CH_3}{|}$$

$$\xrightarrow{H^+}$$

$$CH_3\overset{+}{C}HCH_2CH_2(CH_2)_8CH_3$$

$$-H^+ \nearrow \qquad \qquad \searrow RH \quad R^+$$

$$\boxed{CH_2=CHCH_3} \qquad \boxed{C_3H_8}$$

1. $\sim$H:
2. $\sim CH_3$:
3. $\sim$H:

$$CH_3\overset{+}{C}CH_2CH_2(CH_2)_7CH_3 \xrightarrow[\text{2. }\sim CH_3:]{\text{1. }\sim H:} CH_3CH_2\overset{+}{C}CH_2(CH_2)_7CH_3 \xrightarrow[\text{2. }\sim CH_3:]{\text{1. }\sim H:}$$
$$\underset{CH_3}{|} \qquad\qquad\qquad\qquad \text{3. }\sim H: \qquad \underset{CH_3}{|}$$

$$CH_3CH_2\overset{+}{C}HCH(CH_2)_7CH_3 \xrightarrow[\text{2. }\sim H:]{\text{1. }\sim CH_3:} CH_3\overset{}{C}-CHCH_2(CH_2)_6CH_3 \xrightarrow[\text{3. }\sim H:]{\text{1. }\sim CH_3:}$$
$$\underset{CH_3}{|} \qquad\qquad\qquad\qquad\qquad\quad \underset{H_3C}{|}\overset{+}{\underset{CH_3}{|}} \qquad \text{2. }\sim H:$$

$$\underset{CH_3}{\overset{CH_3}{|}} \qquad\qquad\qquad\qquad\qquad \underset{CH_3}{\overset{CH_3}{|}}$$
$$CH_3\overset{}{C}CH_2\overset{+}{C}H(CH_2)_6CH_3 \xrightarrow{\beta\text{-scission}} CH_3\overset{}{C}{}^+ \quad + \quad CH_2=CH(CH_2)_6CH_3$$
$$\underset{CH_3}{|} \qquad\qquad\qquad\qquad\qquad \underset{CH_3}{|}$$

$$-H^+ \nearrow \qquad \searrow RH \quad R^+$$

$$\boxed{\underset{CH_3}{\overset{}{\underset{|}{CH_3=CCH_3}}}} \qquad \boxed{(CH_3)_3CH}$$

Scheme 1.9. Mechanisms of cracking of hexadecane.

The catalytic cracking of straight-chain alkenes gives products similar to those from the catalytic cracking n-alkanes, although their rate of cracking is faster. The initial cation formation can occur through a direct addition of a proton derived from the Brønsted acid sites of SiO_2–Al_2O_3 to the double bond of the alkene.

Catalytic cracking as compared with thermal cracking affords a very small yield of methane and C_2 hydrocarbons. This can be explained by the fact that through β-scission the more stable secondary and tertiary alkyl cations are preferentially produced, and therefore methane, ethane, and ethylene, which require the generation of the least stable primary cations, are not formed.

b. Alkylaromatic Hydrocarbons. The catalytic cracking of aromatic hydrocarbons proceeds with a high degree of specificity. The aromatic nucleus is virtually inert to fragmentation, and the cracking reaction, in terms of carbon–carbon bond breaking, is confined to substituted alkyl and cycloalkyl groups. In addition to cracking, skeletal isomerization of the substituent alkyl groups, such as reversible isomerization of xylenes to ethylbenzene (Section II,C,3), can occur.

The main characteristic of the cracking of alkylbenzenes is protonation of the nucleus, chiefly at the point of attachment of the substituent to the ring, followed by detachment of the alkyl group as a cation with ultimate formation of benzene and an olefin. The ease of removal of the alkyl group is consistent with the order of stability of the alkyl cations produced as exemplified by $(CH_3)_3C^+ > H_5C_2{-}CH^+{-}CH_3 \gg n\text{-}C_4H_9^+$.

$$C_nH_{2n} + H^+$$
(olefin)

The cleavage of the *tert*-butyl group over SiO_2–Al_2O_3 thus occurs at 280°C, whereas *sec*-butylbenzene or cumene requires temperatures of 380°C. The relative ease of cracking of the respective alkylbenzenes also depends on the intrinsic acidities of the cracking catalyst. The relative rates of dealkylation of cumene were used to determine the relative acidities of silica–alumina catalysts as a function of their temperature of calcination, and as to their content of alumina.

c. Secondary Reactions. A number of secondary reactions accompany catalytic cracking of hydrocarbons, and the most important of these reactions are listed in Table 1.27. Most of the secondary reactions are of the type discussed in previous sections and occur by a carbonium ion mechanism. Aromatic hydrocarbons are formed through a cationic dehydrogenation of polymethylated cyclohexanes. The latter are produced via skeletal isomerization of alkylcyclopentanes originally present in petroleum. The transformation of alkylcyclohexanes to aromatic hydrocarbons can

TABLE 1.27

Secondary Reactions in Catalytic Cracking[a]

Reacting bonds	Secondary reactions	Amenable hydrocarbon types
C—H	Double-bond shift	Olefins
	Double-bond geometrical isomerization	Olefins
	Hydrogen transfer	To olefins, from diverse donors
	Dehydrogenation to aromatics	Cyclohexane naphthenes
	Self-saturation	Olefins
Both C—C and C—H	Polymerization	Olefins
	Condensation	Aromatics
	Aromatization	Olefins
	Skeletal isomerization	Olefins; bicyclic naphthenes and cycloalkylaromatics

[a] From Greensfelder (1955, p. 156).

proceed by a series of cationic reactions as exemplified by the following steps:

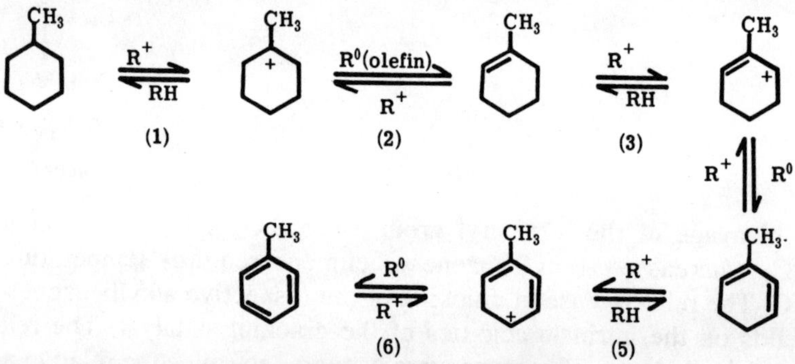

The first step in the reaction, involving the removal of a hydride from methylcyclohexane to form methylcyclohexyl cation, is the most difficult one and is probably rate determining. However, the removal of a hydride from cyclohexene and even more so from a cyclohexadiene is greatly facilitated by the relative stabilities of the resulting allylic and cyclohexadienyl cations (steps 3 and 5, respectively).

Aromatization may also occur via cyclization of conjugated dienyl cations to form alkylcyclopentadienes, which in the presence of SiO_2–Al_2O_3 can undergo skeletal isomerization, leading to the formation of aromatics (Sections II,C,3 and III,C).

Transmethylation is another type of reaction that takes place over silica–alumina catalysts; xylenes undergo disproportionation under cracking conditions to afford toluene and trimethylbenzenes. The mechanism of this reaction is similar to that of transalkylation and involves an intermediate formation of diarylmethanes (Section V,H,3).

$$2\ C_6H_4(CH_3)_2 \longrightarrow C_6H_5CH_3 + C_6H_3(CH_3)_3$$

3. "Coke" Formation

The formation of carbonaceous residue that remains on the catalyst is a characteristic part of all catalytic cracking reactions. This residue is not pure carbon but is composed of a mixture of highly unsaturated hydrocarbons with the average atomic ratio of hydrogen to carbon varying from 0.3 to 1.0 or even higher in carbon, depending on the conditions of formation of the "coke."

"Coke" formation is a complex reaction, and little fundamental information has been published regarding the steps involved. However, on the basis of the hydrocarbon conversion reaction in the presence of acids, it is not unreasonable to suggest that at least part of the carbonaceous material deposited on the catalyst may result from conjunct polymerization of olefins accompanying the cracking reaction. This polymerization may generate highly unsaturated cyclic hydrocarbons that could form complexes with the acid sites of the catalyst, as discussed in Section III,C. A portion of the "coke" might also result from dehydrogenation of polynuclear compounds produced from the oligomerization of cyclopentenes, as described in Section III,E. The highly unsaturated hydrocarbons thus produced, which are rich in electrons, could act as organic bases, neutralize the acid sites of the silica–alumina, and render the catalyst inactive.

Commercially, the catalyst is regenerated at 600°–650°C with air, part of the carbon is removed as CO, and the "regenerated" catalyst, which may still contain 0.2–0.4 wt % of residual coke, is returned to the reactor. The regeneration of the catalyst provides a source of heat for the endothermic cracking reactions.

C. Zeolite-Catalyzed Cracking

1. Introduction

In 1964 Plank et al. (1964) reported a new class of acidic catalysts useful in the catalytic cracking of hydrocarbons. These catalysts consisted of crystalline aluminosilicates, zeolites. These zeolites were modified by replacing a portion of the original sodium ions with hydrogen ions and by replacing the major part of the remaining sodium ions with polyvalent ions or ions of rare earth metals. The resulting catalysts have an activity for gas oil cracking that is more than 100 times as great as that of an amorphous silica–alumina catalyst. In addition, up to 20% more gasoline with a higher octane number has been obtained than with conventional silica–alumina catalysts.

This discovery resulted in the rapid conversion of catalytic cracking by amorphous silica–aluminas to catalytic cracking by zeolites and stimulated much basic research on the application of zeolites to hydrocarbon conversion reactions. The practical significance of this development can be judged from an estimate that catalytic cracking with zeolites rather than SiO_2–Al_2O_3 in the United States alone saved over 200 million barrels of crude oil in 1977.

2. Description of Zeolite Catalysts

Zeolites, whether synthesized or naturally occurring, are crystalline aluminosilicates. Structurally, they comprise a framework based on a three-dimensional network of SiO_4 and AlO_4 tetrahedrons linked together at the corners through common oxygen ions.

In typical syntheses, sodium zeolites result when gels, which are formed from mixing aqueous solutions of sodium aluminate, sodium silicate, and sodium hydroxide, are allowed to crystallize at 25°–100°C. Sodium ions are required to maintain molecular neutrality since each AlO_4 tetrahedron is associated with a unit negative charge. Zeolites can undergo base exchange with a variety of metallic cations and with protons, a procedure that substantially modifies their properties.

The key structural feature of zeolites is the narrow, uniform, continuous channel system which is formed after the water has been eliminated by heating and evacuation. The cations stay near the anions in the structure. The geometry of the individual channel and cavity system is characteristic of the individual zeolite. Entrance to the intracrystalline volume is through orifices located periodically throughout the structure. Access to the intrazeolitic environment is limited to molecules with dimensions less than a certain critical size. The zeolitic orifices have dimensions ranging from 3 to 9 Å.

a. Zeolite Structures. The framework structures of zeolites are composed of a combination of truncated octahedra of which the simplest unit

Fig. 1.5. Truncated octahedron or sodalite unit. Vertices represent Si of Al atoms; lines represent oxygens.

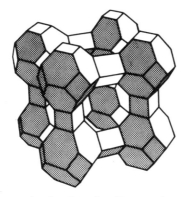

Fig. 1.6. Line drawing of zeolite type A structure.

is sodalite (Fig. 1.5). Each sodalite unit can be linked to its neighbor by four bridging oxygen ions to form zeolite A (Fig. 1.6). This configuration gives a spherical internal cavity with a diameter of 11.4 Å (α cage), which is entered through six circular orifices with a diameter of 4.2 Å. The sodalite units themselves enclose a second set of small cavities (β cage) of 6.6-Å internal cavities. The unit cell composition of zeolite A is $Na_{12}(AlO_2)_{12}\cdot(SiO_2)_{12}\cdot27H_2O$.

Sodalite units can also be linked by hexagonal faces with six bridge oxygen ions to afford faujasite-type zeolites (Fig. 1.7). This results in a series of wide, nearly spherical cavities, each of which opens through a window of 8–9 Å diameter into four identical cavities of 12 Å diameter. The unit cell of faujasite can exist as zeolites X and Y. The unit cell of zeolite X is $Na_{86}(AlO_2)_{86}(SiO_2)_{106}\cdot264H_2O$. The basic framework of zeolite Y is the same as that of zeolite X, but the Si/Al ratio is higher, ranging from 1.5 to 3.0.

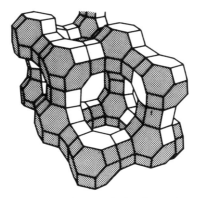

Fig. 1.7. Line drawing of faujasite structure.

In contrast to zeolite type A and faujasite with three-dimensional pore structures that contain cages, mordenite has only a two-dimensional tubular pore system (Fig. 1.8). The crystal structure consists of chains of four- and five-membered rings of Si and Al tetrahedra. The tubes that are formed between the chains by rings of 12 tetrahedra have a major and minor diameter of 6.95 and 5.81 Å, respectively. The unit cell of mineral mordenite is $Na_8(AlO_2)_8(SiO_2)_4 \cdot 24H_2O$.

b. Ion-Exchange Capabilities. Owing to the isomorphic substitution of silicon by aluminum, the three-dimensional oxygen framework contains an excess of negative charge compensated by sodium ions. These ions can be exchanged by cations of different nature and valency or by protons, which permits the introduction into zeolites of catalytically important elements.

Catalytic activity in ion-exchanged faujasites is influenced by cation type, including size and charge, Si/Al ratio, and the presence of proton donors. For a given cation the catalytic activity for a Y-type faujasite is much greater than that for an X-type, and this was attributed to a greater electrostatic field strength of the latter. The catalytic activity of cation-exchanged zeolites depends on the temperature of activation and method of removing water. The optimal temperature is about 500°–625°C; at 700°–750°C the cracking activity of the catalyst deteriorates. This is exemplified by the data in Table 1.28, which were obtained when cumene was used as a probe to relate the cracking abilities of zeolites as a function of exchanged cations.

Hydrogen ion-exchanged zeolites show high activity in hydrocarbon conversion reactions. They are usually produced by heating NH_4^+-exchanged X or Y zeolite at 250°–400°C in an inert atmosphere. This leads to

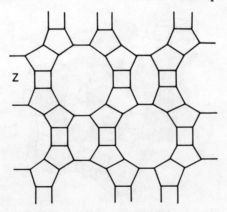

Fig. 1.8. Section through the mordenite structure (Z) perpendicular to the pores.

TABLE 1.28

Cracking of Cumene over Zeolites Y[a,b]

| Catalyst | Activation | | Cumene conversion (%) |
	Temp (°C)	Time (hr)	
NaY	550	16	<0.1
CaY[c]	550	16	>97
CaY[c]	710	2	2
LaY[d]	550	16	>98
LaY[d]	700	3	60
LaY[d]	750	16	20

[a] From Rabo and Poutsma (1971).
[b] Conditions: Temp, 325°C; 0.0157 mol cumene was passed per hour plus 0.39 mol/hr N_2 fed; 16 gm catalyst.
[c] CaY: Si/Al = 2.42; Ca(II) exchange = 83%; equivalent cation/Al = 0.99.
[d] LaY: Si/Al = 2.46; La(III) exchange = 86%; equivalent cation/Al = 1.03.

stoichiometric evolution of NH_3 and to production of a hydrogen ion-exchanged zeolite commonly called HX or HY zeolite, which cannot be achieved by direct ion exchange because of the sensitivity of the zeolite structure to aqueous mineral acids.

3. Catalytic Cracking over Zeolites

Cation- and hydrogen-exchanged zeolites act as acids, and their catalytic behavior is similar to that of silica–alumina catalysts. However, the useful catalytic properties of zeolites which distinguish them from the conventional silica–alumina cracking catalysts hinge mainly on the regular crystalline structure and uniform pore size of zeolites, and since the major catalytic activity occurs within their pores and cavities, only hydrocarbon molecules below a certain dimension can react.

The cracking of gas oil over zeolite catalysts leads to a gasoline richer in paraffins and aromatics than does cracking with silica–alumina catalysts. This can be explained by a more favorable hydrogen transfer reaction over zeolites.

$$\text{Olefins + naphthenes} \longrightarrow \text{paraffins + aromatics}$$

The rate of cracking over modified zeolites is much higher than that over SiO_2–Al_2O_3 catalysts. This is well illustrated by the cracking of hexane over catalysts derived from the parent sodium zeolites in which sodium was exchanged with polyvalent metal ions, rare earths, or ammonium followed by thermal treatment (Table 1.29). Low sodium content REHx (rare earths hydrogen exchanged)-faujasite and H-mordenite zeolite type cracking catalysts were obtained with relative activities more than 10^4 times the activities of conventional $SiO_2-Al_2O_3$ cracking catalysts. This greatly enhanced activity cannot, however, be completely utilized commercially since a catalytic cracking process is a balanced cyclic operation. An integral part of the process is the deposition of "coke" on the catalyst, and the rate of the cracking cannot be increased beyond corresponding "coke" regeneration rates. For that reason the activity of the catalyst is modified by dispersing the zeolite component in a synthetic or natural $SiO_2-Al_2O_3$ matrix. The most important function of the matrix is creating the heat sink in the regenerator, thus protecting the heat-sensitive zeolite.

The product distribution from the cracking of some representative C_{12} and C_{18} hydrocarbons over REHx (rare earths hydrogen exchanged) zeolite is given in Table 1.30. The rate of cracking of dodecene is much greater than that of dodecane, which is in line with the proposed carbonium ion mechanism. The necessary cation formation can be produced by direct protonation of the double bond of the olefin, and in addition the withdrawal of a hydride from an allylic carbon of an olefin occurs more readily than the withdrawal of hydride from a normal alkane. The product distribution of dodecene and that of the naphthenes are similar, which is an indication that the first step in the cracking of cycloparaffins is the cleavage of the ring to form olefins. In conformity with the cationic mechanism of cracking, dodecylbenzene produces benzene as one of the principal products of reaction.

The most significant improvement in the use of zeolite catalysts over

TABLE 1.29

Cracking of n-Hexane over Cation-Exchanged Zeolites[a]

Zeolite	SiO_2/Al_2O_3 molar ratio	Na (wt %)	Relative activity[b]
Silica–alumina	15.29	0	1
Ca-Faujasite	2.58	7.70	1.1
H,RE-Faujasite[c]	2.06	0.22	10,000
H-Faujasite	5.57	0.40	6,400
H-Mordenite	10.16	0.1	10,000
	10.16	0.3	2,500

[a] From Miale et al. (1966).
[b] After 5 min on stream.
[c] RE denotes rare earth.

TABLE 1.30

Comparison of Product Distribution from Cracking of Representative C_{12} and C_{18} Hydrocarbons[a,b]

	Hydrocarbon				
	Dodecane	Dodecene	Cyclododecane	Dodecylcyclohexane	Dodecylbenzene
LHSV	650	1300	1300	1300	650
Cracking conversion (%)	9.0	14.8	8.7	19.9	16.0
Isomerization conversion (%)	—	25.3	3.0	2.3[d]	0.0
Cracking rate constant k (hr⁻¹)	500	1750	1020	1890	765
C on catalyst (wt %)	1.3	7.9	0.8	1.2	1.0
C_3	16.9	9.2	8.3	6.9	12.6
C_4	31.0	17.7	20.7	16.4	26.4
C_5	22.6	19.6	17.5	15.1	24.1
C_6	17.6	18.2	16.5	15.9	18.6
C_7	7.7	16.2	15.1	14.7	9.1
C_8	3.4	12.3	13.9	13.4	4.7
C_9	1.0	5.4	6.7	10.5	2.2
C_{10}	—	1.4	1.0	7.1	2.3
C_{11} or i-C_{12}	—	—	0.5	4.4	5.9
C_{12}	—	—	—	2.8	9.8
C_{13}	—	—	—	1.0	1.7
C_{14}	—	—	—	0.4	2.5
Benzene	—	—	—	2.7	21.8
Toluene	—	—	—	—	6.0
C_8 aromatic	—	—	—	—	3.8

(Bracketed totals: Dodecylcyclohexane column C_7–C_{10} group = 88.7; Dodecylbenzene column C_6–C_{10} group = 48.4)

[a] From Nace (1969).
[b] Catalyst, REHX; temp 482°C, 2-min instantaneous samples.
[c] i-C_{12} and n-C_{12}.
[d] i-C_{18} and n-C_{18}.

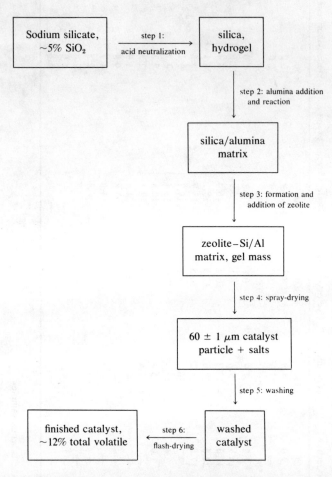

Fig. 1.9. Flow diagram of the preparation of (silica–alumina gel)-based zeolite catalyst (from Magee and Blazek, 1977, p. 617).

that of conventional silica–alumina cracking catalysts is not so much in the increase in catalytic activity as in the selectivity. The zeolites give a higher yield of products from the cracking of paraffins in the C_5 to C_{10} range and fewer in the C_3 to C_4 range. The average molecular weight of products made with RHEX zeolite is always higher than with SiO_2–Al_2O_3.

The fact that zeolite has better selectivity than SiO_2–Al_2O_3 in the gasoline range is ascribed to its greater hydrogen transfer activity relative to the β-carbon scission reaction of the cations. This can be explained by the regular micropore structure of the zeolites, which allows the gas oil molecules to remain in a condensed or pseudoliquid state. The cations, once formed, are in proximity to the adsorbed gas oil and undergo many more collisions with gas oil molecules than they would under strictly vapor-phase conditions. Therefore, it is more likely that the activated ion abstracts hydride from another gas oil molecule and is converted to a gasoline-range hydrocarbon before it unzips by successive β-fissions to light gases.

4. Preparation of Commerical Cracking Catalyst

A typical preparation of a commercial cracking catalyst, in which the zeolite component is added to an amorphous mass of silica–alumina gel that contains kaolin clay diluent, is presented in Fig. 1.9.

VII. REACTIONS CATALYZED BY STRONG ACIDS (SUPERACIDS)

A. Introduction

The interest in strong acids (superacids) as catalysts for hydrocarbon conversion reactions arose with the discovery in 1969 by G. A. Olah that a carbon–hydrogen bond of ethane, neopentane, and even the extraordinarily unreactive methane undergoes ionization in the presence of SbF_5/HSO_3F at 50°C. Methane and ethane underwent hydrogen exchange and polycondensation to produce tertiary alkyl carbonium ions, whereas neopentane underwent protolytic fragmentation and hydrogen abstraction followed by rearrangement to *tert*-pentyl cation. Subsequent study has demonstrated that in the presence of strong acids it is possible to achieve catalytic alkylation of methane and ethane with alkenes.

Most acid catalysts that are used in conjunction with hydrocarbon conversion, as described in previous sections, have an acidic strength equal to $-H_0$ of 11 or lower, with the exception of $AlCl_3/HCl$ or $AlBr_3/HBr$, the acidity function $-H_0$ of which is assumed to be around 17, equal to that of BF_3/HF. The catalytic capabilities of acids depend greatly on their acidities. Using the isomerization of alkanes as an example (Section II), it was observed that $AlCl_3/HCl$ and $AlBr_3/HBr$ act as catalysts for the isomerization of butane to isobutane, whereas HF or H_2SO_4 do not. It is therefore not surprising that by the use of strong acids with $-H_0 > 17$, such as TaF_5/HF, SbF_5/HF, or SbF_5/HSO_3F, which are from 10^2 to 10^6 times stronger than AlX_3/HX, a new magnitude to catalytic reactions was uncovered.

B. Reactions Catalyzed by Strong Acids

1. Polycondensation of Methane in the Presence of SbF_5/HSO_3F

The most striking observation about the strong acid SbF_5/HSO_3F is that in its presence at 80°–150°C methane, which is considered to be the most stable alkane, undergoes hydrogen exchange and self-condensation to a trimethyl cation. Hydrogen gas is liberated in these reactions but in small amounts.

The above-mentioned reactions were explained by assuming that in strongly acidic solutions methane behaves as a base and protonates to CH_5^+, methanium cation, which can lose hydrogen to form the extremely active methyl cation CH_3^+. The latter then reacts with excess methane to form ethanium ion $C_2H_7^+$, which loses H_2 to form $C_2H_5^+$; the latter ion through proton elimination gives ethylene.

Methanium ion should not be considered a carbocation, since eight electrons surround the C, as in methane; hence, the (+) charge must be

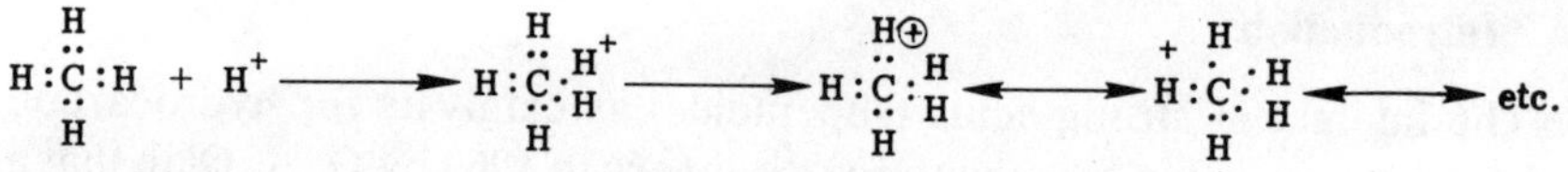

shared among the five hydrogens. Ethylene in turn can react with methyl or ethyl cation to give ultimately higher tertiary cations (Scheme 1.10).

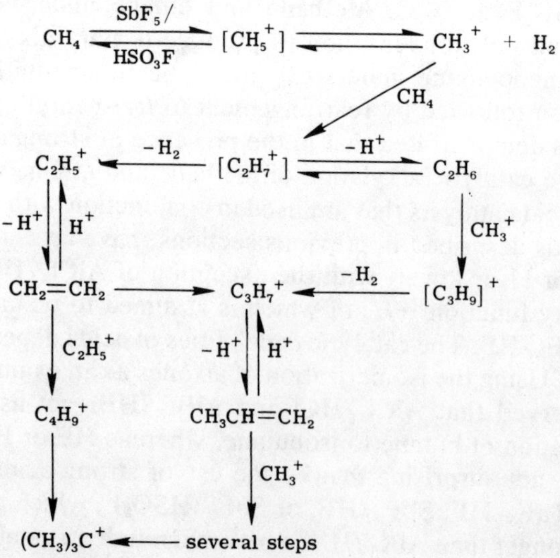

Scheme 1.10. Polycondensation of methane in the presence of SbF_5/HSO_3F (Olah *et al.*, 1969).

The protonation of methane and related alkanes is explained as occurring through the formation of a transition state containing pentacoordinated carbon, and for simplicity this is represented as triangular three-center bonds with dashed lines. Some of the steps in the condensation of methane as given in Scheme 1.10 are represented as follows:

$$CH_4 \xrightarrow{H^+} \left[H_3C\text{---}\overset{H}{\underset{H}{\diagdown}} \right]^+ \text{ or } \left[H_3C\diagdown\overset{H}{\underset{H}{\vdots}} \right]^+ \longrightarrow H_3C^+ + H_2$$

It was also suggested that the formation of cations from saturated hydrocarbons in superacid media is due to powerful oxidizing species of the catalyst such as SO_3 or SbF_5 (Herlem, 1977).

The formation of ethane and higher hydrocarbons is suggested to occur as follows:

$$CH_4 + CH_3^+ \longrightarrow \left[H_3C\text{---}\overset{H}{\underset{CH_3}{\diagdown}} \right]^+ \longrightarrow C_2H_6 + H^+$$

$$\downarrow CH_3^+$$

$$C_3H_8 + H^+ \longleftarrow \left[CH_3CH_2\text{---}\overset{CH_3}{\underset{H}{\diagdown}} \right]^+$$

$$\downarrow$$

$$CH_3CH_2^+ + CH_4$$

2. Cleavage and Rearrangement of Neopentane

Neopentane, 2,2-dimethylpropane, reacts in SbF_5/HSO_3F at room temperature to produce methane and *tert*-butyl cation. However, in the presence of $1:1$ SbF_5/HSO_3F diluted with SO_2ClF, a new pathway is followed, and *tert*-pentyl cation is produced.

$$CH_3\text{-}\overset{\overset{CH_3}{|}}{\underset{\underset{CH_3}{|}}{C}}\text{-}CH_3 \underset{}{\overset{SbF_5}{\underset{HSO_3F}{\rightleftharpoons}}} \left[(CH_3)_3C\text{---}\overset{CH_3}{\underset{H}{\diagdown}} \right]^+ \rightleftharpoons (CH_3)_3C^+ + CH_4$$

$$\Big\updownarrow SbF_5/HSO_3F/SO_2ClF$$

$$\left[(CH_3)_3CCH_2\text{---}\overset{H}{\underset{H}{\diagdown}} \right]^+ \xrightarrow{-H_2} (CH_3)_3CCH_2^+ \xrightarrow[\text{fast}]{\sim CH_3:} (CH_3)_2\overset{+}{C}CH_2CH_3$$

It should be recalled that neopentane remains unaltered in the presence of weaker acids such as $AlBr_3/HBr$ or $AlCl_3/HCl$.

3. Alkylation of Alkanes

In the presence of the conventional acid catalysts, n-alkanes do not undergo reaction with alkenes (Section IV,B,3). However, by the use of a strong acid it is possible to extend the alkylation of alkanes to include C_1 to C_3 alkanes. The catalyst TaF_5/HF was found to be useful for this reaction. The positive nature of the tantalum atom, enhanced by five very electronegative fluorines, allows the tantalum to accept an anion from HF and generate a proton active enough to protonate the weakly basic HF solvent:

$$2\,HF + TaF_5 \rightleftharpoons HF + H \overset{\delta+}{\cdots} F \cdots TaF_5 \rightleftharpoons H_2F^+ + TaF_6^-$$

The TaF_6 is so weakly basic as to be almost inert, which thus allows the formation of stable carbocations in hydrocarbon reactions. In addition TaF_5 is thermally stable and resistant to reduction by molecular hydrogen and hydrocarbon ions, in sharp contrast to SbF_5, which is reduced to SbF_3.

a. Methane–Ethylene. Selective interaction of the title hydrocarbons to produce propane can be achieved by passing a CH_4–C_2H_4 (85.9–11.4%) gas mixture at a rate of 42 cm^3/min through a 300-cm^3 stainless steel autoclave containing 50 cm^3 of 1:10 TaF_5/HF stirred at 1000 rpm at 40°C and maintained at ~2 atm.

b. Ethane–Ethylene. Ethylene, 17.9 wt %, was allowed to react with ethane under conditions specified in Section a above to form butane as the

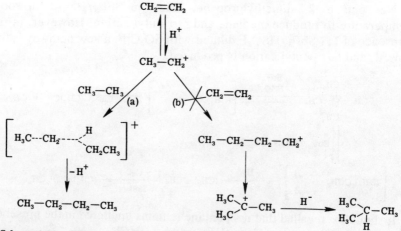

Scheme 1.11. Ethane–ethylene alkylation catalyzed by TaF_5/HF (Siskin *et al.*, 1977).

only product. The absence of isobutane means that the ethyl cation is alkylating a primary ethane position and that there is no free primary butyl
cation formed and thus provides evidence for the existence of pentacoordinated carbon. n-Butyl cation would lead to the production of the more
stable $tert$-butyl cation. The mechanism of the reaction is outlined in
Scheme 1.11.

    c. Methane–Propene. Propene, 3.4%, reacts with methane, 96.6%, at
40°C in the presence of 1:10 TaF$_5$/HF to form isobutane with over 60%
selectivity.

$$CH_3CH{=}CH_2 \; \overset{H^+}{\underset{}{\rightleftharpoons}} \; CH_3\overset{+}{C}HCH_3 \; \xrightarrow{CH_4} \; \left[H_3C\text{----}\overset{H}{\underset{H}{\overset{|}{C}}}\overset{CH_3}{\underset{CH_3}{\overset{}{C}}} \right]^+$$

$$\downarrow{-H^+}$$

$$(CH_3)_3CH$$

4. Hydrogenation of Aromatic Hydrocarbons

    Strong acids offer a novel method of hydrogenating aromatic hydrocar-
bons through protonation and reaction of the resulting carbonium ions
with hydrogen. Molecular hydrogen is not directly involved in this reaction. For the reaction to occur effectively either saturated hydrocarbons
containing a tertiary carbon atom must be present, or hydrocarbons with
a tertiary carbon atom must be produced under the conditions of the reaction. Pentane is a useful solvent and participant in the hydrogenation reaction.

    a. Benzene. The hydrogenation of benzene was carried out in a stirred
300-ml stainless steel autoclave into which were introduced 0.4 mol benzene, 0.9 mol pentane, 0.2 mol TaF$_5$, and 2–2.3 mol of HF. Enough hydrogen was added to give a final pressure of about 33 atm. The solution
was stirred and heated to 50°C. As hydrogen was consumed it was replen-
ished.

    The study of the hydrogenation of benzene has revealed that before an
uptake of hydrogen occurs pentane isomerizes to isopentane. Then the
protonated benzene abstracts a hydride ion from isopentane to form cyclohexadiene and $tert$-pentyl cation. A similar type of hydrogenation also
occurs when AlBr$_3$/HBr, AlCl$_3$/HCl, or SbF$_5$/HF is used as the catalyst.

    The paths involved in hydrogenation of benzene are presented in
Scheme 1.12. Cyclohexadiene, cyclohexene, and methylcyclopentane are
products.

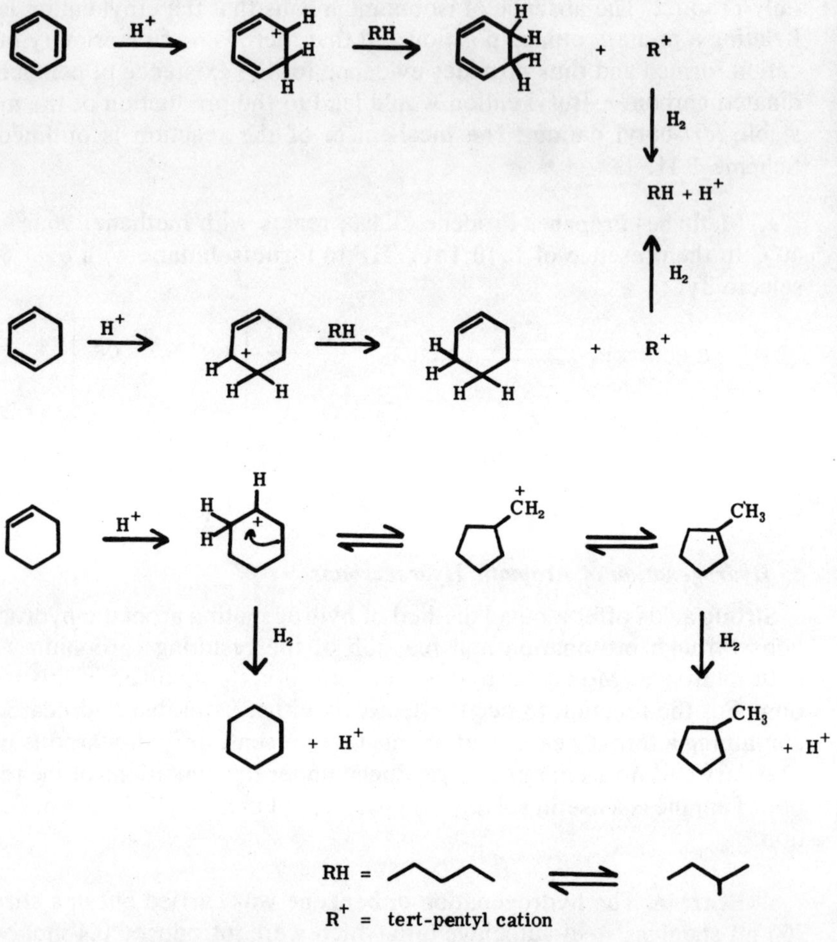

Scheme 1.12. Mechanism of acid-catalyzed hydrogenation of benzene (Wristers, 1976, 1977).

b. Polynuclear Aromatic Hydrocarbons. The hydrogenation of naphthalene at 50°C under conditions specified for benzene takes place rapidly to form tetrahydronaphthalene. On longer contact with TaF_5/HF, tetrahydronaphthalene is slowly hydrogenated to decahydronaphthalene or converted to benzene and isobutane.

Anthracene and phenanthrene undergo a more complex reaction. The octahydro intermediates produced undergo further reaction to form benzene and dimethylcyclohexanes.

VIII. REFORMING AND HYDROCRACKING

A. Introduction

Reforming and hydrocracking are two major catalytic processes used in the petroleum industry. Reforming is used to upgrade the octane number of gasoline in the presence of hydrogen. Typical petroleum naphthas obtained from the distillation of virgin oil and other refinery feed stock, boiling within gasoline range, contain a large proportion of unbranched alkanes with an octane number of less than 50. The purpose of reforming is to rearrange these hydrocarbons to those of higher octane number with-

TABLE 1.31

Octane Number of Pure Hydrocarbons[a,b]

Hydrocarbon	Octane number (0.N)	Hydrocarbon	Octane number (0.N)
Paraffins			
Isobutane	122	n-Heptane	0
n-Pentane	62	2-Methylhexane	41
Isopentane	99	3-Methylhexane	56
n-Hexane	19	2,2-Dimethylpentane	89
2-Methylpentane	83	2,3-Dimethylpentane	87
3-Methylpentane	86	2,4-Dimethylpentane	77
2,3-Dimethylbutane	96	2,2,3-Trimethylbutane	113
Naphthenes			
Methylcyclopentane	107	1,2-Dimethylcyclohexane	85
1,3-Dimethylcyclopentane	98	(cis or trans)	
(cis)		1,3-Dimethylcyclohexane	67
1,3-Dimethylcyclopentane	91	(cis)	
(trans)		1,1,3-Trimethylcyclohexane	85
Cyclohexane	110	1,3,5-Trimethylcyclohexane	60
Methylcyclohexane	104	(cis)	
Ethylcyclohexane	43	Isopropylcyclohexane	62
Aromatics			
Benzene	99	Ethylbenzene	124
Toluene	124	Isopropylbenzene	132
o-Xylene	120	1-Methyl-2-ethylbenzene	125
m-Xylene	145	1-Methyl-3-ethylbenzene	162
p-Xylene	146	1-Methyl-4-ethylbenzene	155
		1,2,4-Trimethylbenzene	171

[a] From American Petroleum Institute Research Project 45, Sixteenth Annual Report (1954) as tabulated in Ciapetta et al. (1958).
[b] Blending research octane number (clear).

out incurring substantial losses. The octane numbers of some representative hydrocarbons are listed in Table 1.31.

In most catalytic reforming processes the reactions are promoted by fixed beds of platinum deposited on acidic aluminas. The main reactions are isomerization, dehydrocyclization, dehydrogenation, and hydrocracking. Alkanes undergo skeletal isomerization and dehydrocyclization to naphthenes, and the C_6 cyclic compounds undergo rapid dehydrogenation to arenes. Alkanes also are hydrocracked to lower homologues. Minor reactions include hydrogenation of arenes to naphthenes, hydrogenolysis of naphthenes to alkanes, and dealkylation of alkylnaphthenes.

Under reforming conditions alkylcyclopentanes and alkylcyclohexanes undergo skeletal and positional isomerization and dehydrogenation to produce aromatic hydrocarbons with the liberation of hydrogen. The arenes with relatively high octane numbers are valuable components of gasoline, or they can be extracted and used as solvents or converted to petrochemicals used in the manufacture of plastics, synthetic rubber, fibers, and detergents.

Hydrocracking, which is practiced extensively in petroleum refining, converts relatively high boiling hydrocarbons to lower boiling products that are used in the production of high-quality gasoline, jet fuel for airplanes, and lubricants.

Reforming and hydrocracking processes are carried out in the presence of dual-function catalysts consisting of hydrogenation–dehydrogenation components deposited on acidic supports and in an atmosphere of hydrogen under pressure. In hydrocracking there is an overall consumption of hydrogen, whereas in reforming hydrogen is produced from the dehydrogenation of saturated hydrocarbons, mainly cycloalkanes, to arenes.

B. Reforming

1. Introduction

Reforming was developed by Vladimir Haensel in the mid-1940's using as catalyst platinum deposited on acidic alumina. In 1978 in the United States alone more than 3 million barrels of reformate gasoline was produced daily.

In reforming hydrocarbons boiling within gasoline range are reconstructed without their carbon numbers being drastically changed. The reactions, which involve isomerization, hydrogenation, dehydrogenation, and dehydrocyclization, lead to an improvement in the fuel quality of the charged stock. At higher temperature or in the presence of more acidic catalysts hydrocracking may accompany reforming.

A variety of catalysts are active for reforming, including platinum–alumina–combined halogen, nickel, cobalt or platinum supported on silica-alumina, and oxides, such as molybdenum oxide and tungsten oxide deposited on silica–alumina. The most effective catalysts are composed of platinum on acidic alumina in combination with another metal, such as Re or Pd. Rhenium addition improves the thermal stability of 0.5% Pt/Al_2O_3. Although the function of Re is not completely understood, it is surmised that Re forms an alloy with Pt and inhibits the growth of the Pt crystallites. The weights of Re and Pt in the catalyst are roughly equal.

2. Platinum-Containing Catalysts

Reforming processes using a platinum-containing catalyst have the widest current application. The catalyst consists of about 0.6% Pt and 0.2–0.6% Re, well dispersed on η- or γ-alumina. This is a dual-function catalyst. Platinum acts as the hydrogenation–dehydrogenation component, and alumina as the acidic component that is principally responsible for the skeletal rearrangement of the hydrocarbons. The acidic properties of the aluminas are greatly enhanced through the incorporation of small amounts, about 0.2%, of Cl^- or F^-, in the form of alkyl or hydrogen halides.

The addition of small amounts of Re to Pt/Al_2O_3 results in major improvements in catalyst performance. These include a smaller decline in yield of product with higher temperature, a lower deactivation rate for the catalyst, and excellent catalyst stability during regeneration. The length of the processing period between regenerations can be greatly extended. The activity of the catalyst declines only one-fifth of that of Pt/Al_2O_3 without the Re, and it was reported that after 3–4 months of use the catalyst loses less than 3–5% activity.

3. Preparation of Catalysts and Processing Conditions

The reforming catalyst Pt/Al_2O_3 can be conveniently prepared by impregnating η- or γ-alumina with a saturated aqueous solution of chloroplatinic acid. The particles are drained, dried, and calcined in air at about 500°C to convert the metal salt to a metal oxide. The oxide is reduced by hydrogen at 400°–450°C to the metal.

Similarly, $Pt/Re/Al_2O_3$ is prepared by impregnating alumina with solutions of chloroplatinic acid and ammonium perrhenate. A typical concentration of the platinum component is 0.6% by weight and that of Re is 0.2–0.6% by weight.

Platinum reforming catalysts are susceptible to irreversible poisoning. Impurities in the naphtha feed stock must be rigorously controlled. Sulfur poisons the metal function of the catalyst and must be maintained at concentrations of less than 1 ppm. Organic nitrogen compounds in the feed stock must be avoided because they are converted to ammonia and poison the acid function of alumina in the catalyst; their concentration must be below 2 ppm.

The processing conditions can be adjusted according to the desirability of the product to be derived. Reaction temperatures are selected to balance the advantages of catalytic activity and the disadvantage of increased deactivation rate as the operating temperature is increased. Temperatures between 470° and 530°C are most suitable. The molar ratio of hydrogen to hydrocarbon may vary from 3 to 10, and ratios of 5 to 8 are preferred. The LHSV ranges from 0.9 to 5, with 1 to 2 being most commonly used. Pressures may vary from 10 to 50 atm.

A low-pressure operation, 10–20 atm, results in a high degree of conversion to aromatic hydrocarbons, a small amount of hydrocracking conversion, and the production of a large amount of hydrogen. It also results in a rapid deactivation of the catalyst and requires frequent regeneration. A high-pressure reaction, 30–50 atm, produces much hydrocracking and low yields of aromatic hydrocarbons and prolongs the activity of the catalyst.

4. Mechanism

a. **Kinetic Evaluation.** The rates of the primary steps in re-forming were obtained for individual hydrocarbons containing six and seven carbon atoms (Table 1.32). This was achieved by limiting the reactions to low conversions in order to minimize secondary reactions.

The main reactions of hexane and heptane do not proceed at the same rate. The rate of isomerization of hexane is more than twice that of its

TABLE 1.32

Reaction Rates in Reforming[a,b]

Reaction type	Hexane	Heptane	Methylcyclo-pentane	· Methylcyclo-hexane
Isomerization				
n-Paraffins to isoparaffins	0.12	0.16	—	—
Five- to six-carbon rings	—	—	0.13	—
Cyclization	0.012	0.05	—	—
Hydrodecyclization	—	—	0.09	—
Hydrocracking	0.05	0.06	—	—
Dehydrogenation	—	—	—	1.4

[a] From Krane et al. (1959) as cited in Gates et al. (1979).
[b] Data expressed as gram-moles per hour per gram catalyst.

hydrocracking, and cyclization proceeds at a rate one-fourth that of hydrocracking. Thus, conversion to aromatic hydrocarbons is low. Heptane, on the other hand, isomerizes and hydrocracks at the same rate; the rate of cyclization is four times that of hexane and leads to much greater conversion to aromatic hydrocarbons.

Methylcyclohexane readily dehydrogenates to toluene, without any other significant reactions occurring, whereas methylcyclopentane undergoes ring opening with hydrogenolysis to alkanes. Methylcyclopentane also isomerizes to a six-carbon ring, and this is followed by a rapid dehydrogenation to benzene. The rate of the competing ring-opening reaction is 0.7 that of isomerization.

The general rate behavior and heat effects of the various reactions in reforming are summarized in Table 1.33.

The principal reaction in the reforming of virgin naphtha feed consists of the formation of branched alkanes from the predominantly present unbranched alkanes. Alkylcyclopentanes, the main components of naphthenes in naphtha, are isomerized to six-membered ring hydrocarbons, and these are rapidly converted to arenes and hydrogen.

The seemingly independent participation in the reforming process of the metal and the acid part of the catalyst was illustrated by a series of reactions in which methylcyclopentane was used as the test hydrocarbon (Table 1.34). In the presence of the catalytically inert silica support, 0.3% Pt/SiO_2 catalyzes only dehydrogenation of methylcyclopentane. If, however, an acidic component such as $SiO_2-Al_2O_3$ is added to Pt/SiO_2, skeletal isomerization with the formation of benzene takes place. The reaction can be explained as occurring by the following steps:

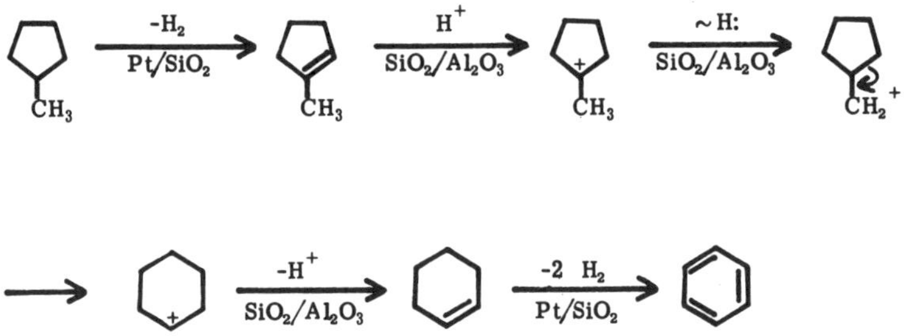

Cations that are generated in reforming processes may undergo a multitude of rearrangements before they are desorbed from the catalyst as electronically neutral hydrocarbons. This was illustrated by experiments in which ^{14}C-labeled ethylcyclohexane having radioactivity in the α or β position of the ethyl group was passed over a re-forming catalyst. The catalyst was composed of 6% Ni deposited on $SiO_2-Al_2O_3$ cracking catalyst. The experiments were carried out at 360°C, 25 atm, LHSV 1, and H_2 to

TABLE 1.33

Rate Behavior and Heat Effects of Important Reforming Reactions[a]

Reaction type	Relative rate[b]	Effect of increase in total pressure	Heat effect
Hydrocracking	Slowest	Increases rate	Quite exothermic
Dehydrocyclization	Slow	None to small decrease in rate	Endothermic
Isomerization of paraffins	Rapid	Decreases rate	Mildly exothermic
Naphthene isomerization	Rapid	Decreases rate	Mildly exothermic
Paraffin dehydrogenation	Quite rapid	Decreases conversion	Endothermic
Naphthene dehydrogenation	Very rapid	Decreases conversion	Very endothermic

[a] From Gates et al. (1979, p. 189).

C_8H_{16} ratio of 4:1. The temperature was lower than the usual reforming temperature of 500°C in order to prevent excessive dehydrogenation of alkylcyclohexanes to aromatic hydrocarbons. Although less than half of the ethylcyclohexane reacted, the relative concentration of dimethylcyclohexanes and C_8 cyclopentanes formed was near equilibrium. Furthermore, the distribution of radioactive carbon was nearly statistical between the ring and side chain in the C_2H_5—C_6H_{11}/$(CH_3)_2$—C_6H_{10} isomers. Such data suggest that the reactant spends a long enough time in

TABLE 1.34

Conversion of Methylcyclopentane Catalyzed by Acid, Metal, and Mixed Catalysts[a,b]

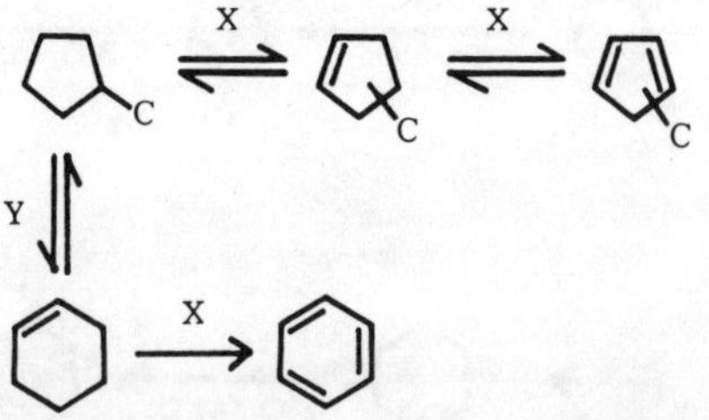

Catalyst[c]	Liquid product analysis (mol %)			

SiO$_2$/Al$_2$O$_3$ (X)	98	0	0	0.1
Pt/SiO$_2$ (Y)	62	20	18	0.8
SiO$_2$/Al$_2$O$_3$ + Pt/Al$_2$O$_3$ (Y + X)	65	14	10	10.0

[a] From Weisz (1961).
[b] Experimental conditions: temp., 500°C; H$_2$ partial pressure, 0.8 atm; methylcyclopentane partial pressure, 0.2 atm; residence time, 2.5 sec.
[c] X = 0.3 wt % Pt on SiO$_2$; Y = SiO$_2$/Al$_2$O$_3$; surface area 420 m^2/gm, 0.8–1.4 mm particle size.

ionic residence in the catalyst for the cations to undergo repeated ring expansion and contraction before they are reconverted to a hydrocarbon. The paths taken by the generated cations are presented in Scheme 1.13.

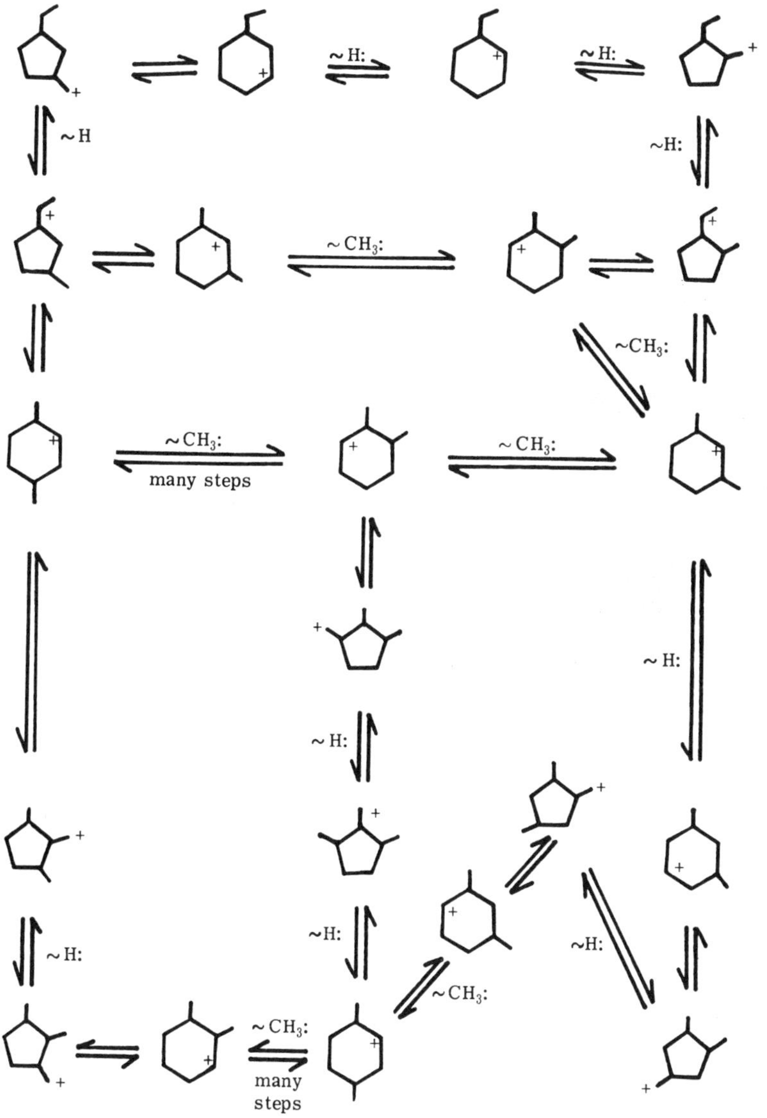

Scheme 1.13. Schematic representation of some of the paths taken by cations in the hydroisomerization of ethylcyclohexane (Pines and Shaw, 1957).

b. Dehydrocyclization. The passage of hexanes at about 275°–300°C over platinum deposited on inert alumina or on silica results in the formation of methylcyclopentane as the only cyclic product, and in addition hexane and 2- and 3-methylpentane undergo reversible isomerization (Table 1.35). Skeletal interconversion of the hexanes can be represented by the following scheme, in which adsorbed methylcyclopentane species is the intermediate:

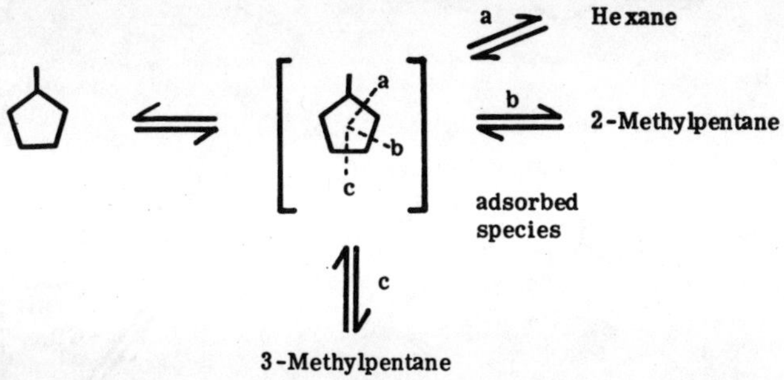

It was proposed that the cyclization path involves a single and a double attachment of 1,5-carbon atoms to two adjacent platinum sites, as follows:

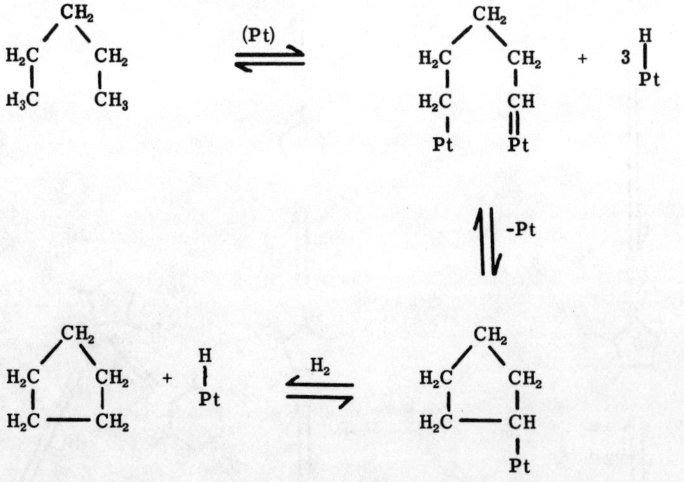

Alternatively, it is possible to formulate this by using only a single platinum.

At higher temperature, 400°C, hexane forms benzene when passed over 1% by weight Pt on inert alumina. A five-membered ring structure is not involved in the conversion of hexane to benzene, and apparently six-membered and five-membered ring closures are parallel reactions.

TABLE 1.35

Distribution of C_5 Reaction Products from the Passage of Hexanes over Platinum Catalysts[a]

Reactant hydrocarbon	Catalyst	Reaction temp (°C)	Initial product distribution[b,c] (mol %)					Proportion of cyclic product (mol %)
			2-Mp	3-Mp	n-H	MCP	CH + B	
Hexane	0.8% Pt/silica	295	46[d]	—	—	54	—	54
	0.2% Pt/alumina	303	55	20	—	25	—	25
	10% Pt/alumina	298	62	31	—	7.3	—	7.3
	Ultrathin Pt film 0.13 µg/cm² (20 Å)[e]	273	8.7	4.0	—	79	8.3	87
2-Methyl-pentane	0.8% Pt/silica	295	—	66[f]	—	34	—	34
	0.2% Pt/alumina	301	—	30	33	37	—	37
	10% Pt/alumina	297	—	47	28	25	—	25
	Ultrathin Pt film, 0.08 µg/cm² (20 Å)[e]	273	—	5.7	14	81	—	81
3-Methyl-pentane	0.8% Pt/silica	295	79[g]	—	—	21	—	21
	0.2% Pt/alumina	303	58	—	22	20	—	20
	10% Pt/alumina	298	76	—	17	6.5	—	6.5
	Ultrathin Pt film, 0.13 µg/cm² (20 Å)[e]	273	12	—	8.5	79	—	79

[a] From Anderson (1973, pp. 44–45).
[b] Products from 2-methylpentane over film catalysts also contain up to 1% 2,3-dimethylbutane.
[c] Abbreviations: 2-MP, 2-methylpentane; 3-MP, 3-methylpentane; n-H, hexane; CH, cyclohexane; B, benzene.
[d] 2-MP plus 3-MP.
[e] Average Pt particle diameter.
[f] 3-MP plus n-H.
[g] 2-MP plus n-H.

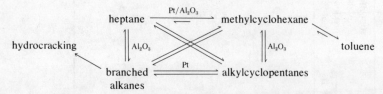

Fig. 1.9. Reaction paths for re-forming heptane over Pt/acidic Al₂O₃ catalyst. The Pt and Al₂O₃ next to the arrows indicate predominant participation. Short arrows indicate low rates of reaction.

Figure 1.10 represents a kinetic model describing the reactions steps that occur between various products obtained from reforming unbranched heptane over a commercial Pt/acidic Al_2O_3 catalyst.

Dehydrocyclization to alkylcyclopentanes and dehydrogenation to toluene are platinum-catalyzed reactions. Skeletal isomerization of cyclic compounds and the direct isomerization of heptanes without the participation of cyclic compounds are catalyzed by the acidic components of the re-forming catalyst. Platinum, however, participates in these reactions in order to initiate the formation of olefins, which are essential for the generation of the intermediate cations.

c. Fission. The products obtained from reforming contain small amounts of lower-boiling hydrocarbons; their formation can be attributed to acid-catalyzed fission of branched hydrocarbons produced in the reaction and not to metal-catalyzed hydrogenolysis of carbon–carbon bonds. Hydrogenolysis by metals would lead to the production of substantial amounts of methane, whereas the lower-boiling fractions from the reforming processes consist principally of isobutane and isopentane and a small amount of propane. Fission is a disproportionation reaction that accompanies many acid-catalyzed hydrocarbon conversions and can be explained by a mechanism similar to that proposed for catalytic cracking (Section VI,B,2,1) and for hydrocracking (Section VIII,C,3).

C. Hydrocracking

1. Introduction

Hydrocracking is a major refining process carried out in the presence of hydrogen under pressure. Because of the specificity of many of the reactions associated with hydrocracking, the process can be used for producing a number of important hydrocarbons as chemicals. Hydrocracking converts relatively high boiling hydrocarbons to lower boiling products with an overall consumption of hydrogen. It is thus a combination of catalytic cracking and hydrogenation.

Hydrocracking catalysts are dual-function catalysts consisting of an acidic component and a hydrogenation–dehydrogenation component. The reactions catalyzed by these two components are different in nature, and the strength of the two components can be varied. The reactions that occur and the products that are formed depend on the balance between the contributions of each of the components.

When hydrocracking is applied to the production of gasoline, catalysts with strong acidic activity are required, and the acidic components usually include amorphous silica–alumina and/or zeolites. The presence of these materials in the catalyst promotes reactions that are usually associated with catalytic cracking (Section VI). The reactions lead to a high ratio of branched to straight-chain alkanes and to the conservation of monocyclic rings. A great deal of skeletal isomerization occurs due to β-scission, leading to the formation of tertiary cations as intermediates. The function of the hydrogenation component is to prevent the formation of olefins from the cations and/or to reduce the olefins, once formed, to the corresponding alkanes. The minimization of olefins in the product prevents the occurrence of conjunct polymerization and consequently also prevents the formation of carbonaceous material, so called coke, which usually deposits on and deactivates the catalyst.

When gas oils are hydrocracked to produce jet fuels, catalysts with milder acidity and stronger hydrogenation activities are used.

In a special category are catalysts containing shape-selective zeolites, in which the pore structure of the zeolites is such that the access of reactants and egress of products are limited to molecules of a particular shape and size.

The major hydrogenating components are platinum, nickel, palladium, molybdenum, and cobalt. Typical catalysts with strongly acidic properties and relatively mild hydrogenation activity are metal sulfides such as nickel, cobalt, or tungsten sulfides on silica–alumina.

2. Hydrocracking of Alkanes

a. **Strongly Acidic Catalysts with Mild Hydrogenation Activity.** Sulfided nickel on $SiO_2–Al_2O_3$ is one of the catalysts frequently used for commercial hydrocracking of gas oil. Table 1.36 contains data on the product distribution obtained from hydrocracking decane over the following catalysts: commercial $SiO_2–Al_2O_3$ cracking catalyst, nickel deposited on $SiO_2–Al_2O_3$, and sulfided nickel deposited on $SiO_2–Al_2O_3$. The catalyst containing sulfided nickel gives much greater conversion of decane than either of the other two catalysts. The product distribution from the sul-

TABLE 1.36

Hydrocracking of Decane[a,b]

	Catalyst		
	Silica–alumina	Ni on silica–alumina	Ni₃S₂ on silica–alumina
Ni in catalyst (%)	0	6.6	6.6
Total conversion (%)	2.5	7.8	52.8
Isomerization	0.3	4.5	5.5
Cracking	2.2	3.3	47.3
Product composition from the converted decane (mol %)			
$CH_4 + C_2H_6$	0.2	3.2	0.1
Propane	6.2	3.2	7.7
Isobutane	24.9	0.7	30.2
Butane	4.2	8.6	6.6
Isopentane	27.0	0.7	20.8
Pentane	2.1	11.5	2.5
"Isohexanes"	20.4	1.1	18.6
Hexane	1.4	9.7	1.5
C_7–C_9 alkanes	7.3	18.0	6.7
Branched decanes	6.2	43.3	5.3

[a] From Langlois et al. (1966).

[b] Conditions: Temp, 288°C; pressure, 82 atm; feed rate, LHSV, 8; H_2/decane mole ratio, 10.

fided nickel catalyst is similar to that obtained from only SiO_2–Al_2O_3 cracking catalyst. Nonsulfided nickel on SiO_2–Al_2O_3 catalyst gives products quite different from the other two catalysts; with the catalyst containing nonsulfided nickel, the primary reaction is isomerization to form branched decanes. Although a significant amount of cracking occurs over this catalyst, the products from cracking are quite different from those from sulfided catalyst inasmuch as they consist mainly of unbranched alkanes. This indicates that the cracking occurs by hydrogenolysis on the nickel metal surface. The products from hydrocracking over sulfided nickel on SiO_2–Al_2O_3 are primarily C_4 to C_7 branched paraffins, with a great increase in the production of isobutane.

These changes all suggest a greatly increased catalyst acidity. It was mentioned in Section VI,B,1 that the Brønsted acidity of SiO_2–Al_2O_3 is believed to be caused by acidic hydroxyl groups attached to the aluminum atoms. During impregnation with a soluble nickel salt, the acidic protons of these hydroxyl groups exchange with the nickel ions to form nickel salts of the silica–alumina acidic sites. The nickel beyond that required to neutralize the acidic sites is converted to nickel oxide on drying and cal-cining. This nickel oxide is reduced with hydrogen to nickel metal under the conditions used before the reaction; the nickel silica–alumina salts, however, remain unaltered. The resulting catalyst is believed to consist of silica–alumina with the acidic sites in the form of nickel salts and with nickel and nickel suboxides dispersed throughout.

When the catalyst is sulfided with hydrogen sulfide, the nickel metal is converted to nickel sulfide. The H_2S also reacts with the nickel salts of the silica–alumina to form nickel sulfide, and at the same time the acidic sites of the SiO_2–Al_2O_3 are regenerated. Thus, the purpose of the hydrogen sul-fide is twofold: to generate the strongly acidic sites of SiO_2–Al_2O_3 origi-nally present and to convert nickel to a mild hydrogenation–dehydroge-nation catalyst. In addition, sulfided nickel as such also has mild catalytic properties.

The mechanism of the hydrocracking of decane in the presence of Ni_3S_2/SiO_2–Al_2O_3 is similar to that of catalytic cracking over SiO_2–Al_2O_3 (Section VI,B,2), with features of hydrogenation and dehydrogenation su-perimposed. The first step in the hydrocracking reaction apparently is de-hydrogenation of alkane to alkene; this is then protonated to form a car-bonium ion, which can initiate the catalytic chain reaction. In hydrocracking, dehydrogenation is the rate-determining step, whereas in catalytic cracking the generation of a cation from an alkane is the most difficult step, and therefore straight cracking requires much higher tem-peratures than hydrocracking for a similar rate of conversion.

b. Catalysts with High Hydrogenation Activity. Catalysts with higher hydrogenation activity relative to acidity give a different product distribu-tion from that obtained with a catalyst of strong acidity. The reactions that occur include isomerization of the reactant without cracking, hydro-genolysis, and, in the presence of noble metals, cyclization.

On nonacidic supports hydrogenolysis is the only reaction of alkanes with nonsulfided metals. A cobalt– or nickel–kieselguhr catalyst under controlled reaction conditions causes the selective demethylation of al-kanes, whereby methyl groups attached to secondary carbons are re-moved more easily than those attached to tertiary or quaternary carbon atoms.

$$CH_3-\underset{\underset{\displaystyle CH_3}{|}}{\overset{\overset{\displaystyle H_3C}{|}}{C}}-\underset{\underset{\displaystyle CH_3}{|}}{CH}-CH_2-CH_3 \xrightarrow[\quad H_2 \quad]{\text{Ni-kieselguhr}} CH_3-\underset{\underset{\displaystyle H_3C}{|}}{\overset{\overset{\displaystyle H_3C}{|}}{C}}-\underset{\underset{\displaystyle CH_3}{|}}{CH}-CH_3 \;+\; CH_4$$

With nonsulfided catalysts, such as nickel, on acidic supports iso-merization is an important reaction, as demonstrated in Table 1.36. At relatively low extended conversion, the product from cracking consists almost entirely of nonbranched alkanes.

Noble metal catalysts frequently mentioned as hydrocracking catalysts are Pt and Pd on SiO_2–Al_2O_3. With increasing Pt content a selective adsorption of the metal by acid sites occurs, which causes a reduction in the acidity of SiO_2–Al_2O_3. Hydrocracking of hexadecane over such a catalyst at 370°C and 69 atm is accompanied by extensive cyclization and isomerization. Cracking gives mainly two intermediate-sized alkanes, with cracking at the center bond preferred. Isoalkanes to normal alkanes are close to equilibrium, whereas on supported sulfided nickel catalysts, the ratio of isoalkanes to normal alkanes is higher than the calculated equilibrium ratio. This suggests that in the presence of supported noble metal catalysts there is a rapid hydrogenation of olefinic product with the prevention of hydroisomerization of alkanes.

3. Hydrocracking of Alkylcyclohexanes

The hydrocracking of C_9 to C_{12} alkylcyclohexanes, such as polymethylated cyclohexanes and butylcyclohexanes, over sulfided nickel on SiO_2–Al_2O_3 catalyst at about 230°–290°C has been studied (Table 1.37). The principal acyclic product is isobutane, and the principal cyclic product contains four carbon atoms less than the original alkylcycloalkane. Very little ring rupture occurs. The small amount of methane in the product indicates that the production of cycloalkanes of lower molecular weight is not due to a simple demethanation reaction. The experimental results suggest that a rapid isomerization step precedes hydrocracking and that the species that undergo hydrocracking are approximately the same regardless of the structure of the original alkylcyclohexane used in the reaction.

The formation of isobutane and methylcyclopentane as the two principal products from the hydrocracking of 1,2,3,4-tetramethylcyclohexane is explained by a mechanism presented in Scheme 1.14. The predominant cracking occurs with tertiary cations, which can produce another tertiary cation by β-scission. This is consistent with the production of isobutane rather than butane and propane. The type of reaction described above, which in effect pares methyl groups from cycloalkane rings and eliminates them as branched alkanes, mainly isobutane, in such a way as to conserve rings, has been called a paring reaction.

4. Hydrocracking of Alkylbenzenes

Products from the hydrocracking of alkylbenzenes exhibit wider variation and greater dependence on the specific structure of the reactant than is the case with a cycloalkane reactant.

TABLE 1.37

Hydrocracking of Alkylcyclohexanes[a,b]

	Alkylcyclohexane used						
	1,2,4,5-Tetramethyl-	Penta-methyl-	Hexamethyl-	Hexamethyl-	Butyl-	1,2-Diethyl-	1,2,4,5-Tetramethyl-[c]
Temp (°C)	291	232	232	288	290	290	289
Cracking (mol %)	39.6	27.0	87.5	99.8	96.0	75.9	19.4
Composition of main product (mol %)							
Propane	2.4	1.2	0.8	3.9	4.2	4.4	2.6
Isobutane	42.9	32.6	33.4	29.9	39.0	35.8	41.7
Butane	0.9	0.4	0.6	1.2	1.6	3.5	1.0
Isopentane	5.1	6.8	12.2	11.4	5.2	5.9	6.5
"Isohexanes"	6.4	2.4	3.8	6.7	6.4	5.4	6.4
Methylcyclopentane	31.5	4.0	1.8	1.8	29.7	27.4	31.3
Cyclohexane	2.2		0.1		2.4	2.6	1.5
Methylcyclohexane	2.4	14.6	6.7	6.6	3.4	3.5	2.6
Dimethylcyclopentanes	1.6	26.4	5.2	4.2	2.6	2.1	2.0
Trimethylcyclopentanes	0.4	2.2	8.7	6.2	0.9	0.8	0.3
Dimethylcyclohexanes	0.7	5.2	18.4	15.5	0.4	2.0	1.0

[a] From Egan et al. (1962).
[b] Catalyst, nickel sulfide on SiO_2–Al_2O_3; pressure, 82 atm; LHSV, 8.0; moles H_2/mole cycloalkane 8.4–9.7.
[c] Catalyst, SiO_2–Al_2O_3.

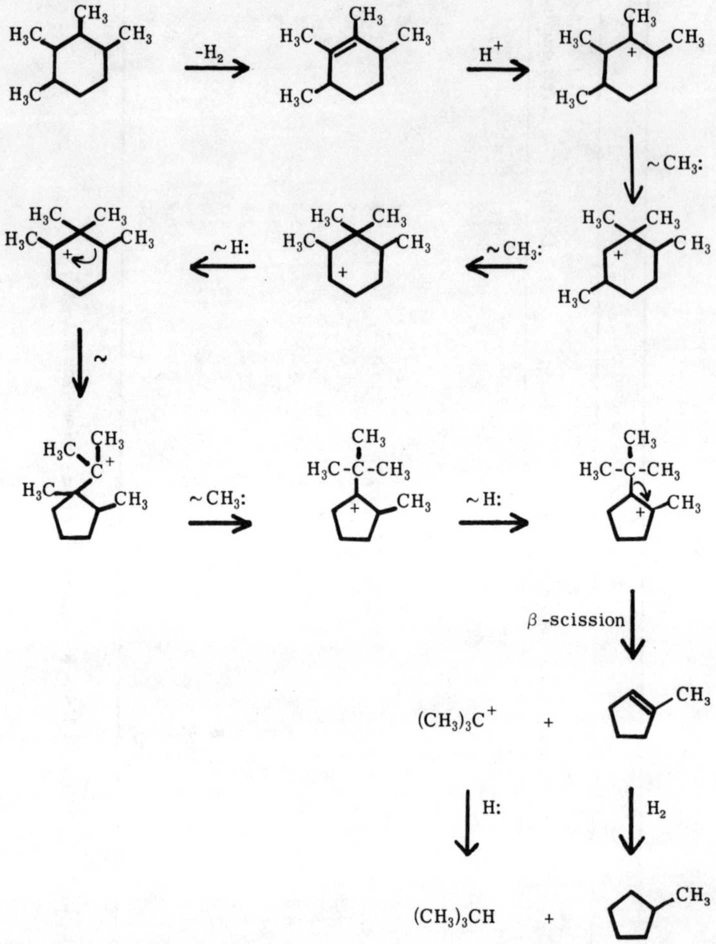

H_3C CH_3 CH_3 H_3C $\xrightarrow{-H_2}$ H_3C CH_3 CH_3 H_3C $\xrightarrow{H^+}$ H_3C CH_3 CH_3 H_3C

$\downarrow \sim CH_3:$

H_3C CH_3 H_3C CH_3 $\xleftarrow{\sim H:}$ H_3C CH_3 H_3C CH_3 $\xleftarrow{\sim CH_3:}$ H_3C CH_3 CH_3 H_3C

$\downarrow \sim$

H_3C CH_3 H_3C CH_3 $\xrightarrow{\sim CH_3:}$ $H_3C-C-CH_3$ CH_3 CH_3 $\xrightarrow{\sim H:}$ $H_3C-C-CH_3$ CH_3 CH_3

β-scission $\downarrow$

$(CH_3)_3C^+$ $+$ $\overset{CH_3}{\bigcirc}$

$\downarrow$ H: $\downarrow$ H$_2$

$(CH_3)_3CH$ $+$ $\overset{CH_3}{\bigcirc}$

Scheme 1.14. Mechanism of the hydrocracking of 1,2,3,4-tetramethylcyclohexane (Egan *et al.*, 1972).

a. Alkylbenzenes with Side Chains of Three or More Carbons. The hydrocracking under pressure of alkylbenzenes containing three to five carbon atoms in the side chain over sulfided Ni on $SiO_2-Al_2O_3$ affords products resulting from a direct dealkylation reaction. *tert*-Pentylbenzene gives benzene and isopentane, and butylbenzene affords benzene and C_4 hydrocarbons, the latter consisting of a mixture of butane and isobutane. With larger alkyl groups attached to the benzene ring the reaction products become more complex, and cyclialkylation to fused bicyclic compounds is also observed. The products from the hydrocracking of decylbenzene are shown in Fig. 1.11. Besides benzene and decane, which

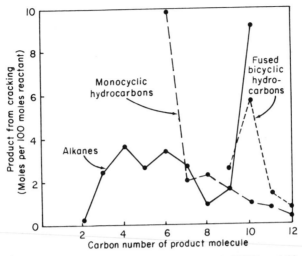

Fig. 1.11. Hydrocracking of decylbenzene at LHSV 8.0, 288°C, and 82 atm (Sullivan *et al.*, 1966).

result from dealkylation, still the most important reaction, all alkanes from C_3 to C_{10} are also produced, as well as alkylbenzenes having from 7 to 12 carbon atoms, indicating that there is also substantial cleavage of bonds in the side chain.

In the hydrocracking of decylbenzene there is considerable formation of C_9 to C_{12} polycyclic hydrocarbons, particularly indans and tetralins. The mechanism of their formation is explained by Scheme 1.15.

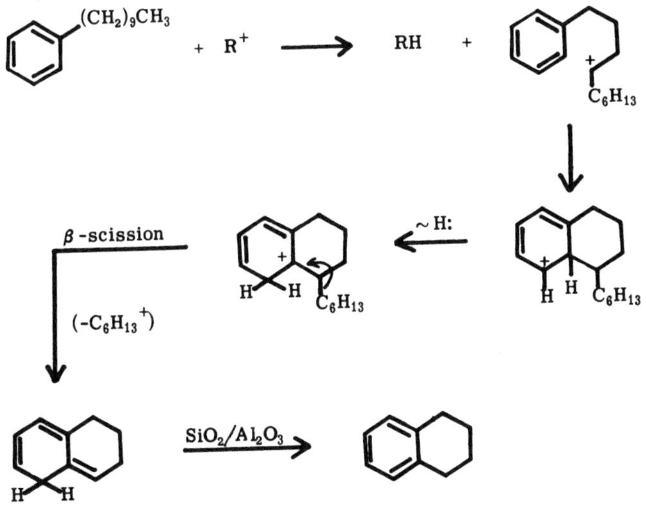

Scheme 1.15. Cyclialkylation of decylbenzene to tetralin (Sullivan *et al.*, 1964).

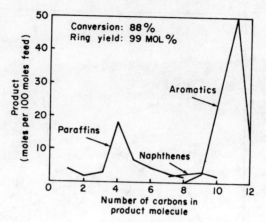

Fig. 1.12. Hydrocracking of hexamethylbenzene: product distribution vs carbon number. Conditions: 349°C; 13.6 atm; moles H_2/mole feed, 1.86; volumes of liquid feed per volume of catalyst per hour, 8.0 (Sullivan *et al.*, 1961).

The "hydrocracking" of alkylbenzenes over nonacidic catalysts is a straightforward hydrogenolysis reaction. Little or no skeletal isomerization occurs. Successive removal of methyl groups from the side chain is the principal reaction over rhodium on neutral alumina.

b. Polymethylbenzene: The Paring Reaction. The product distribution from the hydrocracking of hexamethylbenzene is given in Fig. 1.12. The principal products are isobutane and isopentane and C_{10} and C_{11} methylbenzenes. Essentially no cleavage of the aromatic ring occurs, and the amount of hydrogenolysis to form methane is small. A mechanism to account for the products of reaction is given in Scheme 1.16. This mechanism is similar to the "paring reaction" proposed for tetramethylcyclohexane (Scheme 1.14). The mechanism indicates the formation of a xylene as a primary product. However, little xylene is recovered because of rapid methyl transfer from hexamethylbenzene to form a large amount of C_{10} and C_{11} aromatic hydrocarbons.

A paring reaction occurs also on $SiO_2-Al_2O_3$ as such in the absence of sulfided nickel and hydrogen. Under these conditions, however, the $SiO_2-Al_2O_3$ is deactivated rapidly, and the rate of reaction is much smaller.

REFERENCES

I. Introduction

Gillespie, R. J. (1963). Proton acids and Lewis acids. *In* "Friedel–Crafts and Related Reactions" (G. A. Olah, ed.), Vol. 1, pp. 169–194. Wiley (Interscience), New York.

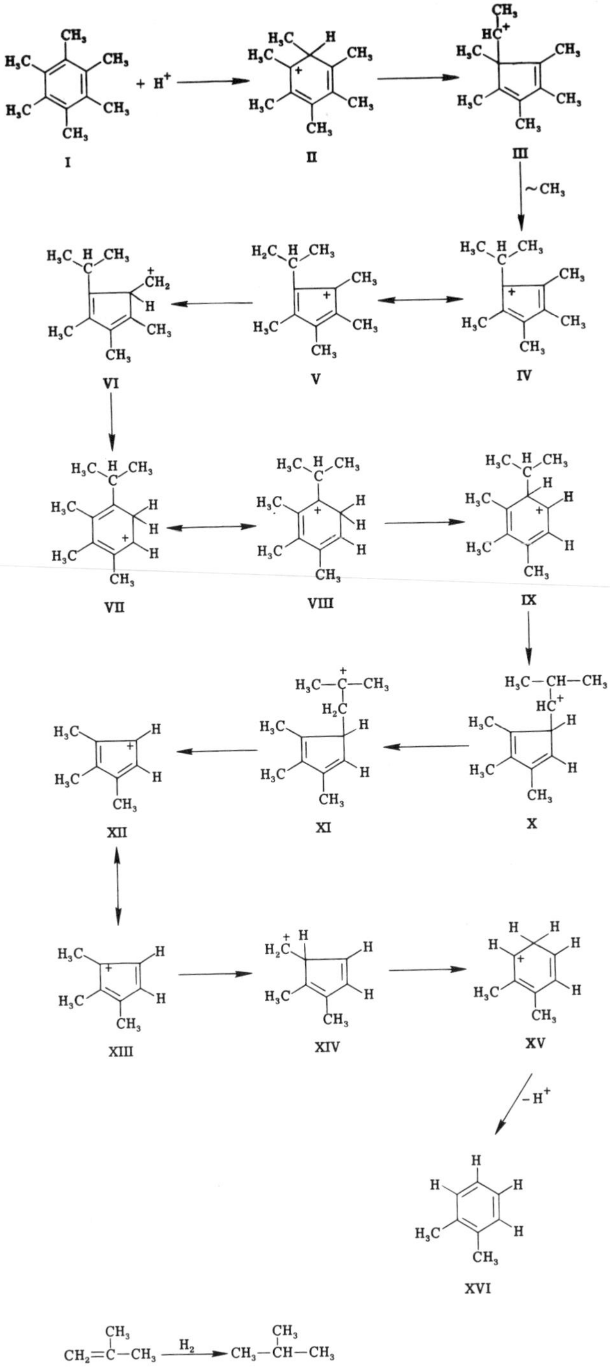

Scheme 1.16. Mechanism of the hydrocracking of hexamethylbenzene (Sullivan *et al.*, 1961).

Gillespie, R. J., and Peel, T. E. (1971). Superacid systems. *Adv. Phys. Org. Chem.* **9**, 1.
Lossing, F. P., and Semeluk, G. P. (1970). *Can. J. Chem.* **48**, 955.
Siskin, M. (1978). *J. Am. Chem. Soc.* **100**, 1838.

II. Isomerization

Allen, R. H., and Yats, L. D. (1959). *J. Am. Chem. Soc.* **81**, 5289.
Allen, R. H., Yats, L. D., and Erley, D. S. (1960). *J. Am. Chem. Soc.* **82**, 4853.
Balaban, A. T., and Farcasia, D. (1967). *J. Am. Chem. Soc.* **89**, 1958.
Buschick, R. D. (1968). *J. Org. Chem.* **33**, 4085.
Condon, F. E. (1958). Isomerization of hydrocarbons. *In* "Catalysis" (P. H. Emmett, ed.), Vol. 6, pp. 43–189. Reinhold, New York.
Evering, B. L., d'Ouville, E. L., Lien, A. P., and Waugh, R. C. (1953). *Ind. Eng. Chem.* **45**, 582.
Haag, W. O., and Pines, H. (1960). *J. Am. Chem. Soc.* **82**, 2486.
Ipatieff, V. N., and Schmerling, L. (1948). *Ind. Eng. Chem.* **40**, 2354.
McCaulay, D. A. (1959). *J. Am. Chem. Soc.* **81**, 6437.
McCaulay, D. A. (1964). Isomerization of aromatic hydrocarbons. *In* "Friedel–Crafts and Related Reactions" (G. A. Olah, ed.), Vol. 2, pp. 1049–1074. Wiley (Interscience), New York.
* McCaulay, D. A., and Lien, A. P. (1952). *J. Am. Chem. Soc.* **74**, 6246.
Mavity, J. M., Pines, H., Wackher, R. E., and Brooks, J. A. (1948). *Ind. Eng. Chem.* **40**, 2375.
Pines, H., and Haag, W. O. (1960). *J. Am. Chem. Soc.* **82**, 2471.
Pines, H., and Hoffman, N. E. (1964). Isomerization of saturated hydrocarbons. *In* "Friedel–Crafts and Related Reactions" (G. A. Olah, ed.), Vol. 2, pp. 1211–1252. Wiley (Interscience), New York.
Pines, H., and Wackher, R. C. (1946). *J. Am. Chem. Soc.* **68**, 599.
Pines, H., Aristoff, E., and Ipatieff, V. N. (1949). *J. Am. Chem. Soc.* **71**, 749.
Pines, H., Pavlik, F. J., and Ipatieff, V. N. (1951a). *J. Am. Chem. Soc.* **73**, 5738.
Roberts, R. M., and Douglas, J. E. (1958). *Chem. and Ind.* 1557.
Roebuck, A. K., and Evering, B. L. (1955). *J. Am. Chem. Soc.* **75**, 1631.
Suld, G., and Stuart, A. P. (1964). *J. Org. Chem.* **29**, 2939.

III. Polymerization of Olefins

Criegee, R., and Riebel, A. (1953). *Agnew-Chem.* **65**, 136.
Ipatieff, V. N., and Pines, H. (1935). *Ind. Eng. Chem.* **27**, 1364.
Oblad, A. G., Mills, G. A., and Heinemann, H. (1958). Polymerization of olefins (to liquid polymers). *In* "Catalysis" (P. H. Emmett, ed.), Vol. 6, pp. 341–406. Reinhold, New York.
Overberger, C. G., Earle, E. M., and Tanner, D. (1958). *J. Am. Chem. Soc.* **80**, 1761.
Pines, H. (1972). *Intra-Sci. Chem. Rep.* **6**(2), 1–41.
* Plesch, P. H. (1963). Isobutene. *In* "The Chemistry of Cationic Polymerization" (P. H. Plesch, ed.). pp. 141–208. Macmillan, New York.
* Schaad, R. E. (1955). Polymer gasoline. *In* "Chemistry of Petroleum Hydrocarbons" (B. T. Brooks, C. E. Boord, S. S. Kurtz, Jr., and L. Schmerling, eds.). Vol. 3, pp. 221–248. Reinhold, New York.
* Schmerling, L., and Ipatieff, V. N. (1950). The mechanism of polymerization of alkenes. *Adv. Catal.* **2**, 21–78.
Whitmore, F. C., and Meunier, P. L. (1941). *J. Am. Chem. Soc.* **63**, 2197.

Whitmore, F. C., Ropp, W. S., and Cook, N. C. (1950). *J. Am. Chem. Soc.* **72**, 1507.

IV. Alkylation of Saturated Hydrocarbons

* Albright, L. F., and Goldsby, A. R., eds. (1977). Industrial and laboratory alkylation. *ACS Symp. Ser.* No. 55.

Hofmann, J. E., and Schriesheim, A. (1962). *J. Am. Chem. Soc.* **84**, 964.

* Kennedy, R. M. (1958). Catalytic alkylation of paraffins with olefins. *In* "Catalysis" (P. H. Emmett, ed.), Vol. 6, pp. 1–41. Reinhold, New York.

McAllister, S. H., Anderson, J., Ballard, S. A., and Ross, W. E. (1941). *J. Org. Chem.* **6**, 647.

Pines, H., and Ipatieff, V. N. (1948). *J. Am. Chem. Soc.* **70**, 531.

* Schmerling, L. (1964). Alkylation of saturated hydrocarbons. *In* "Friedel–Crafts and Related Reactions" (G. A. Olah, ed.), Vol. 2, pp. 363–407. Wiley (Interscience), New York.

V. Alkylation of Aromatic Hydrocarbons

Allen, R. H., and Yats, L. D. (1961). *J. Am. Chem. Soc.* **83**, 2799.

* Boone, D. E. Eisenbraun, E. J., Flanagan, P. W., and Grigsby, R. D. (1971). The acid-catalyzed alkylation and cyclialkylation of the cymenes with isobutylene and related olefins. *J. Org. Chem.* **36**, 2042.

Chen, N. Y., Kaeding, W. W., and Dwyer, F. G. (1979). *J. Am. Chem. Soc.* **101**, 6783.

Csicsery, S. M. (1969). *J. Org. Chem.* **34**, 3338.

Deno, N. C., LaVietes, D., Mockus, J., and Scholl, P. C. (1968). *J. Am. Chem. Soc.* **90**, 6457.

Eisenbraun, E. J., Mattox, J. R., Barsal, R. C., Wilhelm, M. A., Flanagan, P. W. K., Carel, A. B., Laramy, R. E., and Hamming, M. C. (1968). *J. Org. Chem.* **33**, 2000.

Elgavi, A., and Pines, H. (1978). *J. Catal.* **55**, 228.

Friedman, B. S., and Morritz, F. L. (1956). *J. Am. Chem. Soc.* **78**, 2000.

Inukai, T. (1966). *J. Org. Chem.* **31**, 1124.

Ipatieff, V. N., Pines, H., and Olberg, R. C. (1948). *J. Am. Chem. Soc.* **70**, 2123.

Kelly, J. T., and Lee, R. J. (1955). *Ind. Eng. Chem.* **47**, 757.

* Koncos, R., and Friedman, B. S. (1964). Alkylation of aromatics with dienes and substituted alkenes. *In* "Friedel–Crafts and Related Reactions" (G. A. Olah, ed.), Vol. 2, Part I, pp. 289–412. Wiley (Interscience), New York.

Lewis, P. J., and Dwyer, F. G. (1977). *Oil Gas J.* Sept. 26, 55–88.

Nelson, K. L., and Brown, H. C. (1955). Aromatic substitution. *In* "Chemistry of Petroleum Hydrocarbons" (T. T. Brooks, C. E. Boord, S. S. Kurtz, Jr., and L. Schmerling, eds.), Vol. 3, pp. 1465–478. Reinhold, New York.

Olson, A. C. (1960). *Ind. Eng. Chem.* **52**, 833.

* Patinkin, S. H., and Friedman, B. S. (1964). Alkylation of aromatics with alkenes and alkanes. *In* "Friedel–Crafts and Related Reactions" (G. A. Olah, ed.), Vol. 2, Part I, pp. 1–288. Wiley (Interscience), New York.

Pines, H., and Arrigo, J. T. (1958). *J. Am. Chem. Soc.* **80**, 4369.

Pines, H., Huntsman, W. D., and Ipatieff, V. N. (1951b). *J. Am. Chem. Soc.* **73**, 4343.

Pines, H., Huntsman, W. D., and Ipatieff, V. N. (1953). *J. Am. Chem. Soc.* **75**, 2311.

* Ross, L., and Barclay, C. (1964). Cyclialkylation of aromatics. *In* "Friedel–Crafts and Related Reactions" (G. A. Olah, ed.), Wiley (Interscience, New York, Vol. 2, Part II, pp. 785–952.

Schriesheim, A. (1964). Alkylation of aromatics with alcohols and ethers. *In* "Friedel–Crafts and Related Reactions" (G. A. Olah, ed.) Vol. 2, Part I. pp. 477–597. Wiley (Interscience), New York.

VI. Catalytic Cracking

* Benesi, H. A., and Winquist, B. H. C. (1978). Surface acidity of solid catalysts. *Adv. Catal.* **27,** 97–182.

* Greensfelder, B. S. (1955). Theory of catalytic cracking. *In* "Chemistry of Petroleum Hydrocarbons" (B. T. Brooks, C. E. Boord, S. S. Kurtz, Jr., and L. Schmerling, eds.), Vol. 2, pp. 137–164. New York.

Greensfelder, B. S., Voge, H. H., and Good, G. M. (1949). *Ind. Eng. Chem.* **41,** 2573.

* Hansford, R. C. (1952). Chemical concepts of catalytic cracking. *Adv. Catal.* **4,** 1–30.

* Jacobs, P. A. (1977). "Carboniogenic Activity of Zeolites." Elsevier, Amsterdam.

* Magee, J. S., and Blazek, J. J. (1977). Preparation and performance of zeolite cracking catalysts. *In* "Zeolite Chemistry and Catalysis" (J. A. Rabo, ed.), ACS Monograph, No. 171, pp. 615–679. Am. Chem. Soc., Washington, D.C.

Miale, J. N., Chen, N. Y., and Weisz, P. B. (1966). *J. Catal.* **6,** 278.

Nace, D. M. (1969). *Ind. Eng. Chem., Prod. Res. Dev.* **8**(1), 31.

* Oblad, A. G., Milliken, T. H., and Mills, G. A. (1955). The effects of the variables in catalytic cracking. *In* "Chemistry of Petroleum Hydrocarbons" (B. T. Brooks *et al.,* eds.) Vol. 2, pp. 165–188. Reinhold, New York.

Plank, C. J., Rosinski, E. J., and Hawthorne, W. P. (1964). *Ind. Eng. Chem. Prod. Res. Dev.* **3,** 165.

* Rabo, J. A., and Poutsma, M. L. (1971). *Mol. Sieve Zeolites—II; Adv. Chem. Ser.* No. 102, p. 302.

* Venuto, P. B. and Landis, P. S. (1968). Organic catalysis over cystalline aluminosilicates. *Adv. Catal.* **18,** 259–371.

* Voge, H. H. (1958). Catalytic cracking. *In* "Catalysis" (P. H. Emmett, ed.), Vol. 6, pp. 407–493. Reinhold, New York.

VII. Reactions Catalyzed by Strong Acids

Herlem, M. (1977). *Pure Appl. Chem.* **49,** 107–113.

Olah, G. A., Klopman, G., and Schlosberg, R. H. (1969). *J. Am. Chem. Soc.* **91,** 3261.

Siskin, M., Schlosberg, R. H., and Kosci, W. P. (1977). Industrial and laboratory alkylation. *ACS Symp. Ser.* No. 55, 186–204.

Wristers, J. (1976). *J. Am. Chem. Soc.* **97,** 4312.

Wristers, J. (1977). *J. Am. Chem. Soc.* **99,** 5056.

VIII. Reforming and Hydrocracking

* Anderson, J. R. (1973). Metal catalyzed skeletal reactions of hydrocarbons. *Adv. Catal.* **23,** 1–91.

* Ciapetta, F. G., Dobres, R. M., and Baker, R. W. (1958). Catalytic reforming of pure hydrocarbons and petroleum naphthenes. *In* "Catalysis" (P. H. Emmett, ed.), Vol. 6, pp. 495–692. Reinhold, New York.

Egan, C. J., Langlois, G. E., and White, R. J. (1962). *J. Am. Chem. Soc.* **84,** 1204.

* Gates, B. C., Katzer, J. R., and Schuit, G. C. (1979). Reforming. *In* "Chemistry of Catalytic Processes," pp. 184–324. McGraw-Hill, New York.

Krane, H. G., *et al.* (1959). *World Pet. Congr., 5th, New York* Sect. III, p. 39.

Langlois, G. E., and Sullivan, R. F. (1970). Chemistry of hydrocracking. *Adv. Chem. Ser.* No. 97, 38–67.

* Langlois, G. E., Sullivan, R. F., and Egan, C. Y. (1966). *J. Phys. Chem.* **70,** 3666.

Pines, H., and Shaw, A. W. (1957). *J. Am. Chem. Soc.* **79,** 1474.

Sullivan, R. F., Egar, C. J., Langlois, G. E., and Sieg, R. P. (1961). *J. Am. Chem. Soc.* **83,** 1156.

Sullivan, R. F. Egar, C. J. and Langlois, G. E. (1964). *J. Catal.* **3,** 183.

Weisz, P. B. (1961). *Actes Congr. Int. Catal., 2nd* p. 937. Edition's Technip, Paris.

2

Base-Catalyzed Reactions

I. INTRODUCTION

Acid-catalyzed conversions of hydrocarbons have been widely studied because of the many important petrochemical processes involving catalysis by acids. In contrast, base-catalyzed reactions of hydrocarbons received relatively little attention until two decades ago, despite the fact that it has been known for nearly a century that unsaturated hydrocarbons may undergo reactions in the presence of bases. For example, it was reported as early as 1887 that 1-butyne is converted to 2-butyne at 170°C in a sealed tube in the presence of alcoholic KOH and that 2-hexyne is converted to 1-hexyne by reaction with metallic sodium (Faworsky, 1887). A similar isomerization of 3-ethyl-1,2-pentadiene (an allene) to 3-ethyl-1-pentyne by the use of sodium was recorded at the beginning of this century (Mereshkowski, 1913). Such reactions did relatively little to promote research activity in this field. In 1955, however, an observation of the author, that sodium, in the presence of small amounts of organosodium compounds produced *in situ,* acts as an effective catalyst for double-bond isomerization of simple olefins, triggered a great deal of research in this field. It was subsequently discovered that low-molecular-weight olefins can undergo selective oligomerization and that alkylaromatic hydrocarbons can undergo side-chain alkylation and cyclialkylation. Base-catalyzed carbon–carbon addition reactions are of interest in the area of synthesis because they afford hydrocarbons in good yields by a simple one-step procedure.

A. Acidity of Hydrocarbons

Base-catalyzed reactions of hydrocarbons generally involve, as the first step, the formation of a carbanion. This is usually accomplished by the removal of an activated hydrogen from the substrate by a base. Of most interest are the allylic protons of olefins and the benzylic protons of alkylaromatic compounds, both of which are the most active hydrogens in the respective classes of compounds.

The tendency of a hydrocarbon to participate in a base-catalyzed conversion reaction depends to a large extent on the ease of removal of a proton from the hydrocarbon or on the acidity of the hydrocarbon. The classification of weak carbon acids according to acidity is based on the treatment of alkali metal salts of carbon acids in solution with other very weak carbon acids, and the equilibrium constant between the two salts is estimated from colorimetric determinations. The equilibrium involved in the metalation reaction is

$$RH + R'^-M^+ \rightleftharpoons R^-M^+ + R'H$$

$$M = \text{alkali metal}$$

If the acid strength pK_a is defined by the usual equation, then the difference in pK_a values for the two carbon acids can be given by the following equation:

$$pK_a - pK_{a'} = -\log\frac{[R^-]}{[RH]} + \log\frac{[R'^-]}{[R'H]}$$

The acidity scale of various hydrocarbons arranged according to decreasing acidity is presented in Table 2.1. In each class of compounds the primary anion is more stable than the secondary, and the latter is more stable than a tertiary, $RCH_2^- > R_2CH^- > R_3C^-$; this is the opposite of the order encountered in carbonium ion systems.

TABLE 2.1

Acidity Scale of Hydrocarbons[a]

Compound	pK_a	Compound	pK_a
Fluoradene	11	Ethylene	36.5
Cyclopentadiene	15	Benzene	37
9-Phenylfluorene	18.5	Cumene (α position)	37
Indene	18.5	Triptycene (α position)	38
Phenylacetylene	18.5	Cyclopropane	39
Fluorene	22.9	Methane	40
Acetylene	25	Ethane	42
1,3,3-Triphenylpropene	26.5	Cyclobutane	43
Triphenylmethane	32.5	Neopentane	44
Toluene (α position)	35	Propane (sec position)	44
Propene (α position)	35.5	Cyclopentane	44
Cycloheptatriene	36	Cyclohexane	45

[a] From Cram (1965).

B. Catalysts •

The choice of catalysts for base-catalyzed reactions depends to a great extent on the acidity of the hydrocarbons used in the reaction. The most effective catalysts are alkali metals such as Na, K, and Cs. Although these metals do not act catalytically as such, they may interact with some of the reactants present, such as conjugated dienes, to form organoalkali compounds, which then act as initiators of the catalytic reaction. In the absence of compounds with which the alkali metals can interact, small amounts of "promoters" can be added to form organoalkali compounds with the metals.

The function of a promoter is twofold: It forms organoalkali compounds essential for the initiation of the reaction, and in addition it helps to break down the alkali metals into a finely dispersed colloidal powder. With anthracene as a promoter the reaction proceeds as follows:

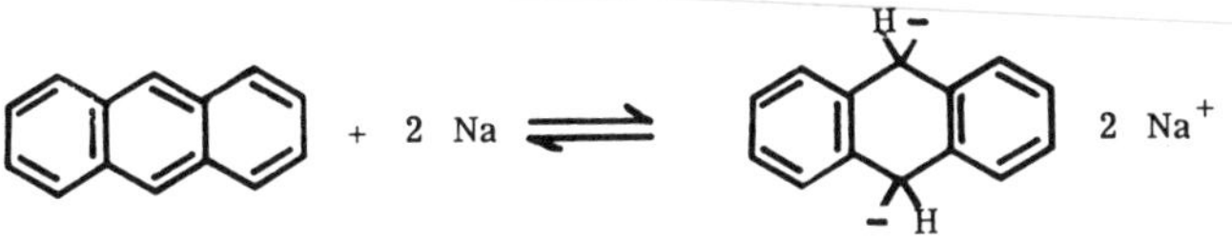

At the end of the experiments much of the anthracene is recovered.

When o-chlorotoluene is used as a promoter, benzylsodium is produced. Then the latter either is converted to toluene (see p. 128) or,

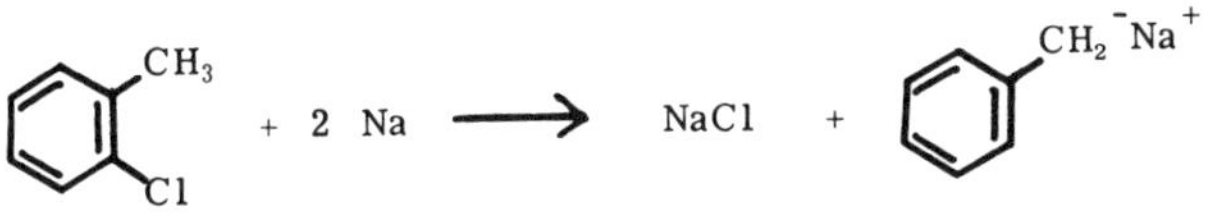

in some special instances, may enter into reaction with the substrate. Both reactions start with Na, which changes stepwise to Na$^+$. Thus, the promoter is a carbanion that is readily formed from metallic Na or K.

Alkali metals as heterogeneous catalysts can be used in a dispersed form or on supports such as alumina, graphite, and silica. Alkali metals as such or in the form of alkoxides, amides, etc., have been used in homogeneous catalytic systems for the double-bond isomerization of olefins and in certain alkenylation reactions of alkyl aromatic hydrocarbons. Most extensively studied have been solutions of potassium *tert*-butoxide in aprotic solvents.

II. ISOMERIZATION OF OLEFINS

A. Olefins

Olefins undergo a reversible double-bond migration in the presence of bases, without an accompanying skeletal isomerization, as is often the case with acid-catalyzed reactions. The ease of isomerization is related to the acidity of the allylic hydrogen of the olefin and to the strength of the base used as catalyst.

1. Heterogeneous Catalysts

Under various conditions all the following bases have shown activity for olefin isomerization: metal amides, metal hydrides, organoalkali metal compounds, and alkali metals supported on inert carriers. At higher temperatures alkali earth metal oxides act as catalysts.

Sodium/organosodium catalysts, which are prepared *in situ* by reacting an excess of sodium with an organic compound such as *o*-chlorotoluene or anthracene, are effective catalysts for the isomerization of olefins. These catalysts require temperatures in the range of 150°–200°C for reasonable rates.

Cyclic unsaturated hydrocarbons, such as 1-*p*-menthene (1), undergo double-bond isomerization in the presence of $Na/C_6H_5CH_2Na$ catalyst. After 20 hr at reflux temperature an equilibrium is established consisting of hydrocarbons 1–3. Similarly, 1-methylcyclopentene, on heating for 80 hr, affords an equilibrium mixture of double-bond isomers consisting of 4–7.

1 (32%)	2 (63%)	3 (5%)

4 (94.9%)	5 (0.7%)	6 (9.7%)	7 (4.5%)

Isomerization of bicyclic alkenes also was noted for the $Na/C_6H_5CH_2Na$ catalyst system in refluxing xylene. After 4 hr the equilibrium between β-pinene (8) and α-pinene (9) was established.

8 (4%)	9 (96%)

The most active catalysts appear to be alkali metals supported on highly surface-activated alumina; they are effective at or near 25°C (Table 2.2). These catalysts can be prepared by adding sodium to dry alumina, which is stirred with a sweep-type stirrer in a glass flask at 150°C under nitrogen atmosphere.

TABLE 2.2

Isomerization of Monoolefins over Na/Al$_2$O$_3$[a]

Starting material	Reaction conditions			Product (mol %)				
	Contact time (min)	ml olefin/ gm catalyst	Temp (°C)	1-	cis-2-	trans-2-	cis-3-	trans-3
cis-2-Butene	15	23	25	3	26	71	—	—
trans-2-Butene	45	23	25	2	26	72	—	—
1-Butene	0.6	20	25	3	26	71	—	—
1-Pentene	5	6.4	0	1.6	17	81	—	—
	150	6.4	0	1.3	16	83	—	—
	950	6.4	25	1.8	18	80	—	—
1-Hexene	6	6.4	0	1.1	17.2	60.2	2.5	18.4
	60	6.4	0	0.6	13.9	62.1	1.7	20.8
1-Octene	7	5.8	25	—[b]	—	—	—	—

[a] From O'Grady et al. (1959).
[b] Product consisted of 1% 1-octene, 4% cis-octenes, and 95% trans-octenes.

A very active catalyst, composed of 5% sodium on alumina, was also prepared by impregnating alumina with an ammoniacal solution of sodium. When passed over it, 1-butene was readily isomerized at 20°C to 2-butene with a ratio of cis to trans equal to 5:1, and 1-heptene isomerized to a mixture of *cis*-2-, *cis*-3-, *trans*-2- and *trans*-3-heptene. A similar type of double-bond isomerization occurs with branched alkenes.

With a cyclic system, the presence of Na/Al₂O₃ led to the conversion of **10** to **11**. Equilibrium was established in 2 hr.

10   (14%)   11   (86%)

The three steps in the isomerization of olefins over strong bases are given in Scheme 2.1. The first step involves abstraction of an allylic pro-

Step 1. B^-M^+ + $RCH_2CH{=}CH_2$ ⟶ BH + $RCHCH{=}CH_2M^+$

(B = H, NH₂, alkyl, aryl, etc. ; M = alkali metal)

Step 2. $R\bar{C}HCH{=}CH_2M^+$ ⟷ $RCH{=}CHCH_2^-M^+$

Step 3. $RCH{=}CHCH_2^-M^+$ + $RCH_2CH{=}CH_2$ ⟶ $RCH{=}CHCH_3$ + $R\bar{C}HCH{=}CH_2M^+$

Scheme 2.1. Mechanism of base-catalyzed olefin isomerization.

ton. The carbanion thus generated is a resonance hybrid (step 2). In the presence of excess olefin, transmetalation occurs, resulting in the isomeric olefin and more of the basic intermediate (step 3). A similar mechanism applies to the isomerization of cycloalkenes.

2. Homogeneous Catalysts

A variety of homogeneous systems were found to catalyze double-bond isomerization, such as Li or NaNH₂ in ethylenediamine and Na or K in hexamethylphosphoramide, (Me₂N)₃PO.

The most widely used base is *tert*-BuOK in aprotic solvents. Changes in the solvent have a great effect on the reaction, and the variation is assumed to be caused by the change in aggregation of the ions in solution. The rate decreases with decreasing dielectric constant of the solvent, Me₂SO >> (Me₂N)₃PO > N-methyl-2-pyrrolidone > tetramethylurea. Since the catalytic system is sensitive to air and water, resublimation of the alkoxide before use is necessary, and the solvents must be anhydrous.

Isomerization of 2-methyl-1-pentene (**12**), using *tert*-BuOK in Me₂SO at 55°C results in an isomeric mixture after 15 hr containing 88% of **12** and 12% of 2-methyl-2-pentene (**13**). Only after 750 hr of contact time do all of the other isomers appear. The relatively rapid interconversion of **12** and

13 in contrast to the slow appearance of 4-methyl-2-pentene (**14**) is readily rationalized in terms of carbanion stability. The mechanism assumes that ionization of the allylic hydrogen determines the rate. Examination of the starting olefins shows that three possible proton abstractions might occur [Eqs. (1)–(3)]. Equations (1) and (2) show that the same mesomeric anion results from either compound **12** or **13**, one a primary and the other a secondary, but Eq. (3) shows that the resonance hybrid consists of a secondary and a tertiary anion (**14a**). Since the stability of a tertiary carbanion is lower than that of a secondary or primary one, the rate of isomerization of **13** to 4-methyl-2-pentene (**14**) should thus be lower than that of **13** to **12**.

$$\text{B}^- + \underset{\textbf{12}}{\diagup\!\!\diagup\!\!\diagdown} \xrightarrow{-\text{BH}} \diagup\!\!\diagup\!\!\diagdown \longleftrightarrow \diagup\!\!\diagup\!\!\diagdown \qquad (1)$$

$$\text{B}^- + \underset{\textbf{13}}{\diagdown\!\!\diagup\!\!\diagdown} \xrightarrow{-\text{BH}} \diagdown\!\!\diagup\!\!\diagdown \longleftrightarrow \diagdown\!\!\diagup\!\!\diagdown \qquad (2)$$

$$\text{B}^- + \diagdown\!\!\diagup\!\!\diagdown \xrightarrow{-\text{BH}} \diagdown\!\!\diagup\!\!\diagdown \longleftrightarrow \underset{\textbf{14a}}{\diagdown\!\!\diagup\!\!\diagdown} \qquad (3)$$

B. Arylolefins

Arylalkenes, being more acidic than the corresponding alkenes, undergo facile isomerization under mild conditions in the presence of both heterogeneous and homogeneous catalytic systems. Bases that are inactive for the isomerization of alkenes, such as C_2H_5ONa/C_2H_5OH or *tert*-BuOK as such, bring about the isomerization of arylalkenes and similar compounds to their equilibrium mixture.

C. Dienes

1. Acyclic Dienes

The isomerization of nonconjugated dienes to the stable conjugated isomers occurs with a variety of catalysts. Over Na/Al_2O_3 at ambient temperature for 10 min the starting hexadiene (**15** or **16**) produced an equilibrium that was 99% in favor of **17** or **18**, respectively. Under the same conditions 1,4-pentadiene yielded 94% of 1,3-pentadiene.

$$CH_2=\underset{\underset{CH_3}{|}}{C}CH_2CH_2CH=CH_2 \qquad CH_3\underset{\underset{CH_3}{|}}{C}=CHCH_2\underset{\underset{CH_3}{|}}{C}=CH_2$$

15   **16**

$$CH_3\underset{\underset{CH_3}{|}}{C}=CHCH=CHCH_3 \qquad CH_3\underset{\underset{CH_3}{|}}{C}=CHCH=\underset{\underset{CH_3}{|}}{C}CH_3$$

17   **18**

Comparison of the relative rate of isomerization of 1,3-, 1,4-, and 1,5-hexadienes in a medium of *tert*-BuOK/*tert*-BuOH indicates the following relative reactivities: 1,4 >> 1,3 > 1,5.

2. Cyclic Dienes

The double-bond isomerization of cyclic hydrocarbons having two double bonds takes place readily. When 1,4-cyclooctadiene is contacted with Na/Al$_2$O$_3$ at 0°C for 10 min, over 95% of 1,3-cyclooctadiene results. Similarly, 1,5-cyclooctadiene is converted to 1,3-cyclooctadiene in quantitative yield upon treatment with *tert*-BuOK/Me$_2$SO at 25°C for 1 hr.

If a cyclic diolefin possesses a six-membered ring, mild reaction conditions must be used to prevent aromatization. Refluxing limonene (**19**) in the presence of Na/C$_6$H$_5$CH$_2$Na catalyst results in initial isomerization to an equilibrium mixture of conjugated dienes **20** and **21** and some aromatization to *p*-cymene (**22**). On further heating, aromatization, with the evolution of hydrogen, is complete. The absence of any endocyclic dienes such as **23** is explained by the rapid rate with which these compounds undergo dehydrogenation.

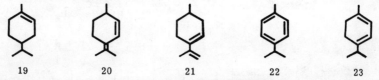

19   **20**   **21**   **22**   **23**

5,5-Dimethyl-3-methylenecyclohexene, (**24**), upon refluxing in the presence of Na/C$_6$H$_5$CH$_2$Na, isomerizes to endocyclic dienes **25** and **26**; this is followed by a rapid aromatization to *m*-xylene with loss of methane.

Methyl anion precedes the formation of methane.

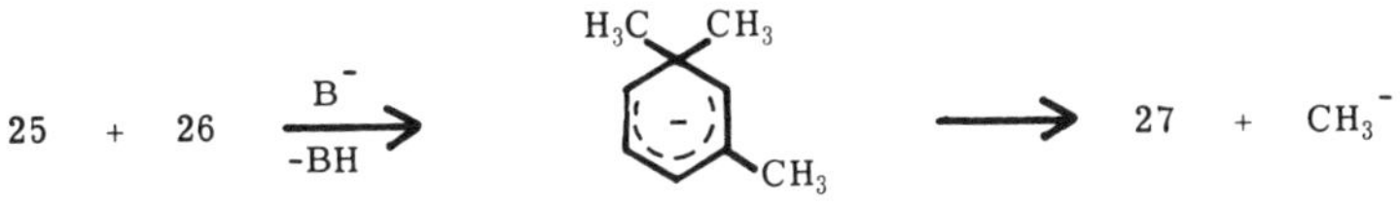

$$25 \;+\; 26 \;\xrightarrow[-BH]{B^-}\; \text{26a} \;\longrightarrow\; 27 \;+\; CH_3^-$$

(structure labeled 26a: cyclohexadiene ring with H₃C, CH₃ at top and CH₃ at right)

$$24 \;+\; CH_3^- \;\longrightarrow\; C_9H_{13}^- \;+\; CH_4$$

D. Acetylenes

The base-catalyzed isomerization of acetylenic hydrocarbons is a much easier process than the isomerization of the corresponding olefins. The catalysts that can be used for this reaction include tert-BuOK/tert-BuOH at 85°–140°C, tert-BuOK/Me₂SO at 20°C, KNH₂/Al₂O₃ at 70°C, NaNH₂/NH₃ at −33°C, and KOH/EtOH at 175°C.

The composition of products obtained from treating alkynes in the presence of various catalysts is presented in Table 2.3. The reversible isomerization of alkynes and the formation of alkadienes (allenes) have been explained by a mechanism that assumes the involvement of resonance-stabilized carbanions as intermediates (Scheme 2.2).

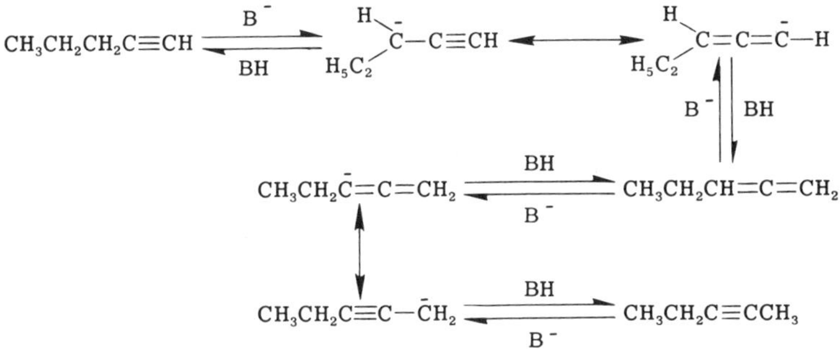

Scheme 2.2. Mechanism of base-catalyzed alkyne isomerization.

The results in Table 2.3 indicate that under the given conditions the 2-alkyne is the predominant isomer, whereas the conjugated dienes, presumably the most stable isomers, are not formed to any appreciable extent. It appears that kinetic factors are of major importance in the isomerization of acetylenes, and conjugated dienes are not favored kinetically.

Which product is obtained during the isomerization of alkynes depends on the conditions of reaction. A mixture of isomerized alkynes including 5-decyne is obtained when 1-decyne is treated with tert-BuONa/BuOH. At 160°C 2-decyne is the predominant component, whereas at 240°C 5-decyne predominates. Treatment of 1-decyne with NaNH₂/Me₂SO at 70°C for 30 hr gives a 94% yield of 2-decyne that is essentially free of other iso-

TABLE 2.3

Equilibration of *n*-Alkynes[a]

Alkyne	Base/solvent	Products (%)						
		1-Alkyne	1,2-Alkadiene	2-Alkyne	2,3-Alkadiene	3-Alkyne	2,4-Alkadiene
Pentyne	KOH/EtOH	1.3	3.5	95.2	—	—	—
Hexyne	NaNH$_2$/EDA	6	—	80	3	11	—
Heptyne	*t*-BuOK/*t*-BuOH	—	0.5	46	7.5	42	4
Octyne	KOH/EtOH	0.2	2.3	97.5	—	—	—

[a] From Pines and Stalick (1977, p. 126).

meric hydrocarbons. The reaction of any of the isomeric decynes with $NaNH_2/(-CH_2NH_2)_2$ results in a mixture of products containing 2-, 3-, 4-, and 5-decyne, with the first two predominating.

E. Cycloisomerization

Linear and cyclic alkenes undergo cycloisomerization in the presence of bases. Treatment of 1,3- or 1,5-cyclooctadiene with the strong bases C_6H_5K or KH at 175°–190°C gives 65–90% yield of bicyclo[3.3.0]oct-2-ene in the following steps:

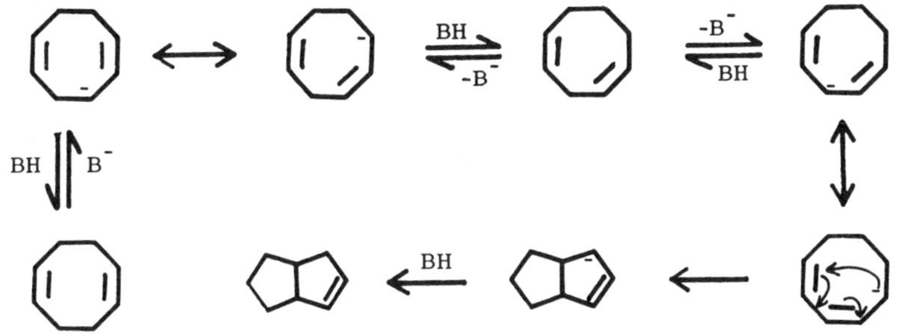

Upon standing in solution of sodium dissolved in isopropylamine, linear trienes are cycloisomerized at 20°C. The treatment of alloocimene (**28**) under these conditions for 8 days yields **29** and **30** in 50 and 18% yields, respectively (Scheme 2.3):

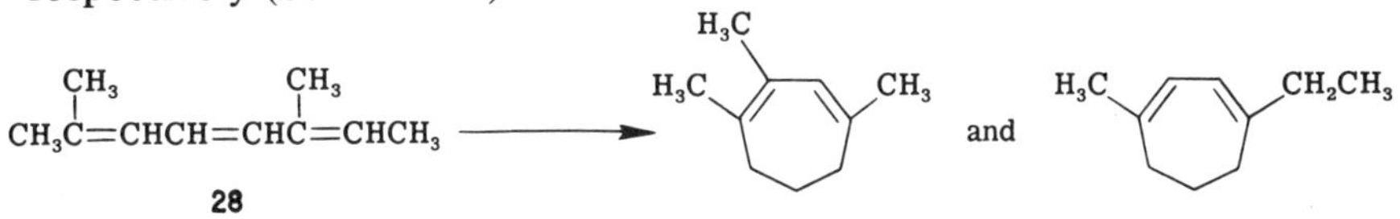

Similarly, compound **31** cyclizes to hexahydroazulene (**32**) after 24 hr in 55% yield.

The cycloisomerization of 3,3-dimethyl-1,4-hexadiene to geminal dimethylcyclohexenes is outlined in Scheme 2.4.

Phenylalkenes **33** and **34** undergo cyclization at 185°C in the presence of catalytic amounts of $K/C_6H_5CH_2K$. Compound **33** affords 57% of **35**, whereas **34** yields 68% of **36**.

The reactions are represented by Scheme 2.5. Under similar conditions $Na/C_6H_5CH_2Na$ causes only double-bond migration.

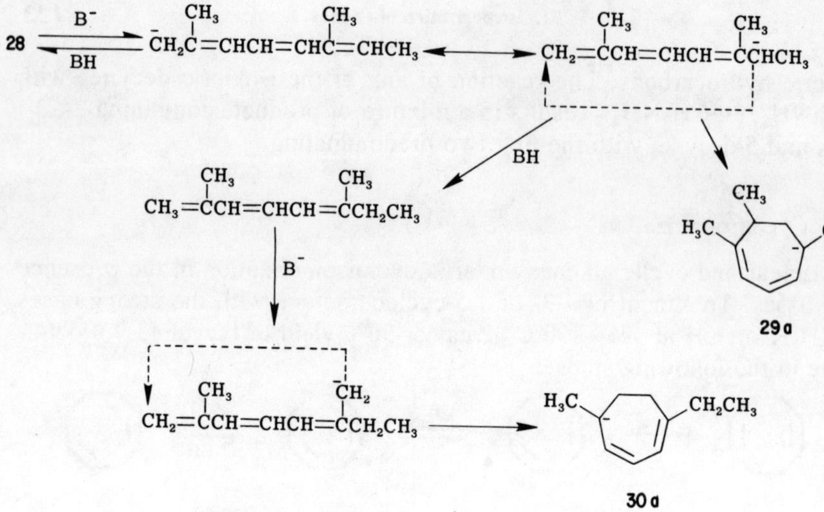

Scheme 2.3. Cycloisomerization of alloocimene.

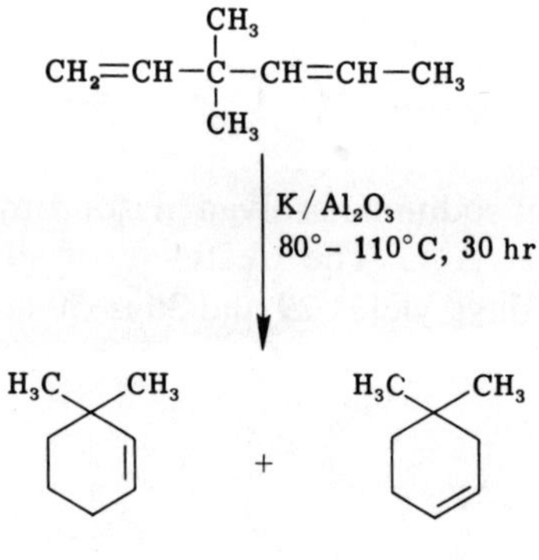

(13%) (40%)

Scheme 2.4. Cycloisomerization of 3,3-dimethyl-1,4-hexadiene.

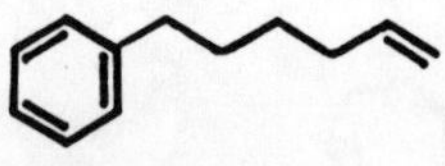

33

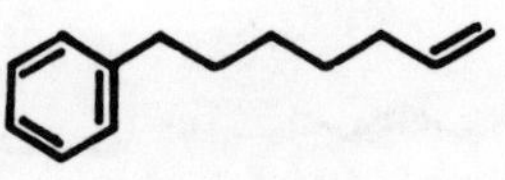

34

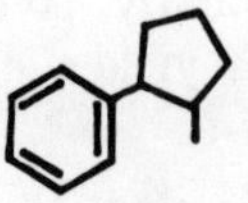

35

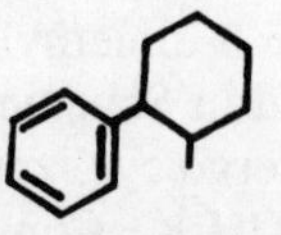

36

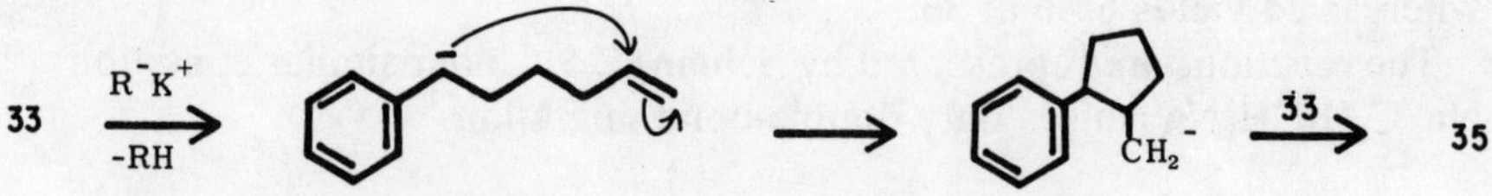

Scheme 2.5. Cyclization of 6-phenylhexene.

III. DIMERIZATION AND OLIGOMERIZATION OF OLEFINS

A. Introduction

The polymerization of conjugated dienes, such as butadiene, isoprene, and styrene, by alkali metals, has been known since the 1910's to lead to high-molecular-weight compounds. It has been found more recently that simple olefins that form allylic carbanions can undergo oligomerization in the presence of an alkali metal catalyst and a promoter. Ethylene can form cooligomers when contacted with these olefins. The reactions are carried out under pressure and with temperatures ranging from 150° to 225°C.

B. Propene

Propene dimerization takes place rapidly under pressure in a stirring type of autoclave and in the presence of an alkali metal catalyst (Table 2.4). The dimerization proceeds through the initial formation of an organoalkali compound, followed by metalation of the propene. The selectivity of conversion of propene to 4-methyl-1-pentene follows the order Cs ~ K >> Na. The rearrangement of this dimer to double-bond isomers could occur by way of a proton transfer reaction generating an allylic anion.

$$CH_2{=}CHCH_3 + R^-M^+ \xrightarrow{\;-RH\;} [CH_2 \cdots CH \cdots CH_2]^-M^+ \xrightarrow{\;CH_2{=}CHCH_3\;}$$

$$\underset{\displaystyle CH_2{=}CHCH_2\overset{\textstyle CH_3}{\underset{|}{C}}HCH_2^-M^+}{} \xrightarrow{\;CH_2{=}CHCH_3\;} \begin{array}{c} CH_2{=}CHCH_2\overset{\textstyle CH_3}{\underset{|}{C}}HCH_3 \\ + \\ [CH_2 \cdots CH \cdots CH_2]^-M^+ \end{array}$$

The dimerization of propene can also be effected over alkali metals dispersed on supports. The reactions are carried out at 150°C and at about 100 atm pressure (Table 2.5). Sodium and potassium deposited on K_2CO_3 are highly selective catalysts for the formation of the dimer 4-methyl-1-pentene.

C. Butenes

Unlike propene, 2-butene and isobutylene undergo dimerization with difficulty in the presence of alkali metal catalysts. In the presence of 1% by weight of Li/K catalyst, 7% of the charged 2-butene is converted at 200°C in 24 hr to octenes and a higher boiling liquid. Under similar condi-

TABLE 2.4

Effect of Alkali Metals on Propene Dimerization[a]

	Experiment			
	1	2	3	4
Conditions				
Temp, °C	196	211	218	200
Time, hr	4.0	4.3	4.0	3.1
Charge[b]				
Catalyst	Na	K	K	Cs
Alkali metal, gm-atom	0.065[c]	0.013	0.100	0.026
Propene conversion (%)	20	34	61	39
Products, wt % of propene converted to:				
Propane	13	10	21	13
Branched C_6 olefins	68	73	38	63
Unbranched C_6 olefins	—	7	4	3
Composition of branched hexenes, %				
4-Methyl-1-pentene	16	70	62	73
4-Methyl-2-pentene	45	29	34	26
2-Methyl-2-pentene	32	1	3	0
2-Methyl-1-pentene	7	0	1	1

[a] Shaw *et al.* (1965).
[b] The charge in each experiment consisted of ~1.52 mol propene and 1 mol benzene.
[c] 1.8 gm anthracene was added as promoter.

TABLE 2.5

Dimerization of Propene over Supported Alkali Metals[a]

Catalyst (wt % alkali metal)	Rate of reaction (gm hexane/gm-atom alkali metal × hr)	Selectivity to dimer (%)	Composition of C_6H_{12} (%)[b]
K/graphite (11.9)	120	98	62:24:4:6:4
K/graphite (5.1)	97	98	30:50:4:12:4
K/K$_2$CO$_3$ (4.4)	150	98	75:16:0:1:8
Na/graphite (3.0)	155	89	21:56:5:15:3
Na/K$_2$CO$_3$ (3.7)	91	98	74:18:0:1:7

[a] Hambling (1969).
[b] Ratio, 3-methyl-1-pentene/4-methyl-2-pentene/2-methyl-1-pentene/2-methyl-2-pentene/*n*-hexene.

tions, but with 3 hr of heating, less than 1% of isobutylene is converted to dimer. The difficulty of dimerizing isobutylene as compared to propene can be ascribed to steric hindrance to the addition of the methallyl carbanion to the olefin and to the inductive effects of the methyl groups.

$$
\underset{\underset{\displaystyle CH_3}{|}}{CH_2{=}C{-}CH_2^-} \;+\; \underset{\underset{\displaystyle CH_3}{|}}{\overset{\overset{\displaystyle CH_3}{|}}{C}{=}CH_2} \longrightarrow \underset{\underset{\displaystyle CH_3}{|}}{CH_2{=}C{-}CH_2{-}\underset{\underset{\displaystyle CH_3}{|}}{\overset{\overset{\displaystyle CH_3}{|}}{C}}{-}CH_2^-}
$$

D. Ethylation of Olefins

Ethylene is not polymerized to any significant extent by alkali metal catalysts below 200°C. Although ethylene is not metallated to form a vinyl anion at these temperatures, it is an excellent anion acceptor.

⁻ The reaction of an equimolar mixture of ethylene and propene passed through an alkali metal supported catalyst bed maintained at 80°–120°C gave 92% unbranched pentenes and 5.2% hexenes. 1-Butene and 2-butene reacted with ethylene at 80°C, whereas the isomeric isobutylene required a higher temperature for copolymerization. Reactions of 1- and 2-butene produced hexenes composed of about 90% 3-methyl-1-pentene, whereas the reaction of isobutylene resulted in 2-methyl-1-pentene with 96% selectivity. The ethylation of simple olefins by ethylene using an equiatomic ratio of lithium/potassium catalyst is presented in Table 2.6.

E. α-Methylstyrene

Styrene undergoes facile polymerization in the presence of alkali metals to produce macromolecules. However, α-methylstyrene, which contains an allylic hydrogen, on refluxing in the presence of $Na/C_6H_5CH_2Na$ catalyst affords a cyclic dimer, 1-methyl-1,3-diphenylcyclopentane (37), in 32% yield, as determined on the basis of reacted olefin. The other products are isopropylbenzene, a trimer, and small amounts of p-terphenyl.

$$
\underset{\underset{\displaystyle C_6H_5\overset{\overset{\displaystyle CH_2}{\|}}{C}CH_3}{}}{} \xrightarrow[-RH]{R^-Na^+} \underset{C_6H_5\overset{\overset{\displaystyle CH_2}{\|}}{C}CH_2^-Na^+}{} \xrightarrow{C_6H_5\overset{\overset{\displaystyle CH_2}{\|}}{C}{-}CH_3} \left[C_6H_5\overset{\overset{\displaystyle CH_2}{\|}}{C}CH_2CH_2\underset{\underset{\displaystyle C_6H_5}{|}}{\overset{\overset{\displaystyle CH_3}{|}}{C}} \right] Na^+
$$

TABLE 2.6

Ethylation of Some Simple 1- and 2-Alkenes over Li/K Catalyst[a,b]

Olefin	Olefin/ethylene (molar ratio)	Temp (°C)	Reaction time (hr)	Conversion (%) Olefin[c]	Conversion (%) Ethylene	Selectivity (wt %) of olefin converted to 3-Ethylated 1-alkene	Selectivity (wt %) of olefin converted to Monoethylated internal $n-4$ alkenes	Selectivity (wt %) of olefin converted to Isomerized monoethylated products	Selectivity (wt %) of olefin converted to Diethylated products	Selectivity (wt %) of olefin converted to Residue
Propene	0.6	200	0.4	89	55	42	42	30	14	13[d]
Isobutylene	1.0	200	0.7	71	89	48	48	19	29	4
1-Butene	0.9	175	0.6	52	67	46	25	13	11	5
2-Butene	0.9	200	0.4	59	64	58	17	10	8	6
1-Pentene[e]	1.0	200	0.3	44	64	43	29	Trace	8	11
2-Pentene[f]	1.0	200	0.2	48	65	37	33	Trace	5	2
3-Hexene	1.0	200	0.3	34	59	46	47	Trace	5	2
2-Hexene	0.7	200	0.3	23	44	38	52	7	Trace	3
1-Heptene	0.6	200	1.1	51	83	—	—	—	7	—

[a] Bush et al. (1965).
[b] Carried out over 1 wt % Li/K in a molar ratio of 1:1.
[c] Exclusive of double-bond position isomerization of starting olefin.
[d] Includes 5% propene dimer.
[e] 4-Methyl-2-hexene (9%) also formed from internal metalation of 2-pentene.
[f] 4-Methyl-2-hexene (13%) also formed from internal metalation of 2-pentene.

F. β-Methylstyrene

This hydrocarbon dimerizes to 1,5-diphenyl-4-methyl-1-pentene with a yield of 80–90% when heated to 100°–155°C in the presence of sodium. The first step in the reaction is metalation of the allylic methyl group.

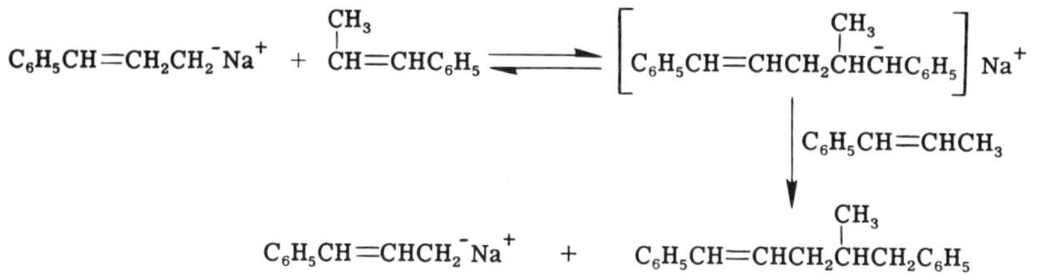

$$C_6H_5CH{=}CH_2CH_2^-Na^+ \; + \; \overset{\overset{\textstyle CH_3}{|}}{CH}{=}CHC_6H_5 \; \rightleftharpoons \; \left[C_6H_5CH{=}CHCH_2\overset{\overset{\textstyle CH_3}{|}}{C}HCHC_6H_5 \right] Na^+$$

$$\Big\downarrow C_6H_5CH{=}CHCH_3$$

$$C_6H_5CH{=}CHCH_2^-Na^+ \quad + \quad C_6H_5CH{=}CHCH_2\overset{\overset{\textstyle CH_3}{|}}{C}HCH_2C_6H_5$$

IV. REACTION OF AROMATIC COMPOUNDS WITH OLEFINS

A. Introduction

The acid-catalyzed reactions of aromatic hydrocarbons with olefins result almost exclusively in the addition of alkyl groups to the aromatic ring (Chapter 1, Section V). The base-catalyzed reaction of alkylaromatic hydrocarbons with olefins is unique in that it allows one to enlarge the alkyl group of an arylalkane. Arylalkanes suitable for this reaction are those that contain a benzylic hydrogen. The olefins used are ethylene, propene, and, to a smaller extent, butenes. Condensations with conjugated dienes, such as butadiene and isoprene, change the alkyl group to a (3-alkenyl)arene. Reactions of styrene and its derivatives, such as α-methyl- and β-alkylstyrenes, produce aralkylated aromatic hydrocarbons.

Catalysts and reaction conditions depend greatly on the aromatic compounds and olefins used in the reaction. Sodium and potassium are the most effective catalysts for the side-chain alkylation of alkylaromatic hydrocarbons. A sodium catalyst usually requires the presence of small amounts of organosodium compounds or the presence of promoters, chain precursors, which can react with sodium to form organosodium compounds. The most frequently used promoters are anthracene and o-chlorotoluene. Sodium hydride, benzylsodium, and supported alkali metals, such as potassium on graphite, are also used as catalysts for the above reaction. In instances in which the substrates react readily with the alkali metal to form organoalkali metal compounds, the presence of promoters is not required.

Sodium and potassium catalysts yield different results when used for side-chain alkylation of alkylaromatic hydrocarbons, whereas lithium compounds are not very effective catalysts. Temperatures of 150°–250°C are usually required for side-chain alkylation, whereas side-chain alkenylation and aralkylation proceed at much lower temperatures.

Of the alkenes, ethylene reacts readily with alkylbenzenes; propene requires higher temperatures, and isobutylene still more drastic conditions. Higher boiling alkenes undergo double-bond isomerization when submitted to side-chain alkylation conditions, and they are less suitable for side-chain alkylation reactions.

B. Alkylation of Aromatic Hydrocarbons

1. Ethylation of Alkylbenzenes

The ethylation proceeds at 150°–200°C, with ethylene pressures ranging from 1 to 70 atm. The reaction involves the replacement of benzylic hydrogens by ethyl groups; i.e., toluene yields n-propylbenzene. Further substitution of the α-carbon yields 3-phenylpentane and 3-ethyl-3-phenylpentane.

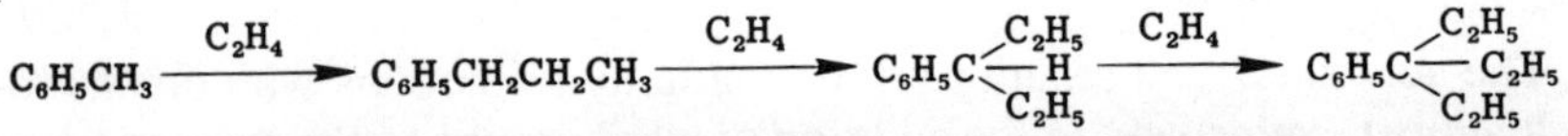

$$C_6H_5CH_3 \xrightarrow{C_2H_4} C_6H_5CH_2CH_2CH_3 \xrightarrow{C_2H_4} C_6H_5C\overset{C_2H_5}{\underset{C_2H_5}{-}}H \xrightarrow{C_2H_4} C_6H_5C\overset{C_2H_5}{\underset{C_2H_5}{-}}C_2H_5$$

The mechanism for this reaction involves the addition of the benzylic carbanion, followed by a transmetalation reaction with more of the aromatic hydrocarbon (Scheme 2.6). The addition of the anion to the ethylene could be considered to be energetically the most difficult of the steps, for it involves the addition of a resonance-stabilized anion to ethylene to

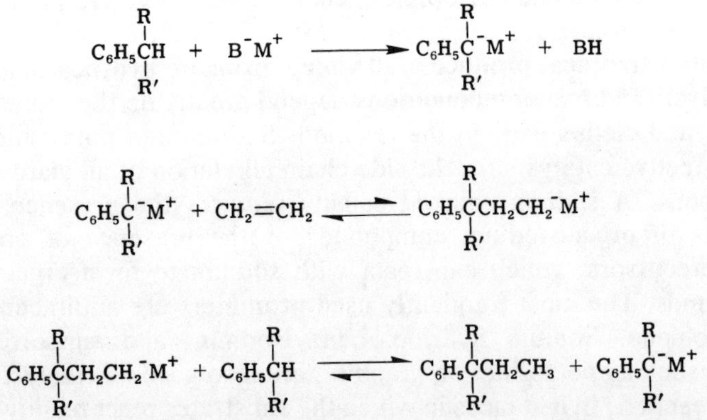

$$C_6H_5\overset{R}{\underset{R'}{C}}H + B^-M^+ \longrightarrow C_6H_5\overset{R}{\underset{R'}{C}}{}^-M^+ + BH$$

$$C_6H_5\overset{R}{\underset{R'}{C}}{}^-M^+ + CH_2{=}CH_2 \rightleftharpoons C_6H_5\overset{R}{\underset{R'}{C}}CH_2CH_2{}^-M^+$$

$$C_6H_5\overset{R}{\underset{R'}{C}}CH_2CH_2{}^-M^+ + C_6H_5\overset{R}{\underset{R'}{C}}H \rightleftharpoons C_6H_5\overset{R}{\underset{R'}{C}}CH_2CH_3 + C_6H_5\overset{R}{\underset{R'}{C}}{}^-M^+$$

Scheme 2.6. Mechanism of base-catalyzed ethylation of alkylbenzenes (M = alkali metal; B⁻M⁺ = an organoalkali metal compound formed by the reaction of an alkyl metal with the promoter or with the hydrocarbon reagent; R = H, alkyl, or aryl; R′ = H or alkyl) (Pines et al., 1955).

form an anion that is not stabilized by resonance. The last step in the reaction is energetically the most favored since the benzylic carbanion is being restored.

a. Sodium Catalyst. The products and yields obtained from the ethylation of a variety of alkylbenzenes in the presence of sodium are presented in Table 2.7. The relative rate of reaction of monosubstituted benzenes is n-alkyl > sec-alkyl > cyclohexyl.

The ethylation of isomeric xylenes and of mesitylene takes place under experimental conditions specified in Table 2.7, footnote b. The monoethylated products consisted of the corresponding n-propyltoluenes and n-propyl-m-xylene, respectively. The diethylated hydrocarbons were composed of di-n-propylbenzenes and 3-tolylpentanes.

b. Potassium Catalyst. Potassium, unlike sodium, also catalyzes a cyclialkylation reaction resulting in the formation of indans (Table 2.8). The ratio of indans to monoethylalkylbenzenes produced depends on the extent

TABLE 2.7

Ethylation of Monoalkylbenzenes and Related Hydrocarbons[a,b]

Aromatic compounds C_6H_5R R =	Sodium, gm (gm-atom)	Anthracene, gm (mol)	Aromatic compounds reacted, mol %	Mono-ethylated aromatic compounds, mol %	Diethylated aromatic compounds, mol %
Methyl	10 (0.43)	4 (0.022)	65	57[c]	22[d]
Ethyl	10 (0.43)	4 (0.022)	68	75[e]	10[f]
Isopropyl	10 (0.43)	4 (0.022)	37	78[g]	—
sec-Butyl	7 (0.30)	2.5[h] (0.019)	41	90[f]	—
tert-Butyl	5 (0.22)	2 (0.011)	2	—	—
Cyclohexyl	6 (0.26)	2 (0.011)	18	50[i]	—
Indan	6 (0.26)	2 (0.011)	14	26[j]	—

[a] Pines *et al.* (1955).

[b] The ethylation was carried out in an 850-ml rotating autoclave using 1 mol arene and 30 atm ethylene pressure. The experiments were done at 200°–225°C for 4 hr.

[c] n-Propylbenzene.

[d] 3-Phenylpentane.

[e] sec-Butylbenzene.

[f] 3-Methyl-3-phenylpentane.

[g] $tert$-Pentylbenzene.

[h] o-Chlorotoluene was used as a prometer.

[i] 1-Ethyl-1-phenylcyclohexane.

[j] Ethylindan.

TABLE 2.8

Potassium-Catalyzed Ethylation of Aromatic Hydrocarbons[a,b]

C_6H_5R, R =	Temp (°C)	Reaction time (hr)	Yield of monoadduct (mol %)	Products	Distribution (mol %)
CH_3	190 ± 5	11	53	n-Propylbenzene	98
				Indan	2
C_2H_5	190 ± 2	3	64	sec-Butylbenzene	86
				1-Methylindan	14
$i\text{-}C_3H_7$	185 ± 1	3.5	33	tert-Pentylbenzene	66
				1,1-Dimethylindan	34
$i\text{-}C_3H_7$	245 ± 3	2.5	22	tert-Pentylbenzene	49
				1,1-Dimethylindan	51
$c\text{-}C_6H_{11}$	186 ± 4	5	13	1-Ethyl-1-phenyl-cyclohexane	63
				Spiro(cyclohexane-1,1'-indan)	37

[a] Schaap and Pines (1957).
[b] One mole of aromatic hydrocarbon and 0.2 mol of ethylene were used in most of the experiments. The catalyst consisted of 1.7 gm potassium and 1.0 gm anthracene. The yields were based on the amount of ethylene charged. The recovered gases consisted of ethylene and ethane.

of substitution of the α-carbon of the alkylarene; it increases from toluene to ethylbenzene to isopropylbenzene.

The mechanism of reaction between alkylbenzenes and ethylene is similar to that suggested for sodium. The adduct of the benzylic carbanion with ethylene, however, is a very reactive alkyl carbanion, and cyclization can take place by the attack of this carbanion on the aromatic ring followed by elimination of a hydride ion:

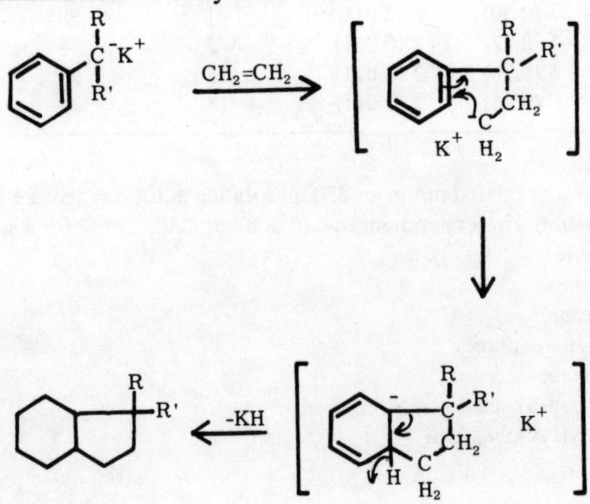

Potassium hydride, which is eliminated in the cyclization reaction, may add to ethylene to form ethylpotassium, which then may react with the alkylbenzene to form ethane and a benzylic carbanion.

$$\text{KH} \ + \ \text{CH}_2{=}\text{CH}_2 \longrightarrow \text{CH}_3\text{CH}_2^-\text{K}^+$$

$$\underset{\underset{R'}{|}}{\overset{\overset{R}{|}}{\text{C}_6\text{H}_5{-}\text{CH}}} \ + \ \text{CH}_3\text{CH}_2^-\text{K}^+ \longrightarrow \underset{\underset{R'}{|}}{\overset{\overset{R}{|}}{\text{C}_6\text{H}_5{-}\text{C}^-\text{K}^+}} \ + \ \text{CH}_3\text{CH}_3$$

2. Ethylation of Alkylnaphthalenes

Sodium or potassium dispersed in 1- or 2-methylnaphthalene is an active catalyst for the reaction of these hydrocarbons with ethylene at a temperature from 90° to 210°C. To form an active sodium catalyst, a promoter such as o-chlorotoluene is needed, whereas in the case of potassium an initiator is not always required (Table 2.9).

Sodium is a selective catalyst for the side-chain ethylation of alkylnaphthalenes. With 1-methylnaphthalene, 1-propyl- and 1-(1-ethylpropyl)naphthalene are the only products obtained. Only on prolonged reaction are small amounts of 1-(*tert*-heptyl)naphthalene produced. 2-Methylnaphthalene, on the other hand, readily forms the 2-*tert*-heptylnaphthalene.

A complex mixture of products is obtained from the ethylation of methylnaphthalenes when potassium is used as the catalyst. Besides the side-chain alkylation, cyclialkylation and nuclear alkylation also occur. In addition, all of these compounds undergo further ethylation. Ethane is also found as a by-product of the cyclialkylation reaction.

3. Propylation of Alkylbenzenes

The reaction of alkylbenzenes with propene in the presence of alkali metals is similar to that of alkylbenzenes with ethylene. The alkylation of toluene with propene is slower than that with ethylene and requires higher temperatures (Table 2.10). With a sodium catalyst temperatures of 230°–280°C are required to make the reaction proceed satisfactorily. Propylation of toluene occurs satisfactorily at about 150°C with potassium, whereas with lithium the alkylation is sluggish even at 300°C.

The main product of side-chain propylation of toluene is isobutylbenzene. *n*-Butylbenzene in amounts ranging from 6 to 10% is also found in the alkylated fractions. The mechanism of the reaction is similar to that of the ethylation of toluene, and it involves the addition of the benzyl anion to propene:

$$\text{C}_6\text{H}_5\text{CH}_2^- \ + \ \text{H}_2\text{C}{=}\text{CHCH}_3 \rightleftharpoons \underset{\underset{|}{\overset{|}{}}}{\text{C}_6\text{H}_5\text{CH}_2\overset{\overset{\text{CH}_3}{|}}{\text{CHCH}_2^-}}$$

TABLE 2.9

Ethylation of Alkylnaphthalenes Catalyzed by Alkali Metals[a,b]

Starting material[c]	Temp (°C)	Reaction time (hr)	Catalyst[a]	Conversion (%)	Yield of products (%)[c]
A	210	8	Na	53	B (50%), C (48%), residue (3.1%)
A	208	48	Na	100	C (92%), D (2.1%), residue (6.2%)
B	195	36	Na	99	C (96%), residue (3.7%)
E	204	40	Na	100	F (94%), residue 5.8%
G	178	10	Na	76	H (26%), I (65%), J (3%), residue (6.3%)
G	175	48	Na	100	H (trace), I (31%), J (63%), residue (6.1%)
A	165	4	K	42	B (18%), C (2.2%), K (14%), L (9.7%), M (12.3%), residue (53%)
A	105	18	K	26	B (25%), C (2.6%), K (6.2%), L (19%), M (12%), residue (35%)
B	160	12	K	49	C (20), M (26%), N (9%), residue (46%)
C	140	8	K	59	N (63%), O (2.3%), residue (35%)
G	90	36	K	49	H (22%), I (2.4%), R (10%), P (42%), Q (3.1%), residue (20%)

[a] From Stipanović and Pines (1969).
[b] The reactions were carried out in a 100-ml-capacity Magne–Dash autoclave using alkylnaphthalene (0.15 mol), 30–35 atm of initial ethylene pressure, and o-chlorotoluene (0.2–0.4 ml) as a promoter in most cases.

[c] Starting materials and products are as follows:

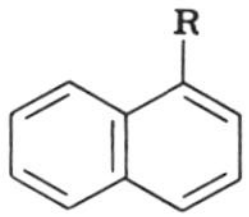

A $R = CH_3$
B $R = n\text{-}C_3H_7$
C $R = CH(C_2H_5)_2$
D $R = C(C_2H_5)_3$

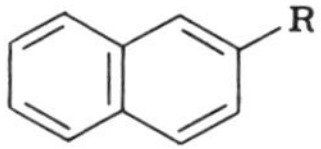

G $R = CH_3$
H $R = n\text{-}C_3H_7$
I $R = CH(C_2H_5)_2$
J $R = C(C_2H_5)_3$

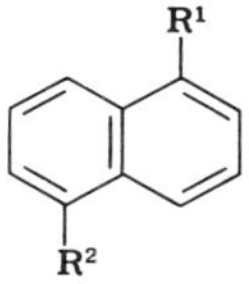

E $R^1 = R^2 = CH_3$
F $R^1 = CH_3;\ R^2 = n\text{-}C_3H_7$
K $R^1 = CH_3;\ R^2 = C_2H_5$
N $R^1 = CH(C_2H_5)_2;\ R^2 = CH(CH_3)(C_2H_5)$
O $R^1 = CH(C_2H_5)_2;\ R^2 = C(CH_3)(C_2H_5)_2$

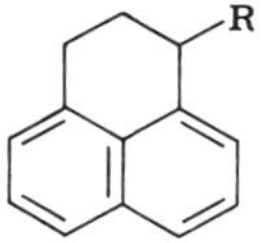

L $R = H$
M $R = C_2H_5$

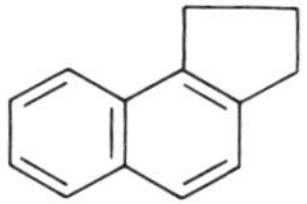

P

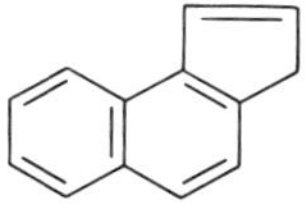

Q

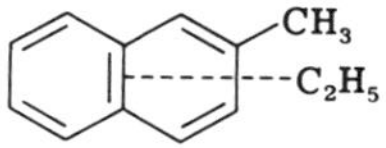

R

[d] Amount of sodium catalyst was 0.2–0.3 gm, potassium 0.3–0.5 gm.

TABLE 2.10

Reaction of Toluene with Propene in the Presence of Alkali Metals[a,b]

Alkali metal (gm-atom)	Toluene (mol)	Propene (mol)	Temp (°C)	Reaction time (hr)	Propene reacted (%)	Products and yield[c] (%)			Ratio isobutyl/ n-butyl- benzene
						Propane	Hexenes	Butyl- benzenes	
Li (0.36)	1.0	0.493	273	18	45	18	22	34	10.3:1
Li (0.36)	1.0	0.485	307	20	78	41	10	14	6.6:1
Na (0.1)	1.0	0.485	232	20	74	16	26	59	16.3:1
Na (0.1)	0.5	0.485	307	4	73	23	17	32	13.2:1
K (0.06)	1.0	0.493	107	29	58	5	10	39	9.2:1
K (0.06)	1.0	0.485	149	6	86	2	10	79	7.9:1
K (0.06)	1.0	0.485	204	2	79	7	8	72	9.1:1

[a] From Schramm and Langlois (1960).
[b] The experiments were performed under pressure in a 250-ml-capacity Magne–Dash autoclave.
[c] Yield is based on reacted propene.

The formation of the more stable primary anion provides the best evidence that ionic intermediates are involved, but the fact that some n-butylbenzene is a product shows that the benzyl anion does not add exclusively at C-2 of propene.

A major side reaction, particularly at higher temperatures, involves the elimination of hydride or methide anions, resulting in the formation of methane, propane, and uncharacterized hydrogen-deficient carbonaceous solid material. The reaction can be explained as follows:

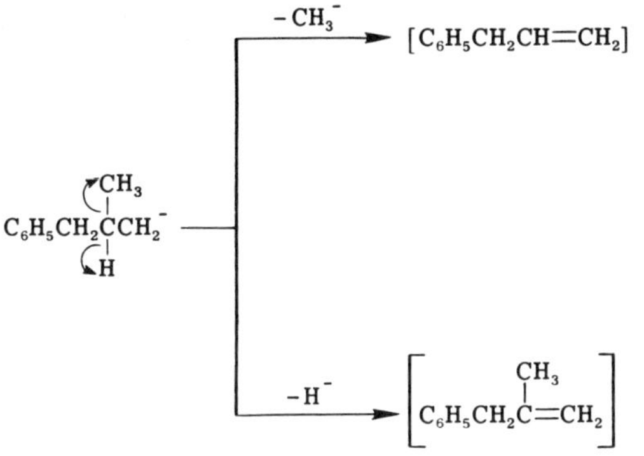

$$\begin{array}{c}
\xrightarrow{-CH_3^-} \quad [C_6H_5CH_2CH{=}CH_2] \\[2em]
C_6H_5CH_2\overset{\displaystyle CH_3}{\underset{\displaystyle H}{\overset{|}{\underset{|}{C}}}}CH_2^- \\[2em]
\xrightarrow{-H^-} \quad \left[C_6H_5CH_2\overset{\displaystyle CH_3}{\overset{|}{C}}{=}CH_2 \right]
\end{array}$$

$$H^- + CH_2{=}CHCH_3 \longrightarrow CH_2CH_2CH_3 \xrightarrow{C_6H_5CH_3} \boxed{CH_3CH_2CH_3} + C_6H_5CH_2^-$$

$$CH_3^- + C_6H_5CH_3 \longrightarrow \boxed{CH_4} + C_6H_5CH_2^-$$

The propylation products of the alkylbenzenes are those expected from the addition of a benzylic anion to propene to form the more stable primary anion. The reaction of ethylbenzene yields **38**, and that of isopropylbenzene affords **39**.

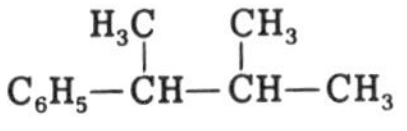

$$C_6H_5{-}\overset{\displaystyle H_3C}{\overset{|}{C}H}{-}\overset{\displaystyle CH_3}{\overset{|}{C}H}{-}CH_3 \qquad\qquad C_6H_5{-}\overset{\displaystyle H_3C}{\underset{\displaystyle H_3C}{\overset{|}{\underset{|}{C}}}}{-}\overset{\displaystyle CH_3}{\overset{|}{C}H}{-}CH_3$$

$$\textbf{38} \qquad\qquad\qquad \textbf{39}$$

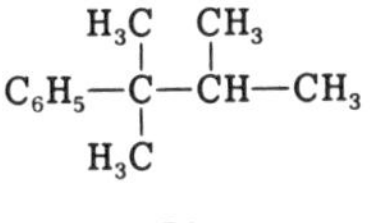

C. Alkenylation of Alkylbenzenes

The base-catalyzed side-chain alkenylation of alkylbenzenes with conjugated dienes is a facile reaction. The stabilization of the intermediate anion adducts makes the reaction proceed at a much lower temperature than that required for side-chain ethylation reactions.

The mechanism involves the addition of a benzyl anion to the diolefins:

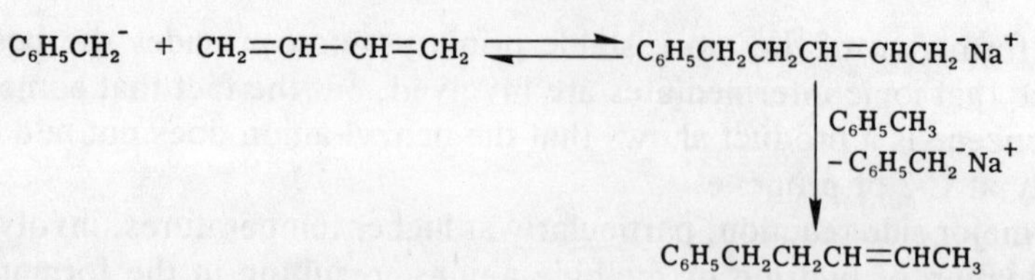

$$C_6H_5CH_2^- + CH_2=CH-CH=CH_2 \rightleftharpoons C_6H_5CH_2CH_2CH=CHCH_2^- Na^+$$

$$\Big\downarrow \begin{array}{l} C_6H_5CH_3 \\ -C_6H_5CH_2^- Na^+ \end{array}$$

$$C_6H_5CH_2CH_2CH=CHCH_3$$

The catalysts most often used for this reaction consist of either metallic sodium or potassium dispersed or as a deposit in 2.5% concentration on calcium oxide. The presence of o-chlorotoluene in the preparation of dispersed alkali metal catalysts facilitates the reaction and eliminates the induction time.

1. Butadiene

The formation of monoadducts in high yields can be achieved by using Na/CaO or K/CaO as a catalyst and bubbling butadiene through a stirred dispersion of the catalyst in liquid alkylbenzene for 3 hr (Table 2.11). Yields of monoadducts obtained from toluene, xylenes, and ethylbenzene range from 80 to 91%, based on the amount of butadiene used.

2. Isoprene

The monoaddition of alkylbenzenes to isoprene sheds light on the mechanism of base-catalyzed side-chain alkenylation reactions because of the unsymmetric structure of isoprene. The addition of a benzyl anion

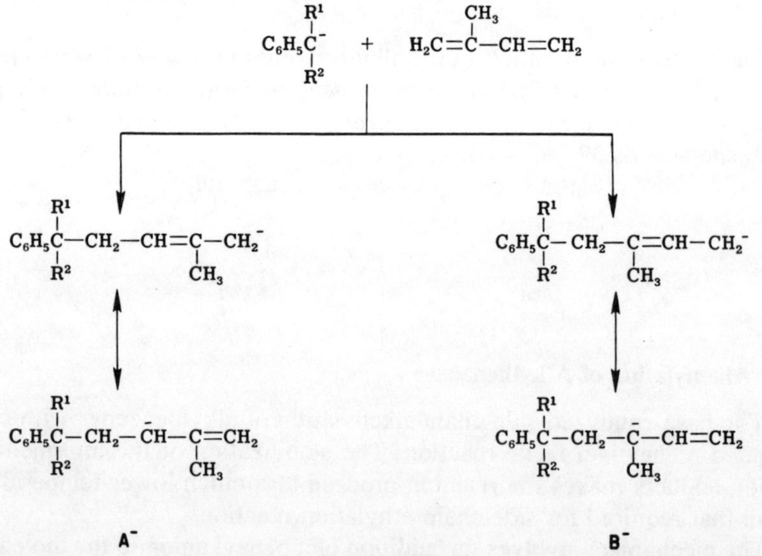

$$C_6H_5\overset{R^1}{\underset{R^2}{C^-}} + H_2C=\overset{CH_3}{\underset{}{C}}-CH=CH_2$$

$$C_6H_5\overset{R^1}{\underset{R^2}{C}}-CH_2-CH=\overset{}{\underset{CH_3}{C}}-CH_2^- \qquad C_6H_5\overset{R^1}{\underset{R^2}{C}}-CH_2-\overset{}{\underset{CH_3}{C}}=CH-CH_2^-$$

$$\updownarrow \qquad\qquad\qquad\qquad \updownarrow$$

$$C_6H_5\overset{R^1}{\underset{R^2}{C}}-CH_2-\overset{-}{C}H-\overset{}{\underset{CH_3}{C}}=CH_2 \qquad C_6H_5\overset{R^1}{\underset{R^{2\cdot}}{C}}-CH_2-\overset{}{\underset{CH_3}{C}}-CH=CH_2$$

$$\mathbf{A^-} \qquad\qquad\qquad\qquad\qquad \mathbf{B^-}$$

leads to the formation of two monoadducts ($\mathbf{A^-}$ and $\mathbf{B^-}$). The results demonstrate that $\mathbf{A}$ is the predominant monoadduct (Table 2.12). This can be

TABLE 2.11

Reaction of Butadiene with Aromatic Hydrocarbons Using as Catalyst Potassium on Calcium Oxide[a,b]

Alkylaromatic	Alkylaromatic (mol)	Butadiene (mol)	Temperature (°C)	Duration (hr)	Yield (%)[c]	Monoadduct structure[d]
Toluene	3.8	1.03	92	6.5	80	**A**
o-Xylene	3.22	0.96	109	3.0	90	**B**
p-Xylene	2.73	0.77	107	4.0	91	**C**
Ethylbenzene	1.74	0.95	109	3.0	87	**D**
1,2,4,5-Tetramethylbenzene	1.0	0.54	109	2.5	35	**E**
2-Methylnaphthalene	1.61	0.60	124–130	3	19	**F**

[a] From Eberhardt and Peterson (1965).
[b] Potassium, 2.5 gm, deposited on CaO, 100 gm.
[c] Based on butadiene used.
[d] Monoadduct structures:

$C_6H_5CH_2CH_2CH{=}CHCH_3$

A

$o\text{-}C_6H_4(CH_3)(CH_2CH_2CH{=}CHCH_3)$

B

$p\text{-}C_6H_4(CH_3)(CH_2CH_2CH{=}CHCH_3)$

C

$C_6H_5CH(CH_3)CH_2CH{=}CHCH_3$

D

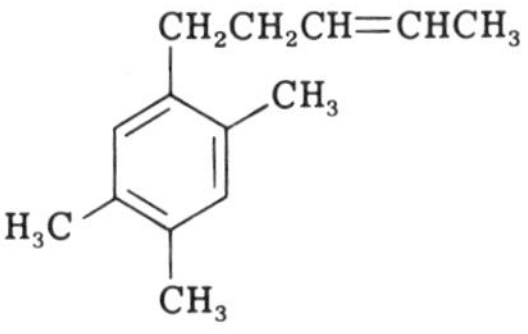

E

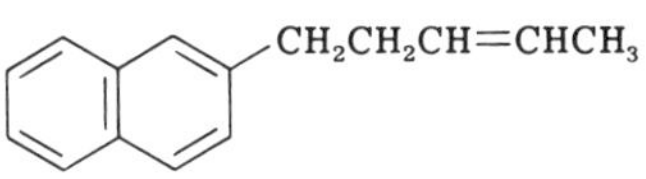

F

attributed to the fact that the stability of **A**$^-$, having resonance-stabilized primary and secondary anionic forms, is greater than that of **B**$^-$, the resonance-stabilized forms of which are primary and tertiary; the relative stabilities of alkyl anions are primary > secondary > tertiary.

Sodium–naphthalene in THF is also an effective catalyst for side-chain alkenylation of alkylbenzenes. A 63–77% yield of pentenylated product was obtained from toluene and isoprene at 25°–60°C.

TABLE 2.12

Reaction of Alkylbenzenes with Isoprene[a,b]

Catalyst	Alkylbenzene C_6H_5R, R =	Yield (%)[c]	Ratio A/B[d]	R_1 =	R_2 =
Na	CH_3	32.2	2.77	H	H
	C_2H_5	31.4	1.88	H	CH_3
	$CH(CH_3)_2$	17.4	1.98	CH_3	CH_3
K	CH_3	38.2	3.04	H	H
	C_2H_5	58.0	2.54	H	CH_3
	$CH(CH_3)_2$	15.7	3.30	CH_3	CH_3

[a] From Pines and Sih (1965).

[b] Reagents: alkali metal, 0.050 gm-atom; o-chlorotoluene, 0.0043 mol; arylalkane, 1.0 mol; isoprene, 0.50 mol. Temp: 125° ± 8°C. Duration of emperiment: 1 hr. Aparatus used: 250-ml Magne–Dash autoclave.

[c] Based on isoprene used.

[d] Structures of **A** and **B**:

$$C_6H_5\overset{\overset{\displaystyle R_1}{|}}{\underset{\underset{\displaystyle R_2}{|}}{C}}-CH_2-CH=\overset{\overset{\displaystyle}{}}{\underset{\underset{\displaystyle CH_3}{|}}{C}}-CH_3 \qquad C_6H_5\overset{\overset{\displaystyle R_1}{|}}{\underset{\underset{\displaystyle R_2}{|}}{C}}-CH_2-\overset{\overset{\displaystyle}{}}{\underset{\underset{\displaystyle CH_3}{|}}{C}}=CH-CH_3$$

A   B

D. Aralkylation of Alkylbenzenes

The reaction of alkylbenzenes with vinylarenes such as styrene, α-methylstyrene, and β-methylstyrene in the presence of a sodium or potassium catalyst affords addition compounds. Mono- and diadducts are obtained in good yields, depending on the experimental conditions and the structure of the hydrocarbons used in the reactions.

The mechanism of side-chain aralkylation is similar to that of alkenylation of alkylbenzenes. Owing to the resonance stabilization of the anion monoadduct by the benzene ring, the reaction proceeds with ease at about 100°–120°C.

1. Styrene

Side-chain aralkylation of alkylbenzenes with styrene occurs satisfactorily at 110°–125°C. The experiments were carried out with efficient stirring and using as catalyst sodium promoted by the addition of small amounts of benzylsodium or anthracene. In the presence of promoters the reaction proceeds more selectively, and the yield of polymeric product is somewhat smaller (Table 2.13).

TABLE 2.13

Reaction of Styrene with Alkylbenzenes[a,b]

Alkylbenzene C_6H_5R, R =	Promoter[c]	Product formed				Residue (gm)
		Monoadduct		Diadduct		
		Yield (%)[d]	Structure[e]	Yield (%)[d]	Structure[e]	
CH_3	—	21	A	23	B	15
CH_3	a	40	A	29	B	7
CH_3	b	33	A	26	B	12
C_2H_5	—	12	C	16	D	22
C_2H_5	a	22	C	35	D	13
$i\text{-}C_3H_7$	—	—	—	—	—	35
$i\text{-}C_3H_7$	a	9	E	29	F	16

[a] From Pines and Wunderlich (1957).
[b] The experiments were conducted using 2.0 gm (0.09 gm-atom) of sodium, 0.01 mol or promoter, 34.5 gm (0.33 mol) of styrene, and about 1 mol of alkylbenzene. Temperature 110°–125°C, duration, 1.0–1.5 hr.
[c] Promoters: a, o-chlorotoluene; b, anthracene.
[d] The yields were based on moles of styrene used in the reaction.
[e] Mono- and diadduct structures:

$$C_6H_5CH_2CH_2CH_2C_6H_5$$

A

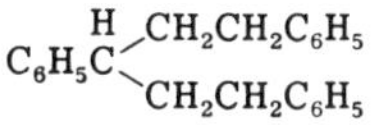

$$C_6H_5C\overset{\displaystyle H}{\underset{\displaystyle CH_2CH_2C_6H_5}{\diagup}}_{CH_2CH_2C_6H_5}$$

B

$$\underset{\displaystyle C_6H_5\overset{CH_3}{\underset{|}{C}}HCH_2CH_2C_6H_5}{}$$

C

$$C_6H_5C\overset{CH_2CH_2C_6H_5}{\underset{CH_2CH_2C_6H_5}{\diagup}}_{CH_3}$$

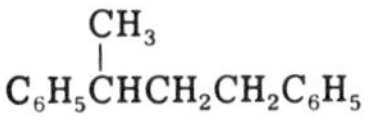

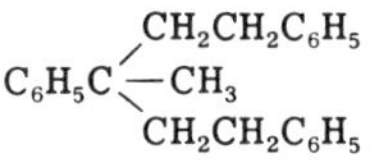

D

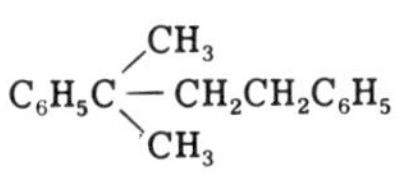

$$C_6H_5C\overset{CH_3}{\underset{CH_3}{\diagdown}}{\!-}CH_2CH_2C_6H_5$$

E

$$C_6H_5C\overset{CH_3}{\underset{CH_3}{\diagdown}}{\!-}CH_2\!-\!\overset{C_6H_5}{\underset{|}{C}}HCH_2CH_2C_6H_5$$

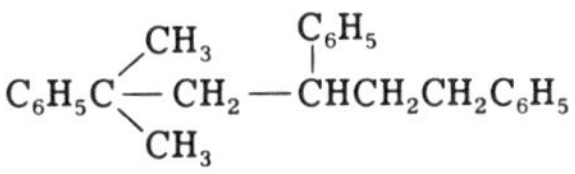

F

2. α-Methylstyrene

α-Methylstyrene reacts readily with n-alkylbenzenes at 100°C in the presence of dispersed sodium to form monoadducts in 73–85% yields. The main competing reaction is the dimerization of α-methylstyrene, and this is especially pronounced when isopropylbenzene is used as the alkylbenzene (Section III,E).

The monoadducts are formed through the addition of the benzylic carbanion to the styrene followed by protonation:

$$C_6H_5\overset{\underset{\displaystyle R}{|}}{CH^-} + C_6H_5\overset{\underset{\displaystyle CH_3}{|}}{C}=CH_2 \longrightarrow C_6H_5\overset{\underset{\displaystyle R}{|}}{CH}CH_2\overset{\underset{\displaystyle CH_3}{|}}{\underset{-}{C}}C_6H_5 \underset{B^-}{\overset{BH}{\rightleftharpoons}} C_6H_5\overset{\underset{\displaystyle R}{|}}{CH}CH_2\overset{\underset{\displaystyle CH_3}{|}}{CH}C_6H_5$$

R = H, CH$_3$, C$_2$H$_5$;   BH = alkylbenzene

3. β-Alkylstyrenes

β-Methyl-, β-ethyl-, and β-isopropylstyrene react with toluene at 110°C in the presence of dispersed potassium to form the corresponding 1,3-diphenyl-2-alkylpropanes in 60–95% yields. The aralkylation with β-isopropylstyrene is relatively slow, probably as a result of steric hindrance due to the isopropyl group. The reaction can be presented as follows:

$$C_6H_5CH_2^- + C_6H_5CH{=}CHR \longrightarrow C_6H_5CH_2\overset{\underset{\displaystyle R}{|}}{\underset{-}{CH}}CHC_6H_5 \overset{BH}{\underset{-B^-}{\longrightarrow}} C_6H_5CH_2\overset{\underset{\displaystyle R}{|}}{CH}CH_2C_6H_5$$

R = CH$_3$, C$_2$H$_5$, i-C$_3$H$_7$

V. HYDROGENATION

A. Introduction

A wide variety of catalysts can be used for the base-catalyzed hydrogenation of hydrocarbons. These include alkali metals and their hydrides as such or deposited on solid supports, and metal–graphite intercalation compounds.

The hydrogenation of alkenes usually requires a temperature of 170°–250°C. Conjugated alkadienes and cycloalkadienes, however, undergo hydrogenation at much lower temperatures. Probably the majority of these hydrogenation reactions proceed through the addition of alkali metals or their hydrides to the multiple carbon–carbon bonds, followed by hydrogenolysis with molecular hydrogen. Polynuclear aromatic hydrocarbons undergo hydrogenation at 250°C and superatmospheric pressure. However, if amines are used as solvents, hydrogenation occurs at 120°C.

Hydrogenation in the presence of base catalysts proceeds best at superatmospheric pressures. Alkali metals deposited on alumina selectively hydrogenate alkadienes to monoolefins at 25°–90°C. However, since these catalysts are very effective for the isomerization of diolefins to conjugated dienes, the position of the double bonds in the original dienes is not of great consequence.

B. The Hydrogenation of Polynuclear Aromatic Hydrocarbons under Superatmospheric Pressure

The hydrogenation of naphthalene in the presence of sodium occurs at 172°–215°C under an initial hydrogen pressure of 68–136 atm. After 7.5 hr

TABLE 2.14

Hydrogenation of Polynuclear Aromatic Hydrocarbons[a]

Compound[b]	Catalyst	Temp (°C)	Solvent (other than benzene or toluene)	Principal products (%)[b,c]
Naphthalene	Na	300		**Aa** (62)
(**A**)	NaK	250		**Aa** (91)
	NaRb	180		**Aa** (99)
	NaCs	200		**Aa** (88)
Anthracene	Na	250		**Ba** (53), **Bb** (38)
(**B**)	NaK	250		**Bb** (37), **Bc + Bd** (53)[d]
	NaK	350		**Bc + Bd** (85)[d]
Phenanthrene	Na	250		**Ca** (23), **Cb** (72)
(**C**)	NaK	250		**Ca** (10), **Cc + Cd** (80)[d]
	NaRb	180		**Ca** (82), **Cb** (5)
	NaRb	200		**Ca** (10), **Cc + Cd** (78)[d]
	NaCs	220		**Ca** (50), **Cc + Cd** (36)[d]
	NaRb	120	Ethylenediamine	**Cc + Cd** (91)
	NaRb	120	Butylamine–benzene (1:1)	**Cc + Cd** (97)
	NaRb	120	Butylamine–benzene (1:1)[e]	**Ca** (12), **Cb** (80)

[a] From Friedman *et al.* (1971).
[b] Structure of starting and hydrogenated compounds:

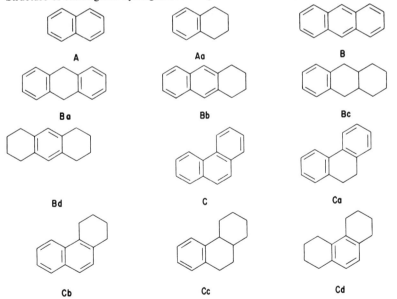

A   Aa   B

Ba   Bb   Bc

Bd   C   Ca

Cb   Cc   Cd

[c] Percent yield, based on recovered material.
[d] Approximately 2:1 mixture of symmetric and unsymmetric isomers.
[e] Gas pressure 7 atm.

a 40% yield of tetrahydronaphthalene is obtained. Liquid sodium/potassium alloy, and sodium alloy with alkali metals of higher atomic weight, are efficient catalysts for the hydrogenation of polynuclear hydrocarbons (Table 2.14).

The procedure for all experiments, which were carried out in a 500-ml-capacity rocking autoclave, was approximately the same. Benzene or toluene was the solvent for reactions at up to 250°C. For higher temperatures, decalin and tetralin served as solvents. All runs were held at a predetermined temperature for 4.5–5 hr. In most experiments 5 gm reactant in 80 ml solvent with 2 gm Na/K, or with 1.5 gm sodium and 2.0 gm rubidium carbonate, was placed in the autoclave. Hydrogen was then introduced up to about 85 atm.

Solvents exert an effect on the rate of reaction. Whereas in benzene the lowest temperature at which measurable hydrogen uptake occurs is 180°C, in solvents such as ethylenediamine or butylamine–benzene, hydrogenation of phenanthrene to octahydrophenanthrene takes place at 120°C. By lowering the initial hydrogen pressure to 7 atm it is possible to produce 1,2,3,4-tetrahydrophenanthrene in 80% yield.

Polynuclear hydrocarbons other than those listed in Table 2.14, such as chrysene, naphthacene, and pyrene, undergo a similar hydrogenation.

Polyaryl hydrocarbons undergo hydrogenation under experimental conditions described in Table 2.13, footnote b, but at a much higher temperature, namely, at 350°C for NaK and 250°C for NaRb. Biphenyl yields phenylcyclohexane, and o-terphenyl produces sym-dodecahydrotriphenylene and other hydrogenated compounds. The hydrogenation of polyaryl compounds is accompanied by extensive hydrogenolysis.

REFERENCES

Bush, W. V., Holzman, G., and Shaw, A. W. (1965). *J. Org. Chem.* **30,** 3290.
* Cram, D. J. (1965). "Fundamentals of Carbanion Chemistry." Academic Press, New York.
Eberhardt, G. G., and Peterson, H. J. (1965). *J. Org. Chem.* **30,** 82.
Faworsky, A. (1887). *J. Russ. Phys. Chem. Soc.* **19,** 414.
Friedman, S., Kaufman, M. L., and Wender, I. (1971). *J. Org. Chem.* **36,** 694.
Hambling, J. K. (1969). *Chem. Br.* **5,** 354.
* Hubert, A. J., and Reimlinger, H. (1969). The isomerization of olefins. Part I. Base-catalyzed isomerization of olefins. *Synthesis* pp. 97–112.
Mereshkowski, B. K. (1913). *J. Russ. Phys. Chem. Soc.* **19,** 414.
O'Grady, T. M., Alm, R. M., and Hoff, M. C. (1959). *Am. Chem. Soc., Div. Pet. Chem., Prepr., 136th Meet., Atlantic City, N.J.* **4,** B65.
* Pines, H. (1974). Base-catalyzed carbon–carbon additions of hydrocarbons. *Synthesis* pp. 309–327.
Pines, H., and Sih, N. C. (1965). *J. Org. Chem.* **30,** 280.

* Pines, H., and Stalick, W. M. (1977). "Base-Catalyzed Reactions of Hydrocarbons and Related Compounds." Academic Press, New York.
Pines, H., and Wunderlich, D. (1957). *J. Am. Chem. Soc.* **79,** 5820.
Pines, H., Veseley, J. A., and Ipatieff, V. N. (1955). *J. Am. Chem. Soc.* **77,** 554.
* Rudledge, T. F. (1969). "Acetylenes and Allenes." Van Nostrand-Reinhold, Princeton, New Jersey.
Schaap, L., and Pines, H. (1957). *J. Am. Chem. Soc.* **79,** 4967.
Schramm, R. M., and Langlois, G. E. (1960). *J. Am. Chem. Soc.* **82,** 4912.
Shaw, A. W., Bittner, C. W., Bush, W. V., and Holzman, G. (1965). *J. Org. Chem.* **30,** 3286.
Stipanović, B., and Pines, H. (1969). *J. Org. Chem.* **34,** 2106.
* Wotiz, J. H. (1969). Propargylic rearrangements. *In* "Chemistry of Acetylenes" (H. G. Viehe, ed.), pp. 365–424. Dekker, New York.

3

Heterogeneous Hydrogenation

I. INTRODUCTION

Although catalytic hydrogenation was discovered at the turn of the last century by Sabatier and Senderens, not until about four decades later, when ortho- and para- and isotopes of hydrogen became available, were the mechanistic aspects of hydrogenation intensively studied. In view of the fact that volumes of books and scores of review papers have appeared on the topic of hydrogenation, this chapter is limited to a description of general and mechanistic aspects as applied to hydrocarbon chemistry.

The hydrogenation of hydrocarbons can be divided into three groups: (a) the hydrogenation of alkenes and cycloalkenes, (2) that of alkynes and dienes, and (3) that of arenes. Several catalyst systems have been used for these hydrogenations, the choice depending on a multitude of factors: the reactivity of the hydrocarbon selected; the experimental conditions, such as a liquid-phase or vapor-phase system; the pressure, whether atmospheric or higher; and the temperature. If a hydrocarbon with several multiple bonds is to be hydrogenated, a catalyst may be chosen which selectively hydrogenates one multiple bond and not the others. Occasionally a catalyst is selected because of its stereospecificity (for disubstituted alkynes, cycloalkenes, or arenes).

II. CATALYSTS

Group VIII of the transition elements includes those of atomic numbers 26–28 (Fe, Co, Ni), 44–46 (Ru, Rh, Pd), and 76–78 (Os, Ir, Pt). Of these, nickel, platinum, palladium, rhodium, ruthenium, and iridium have been used most frequently as hydrogenation catalysts. Other transition elements (or their oxides or sulfides) that have been found to be useful are those of atomic number 24 (Cr), 42 (Mo), 74 (W), 29 (Cu), and 30 (Zn). For

156

example, specially prepared copper and chromium oxide, "copper chromite," is useful for the selective hydrogenation of hydrocarbons since olefins are hydrogenated with it whereas benzenoid rings remain unchanged. Molybdenum and tungsten oxides and sulfides at elevated temperatures and pressures are catalysts for the hydrogenation of fused polycyclic aromatic hydrocarbons.

A. Nickel

Nickel is one of the oldest hydrogenation catalysts and probably the most common. It is usually associated with hydrogenations that are carried out at superatmospheric pressures and at temperatures in the range of 50°–200°C. It is especially suitable for the hydrogenation of all of the multiple bonds of hydrocarbons including benzenoid rings.

Nickel catalysts can be used as such in finely divided states or deposited on solid supports. The finely divided catalysts are commonly produced either by the reduction of freshly precipitated nickel oxide or nickel carbonate or from nickel–aluminum alloys.

1. Catalysts Prepared from Nickel Salts

a. **Precipitation.** A convenient method for preparing an active nickel catalyst is to treat an aqueous solution of nickel nitrate with an aqueous solution of ammonium carbonate. The nickel carbonate thus formed is removed by filtration, dried, and decomposed to nickel oxide on heating to about 175°C. The latter is then reduced cautiously, to avoid hot spots, in a stream of hydrogen, with a gradual increase in temperature from 100° to 250°C.

b. **Reduction with Sodium Borohydride.** The reaction of sodium borohydride with aqueous solutions of nickel salts produces a finely divided black precipitate, which is a highly active catalyst for hydrogenations of olefins under atmospheric pressure.

The catalyst is prepared by dissolving 1.24 gm (5.0 mmol) of nickel acetate tetrahydrate in 50 cm³ of water in a 125-ml Erlenmeyer flask (modified for high-speed magnetic stirring). This flask is attached to the borohydride hydrogenator. With vigorous stirring, 10.0 cm³ of 1.0 M NaBH$_4$ solution in 0.1 M NaOH is injected over 30–45 sec. When gas evolution ceases, a further 5.0 cm³ of solution is added. After gas evolution has subsided, stirring is discontinued, and the supernate is decanted from the catalyst. The catalyst is then washed with ethanol and used for hydrogenation (Brown, 1970).

2. Raney Nickel

The most popular highly dispersed nickel hydrogenation catalyst is Raney nickel, which was named after its inventor, Murray Raney, born in 1885. It is prepared by the action of aqueous sodium hydroxide on a powdered nickel–aluminum alloy. The finely divided nickel is washed from sodium aluminate and excess base by decantation with water. The water is then replaced by ethanol. The active catalyst contains from 25 to 150 ml of adsorbed hydrogen per gram of nickel; the amount of hydrogen adsorbed depends on the method of preparation. The more hydrogen that is adsorbed, the more active is the catalyst. The final catalyst may contain 10–15% of aluminum and aluminate anion. The catalyst loses some of its activity during storage. Various preparations of Raney nickel are given in Table 3.1.

3. Nickel–Kieselguhr

Nickel–kieselguhr is the most common supported nickel catalyst used. Unlike Raney nickel, it shows little activity at 20°C and 1 atm. Under high pressures, however, it exhibits an activity approaching that of Raney nickel. The slightly decreased activity as compared with that of Raney nickel is offset by the fact that this catalyst can be stored, in the absence of air and moisture, without any loss of activity. Nickel–kieselguhr catalyst is available commercially in pellet sizes of various dimensions.

In the laboratory the catalyst is prepared by the impregnation of kieselguhr with a soluble nickel salt, preferably nickel nitrate. Nickel hydroxide or carbonate is precipitated by the addition of an aqueous solution of sodium hydroxide, or sodium or ammonium carbonate. The product is filtered off, washed thoroughly, and dried by heating at 200°–250°C in a stream of nitrogen. The supported nickel oxide thus formed is subsequently reduced to nickel with hydrogen. The reduction is carried out gradually at 125°–250°C in a glass tube, first by passing hydrogen diluted with nitrogen over the catalyst to avoid hot spots, which may deactivate the catalyst. The temperature is gradually raised to about 375°C, and the gases are progressively enriched with pure hydrogen. When the reduction is complete, the hydrogen is replaced by nitrogen and the temperature is lowered to 20°C. The reduced catalyst is pyrophoric and must be kept in a tight container.

The content by weight of nickel as nickel oxide in commercial nickel–kieselguhr catalysts is of the order of 40–60%.

B. Copper–Chromium Oxide ("Copper Chromite")

"Copper chromite" catalyst was developed by Homer Adkins in 1931. It is prepared as follows. A solution of 5 gm of barium nitrate and 42.5 gm

TABLE 3.1

Preparations of Raney Nickel[a]

Catalyst type	Temp (°C)[b]	NaOH–alloy ratio (w/w)	Temp (°C)/ time of digestion	Washing process	Relative activity
W1	0	1:1	115–120/4 hr	Filter; wash to neutral with H_2O; wash with EtOH by decantation	Least active, ≈W8
W2	25	19:15	Steam bath/8–12 hr	H_2O wash to neutral by decantation; EtOH wash by decantation	<W3; >W1; most common type
W3	–20	32:25	⎰50/50 min	H_2O wash by decantation several times; continuous wash with large volume of H_2O; EtOH wash done without contact of catalyst with air	Quite active, >W2; <W7
W4	50	32:25	⎱		
W5	50	4:3			
W6	50	4:3	50/50 min	Continuous H_2O wash under H_2 atmosphere; EtOH wash without contact with air	Most active
W7	50	4:3	50/50 min	Three decantations with H_2O; EtOH wash without contact with air	<W6; >W3
W8	0	1:1	100–105/4 hr	Continuous H_2O wash to remove lighter Ni particles; dioxane wash by decantation; distill portion of dioxane from catalyst	Quite inactive, ≡W1; useful for deuterations

[a] From Augustine (1965, p. 27).
[b] At addition of alloy to base.

of cupric nitrate trihydrate in 150 ml of water at 80°C is added at ambient temperature to 150 ml of solution at 25°C containing 25 gm of ammonium dichromate and 38 ml of 28% ammonium hydroxide. The precipitate is removed by filtration under suction, dried, and decomposed in a casserole over an open flame. The heating must be regulated so that the reaction does not become violent. When no more gas is evolved, the powder is allowed to cool, leached with 10% acetic acid solution, and filtered. The catalyst is then washed with six 50-ml portions of water, dried at 125°C, and pulverized. The finely divided black catalyst is stable during storage.

C. Adams Platinum Oxide

The title catalyst was developed by Roger Adams and V. Vorhees in 1922. An improved method of preparation is as follows. To 9 gm of sodium nitrate are added 10 ml of 10% chloroplatinic acid. The mixture is evaporated to dryness, with caution being taken not to fuse the sodium nitrate. All of the salt is added at once to 100 gm of sodium nitrate that has been heated to 520°C in a 400-ml beaker. The heat is then removed and the olive brown mass is cooled and added to about 2 liters of water. The suspended platinum oxide is collected on a filter and then washed thoroughly with water; it is not permitted to become dry until the washing is completed. A catalyst that is free of the sodium cation shows a great improvement in activity (Frampton *et al.*, 1951).

D. Supported Platinum or Palladium Catalysts

These catalysts are prepared by varying a few basic procedures. A common procedure involves the precipitation of a metal from an aqueous solution of the metal salt onto a support. A 5% palladium on barium sulfate can be prepared by reducing palladium chloride to palladium by alkaline formaldehyde in the presence of finely divided barium sulfate. Hydrogen is often used as a reducing agent.

III. HYDROGENATION OF OLEFINS

A. Introduction

In general, the saturation of simple olefinic bonds occurs readily in the presence of any typical hydrogenation catalyst. Only a few highly hindered olefins are resistant to hydrogenation.

Platinum metal catalysts are active at low pressures and temperatures.

Palladium and platinum make extremely active catalysts for the hydrogenation of olefins, but palladium promotes the migration of the double bond during hydrogenation. The tendency of platinum metals to cause double-bond migration of 1-pentene in the presence of hydrogen decreases in the order Pd >> Ru > Rh > Pt >> Ir.

Nickel–kieselguhr is active only at superatmospheric pressures, and temperatures of 40°–50°C will generally suffice for the hydrogenation of most olefins. The W-6 Raney nickel, which is the most active of all the Raney nickel catalysts, is comparable in activity to platinum and palladium for the hydrogenation of alkenes.

The more active modifications of "copper chromite" catalysts, namely, those that have been pretreated with hydrogen at 100°C and 250 atm, can be used for the hydrogenation of olefins at 25°C and 100 atm pressure. With the exception of certain selective hydrogenations, "copper chromite" catalysts are not used extensively for the hydrogenation of olefins.

The ease of hydrogenation of double bonds depends on the number and nature of the attached groups. Generally speaking, increasing substitution at the double bond is accompanied by decreasing ease of hydrogenation. Terminal olefins react most readily, whereas tetrasubstituted olefins often hydrogenate with difficulty, particularly if the attached groups are bulky. The rate of hydrogenation decreases in the following order: $RCH{=}CH_2 > R_2C{=}CH_2 > RCH{=}CHR > R_2C{=}CHR > R_2C{=}CR_2$.

Cis isomers are hydrogenated in preference to the corresponding trans modification. For example, in the presence of platinum, *cis*-stilbene (1) is hydrogenated more rapidly than *trans*-stilbene (2). In a nonconjugated

diolefin the less hindered double bond is preferentially hydrogenated.

Solvents may influence the rate of hydrogenation of olefins. Vinylcyclooctatetraene undergoes complete hydrogenation to ethylcyclooctane in acetic acid solution and in the presence of Adams platinum catalyst at 3 atm of hydrogen pressure; however, in the presence of methanol solvent a complete saturation is not achieved.

The effect of acetic acid may be attributed to the possible presence of sodium ions that were not completely washed out during preparation of the platinum oxide catalyst but were neutralized by the acetic acid.

B. Mechanism

The addition of hydrogen to olefins is a surface reaction and involves the addition of chemisorbed hydrogen to chemisorbed olefin. There are a large number of experimental data in support of the stepwise addition of hydrogen atoms as proposed in the Horiuti–Polanyi mechanism in 1933. The dissociative addition explains the frequent occurrence, during hydrogenation, of processes other than addition, namely, bond migration and cis–trans isomerization. It also explains why the dideuteriated paraffin is rarely the sole product of reaction when the reactions are conducted with deuterium.

1. Hydrogen Addition

The generally accepted Horiuti–Polanyi mechanism can be summarized as follows:

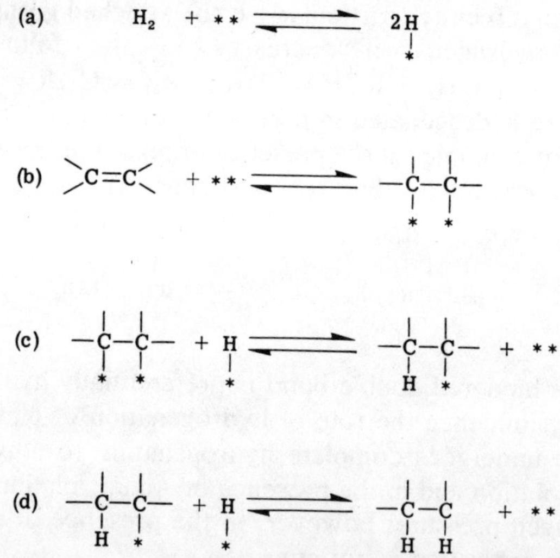

(a) H_2 + ✶✶ ⇌ 2H
       |
       ✶

(b) $\diagup C{=}C\diagdown$ + ✶✶ ⇌ —C—C—
                                  | |
                                  ✶ ✶

(c)   —C—C— + H ⇌ —C—C— + ✶✶
        | | | | |
        ✶ ✶   ✶   H ✶

(d)   —C—C— + H ⇌ —C—C + ✶✶
        | | | | |
        H ✶   ✶   H H

Under hydrogenation conditions, step (d) is almost irreversible at room temperature. Steps (c) and (d) correspond to the addition of two hydrogen atoms (2H·) to the same side of chemisorbed olefin (cis addition).

2. Double-Bond Migration

The double-bond migration accompanying the hydrogenation of an olefin such as 1-butene can also be explained by means of the Horiuti–Polanyi mechanism:

$$
\underset{\overset{|}{H}}{\overset{H}{\underset{|}{CH_3-C}}}-\overset{H}{\underset{}{C}}=\overset{H}{\underset{}{C}}-H \quad + \quad ** \quad \rightleftharpoons \quad \underset{\overset{|}{H}}{\overset{H}{\underset{|}{CH_3-C}}}-\underset{\overset{|}{*}}{\overset{H}{C}}-\underset{\overset{|}{*}}{\overset{H}{C}}-H
$$

$$
CH_3-\underset{\overset{|}{H}}{\overset{H}{C}}-\underset{\overset{|}{*}}{\overset{H}{C}}-\underset{\overset{|}{*}}{CH} \quad + \quad \underset{\overset{|}{*}}{H} \quad \rightleftharpoons \quad CH_3-\underset{\overset{|}{H}}{\overset{H}{C}}-\underset{\overset{|}{*}}{\overset{H}{C}}-CH_3 \quad + \quad **
$$

$$
CH_3-\underset{\overset{|}{H}}{\overset{H}{C}}-\underset{\overset{|}{*}}{\overset{H}{C}}-CH_3 \quad + \quad ** \quad \rightleftharpoons \quad CH_3-\underset{\overset{|}{*}}{\overset{H}{C}}-\underset{\overset{|}{*}}{\overset{H}{C}}-CH_3 \quad \rightleftharpoons \quad \underset{H_3C}{\overset{H}{>}}C=C\underset{CH_3}{\overset{H}{<}} \quad + \quad **
$$

3. Hydrogen Exchange

When deuterium is used in place of hydrogen, the formation of HD is sometimes observed. In addition, hydrogen in the olefin can be exchanged for deuterium, and the hydrogenated olefin molecule often contains more than two deuterium atoms. Return from the half-hydrogenated state of the olefins accounts for olefins exchanged with the deuterium and for the presence of HD:

$$
-\underset{\overset{|}{*}}{\overset{H}{C}}=CH \rightleftharpoons -\underset{\overset{|}{*}}{\overset{H}{C}}-\underset{\overset{|}{*}}{\overset{H}{CH}} \xrightarrow{\overset{|}{\underset{*}{D}}} -CHD \underset{\overset{|}{*}}{\overset{H}{\underset{}{C}}}\overset{H}{} \xrightarrow{} -CD=CH
$$

$$
\underset{\overset{|}{*}}{\overset{D}{|}} + \underset{\overset{|}{*}}{\overset{H}{|}} \quad \rightleftharpoons \quad HD \quad + \quad **
$$

The presence of alkane-d_3, i.e., an alkane molecule containing three deuterium atoms, also can be explained by a dissociative mechanism:

$$
RCH_2CH=CH_2 \underset{**}{\rightleftharpoons} RCH_2\underset{\overset{|}{*}}{CH}-\underset{\overset{|}{*}}{CH_2} \rightleftharpoons RCH_2\underset{\overset{|}{*}}{CH}-CH_2D
$$

$$
RCHDCHDCH_2D \rightleftharpoons RCH\underset{\overset{|}{*}}{CHDCH_2D} \rightleftharpoons R\underset{\overset{|}{*}}{CH}\underset{\overset{|}{*}}{CH}CH_2D
$$

4. Stereochemistry

Stereochemical studies have made a major contribution to the understanding of the mechanism of hydrogenation.

a. Open-Chain Olefins. With open-chain olefins containing two different groups attached to each carbon atom of the double bond, one-sided addition of hydrogen to the cis isomer would lead to a meso saturated product, whereas the addition to the trans isomer would lead to the racemic saturated compound. This fact was demonstrated with *cis*- and *trans*-dimethylstilbene, which produced over platinum in acetic acid solution the meso and racemic 2,3-diphenylbutanes, respectively, in 99% purity (von Wessely and Welleba, 1941).

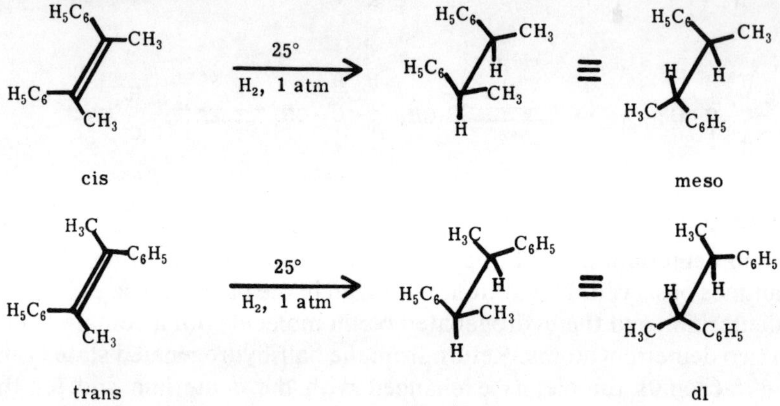

cis   meso

trans   dl

b. 1,2-Dimethylcycloalkenes. The study of the hydrogenation of 1,2-dimethylcycloalkenes has provided additional understanding of the mechanism of hydrogenation. Cis addition is expected to occur during the hydrogenation of 1,2-dimethylcyclopentene and 1,2-dimethylcyclohexene. However, both cis and trans isomers are formed from either compound when hydrogenated in acetic acid solution over reduced platinum oxide catalyst, which is one of the more selective hydrogenation catalysts (Table 3.2). The stereoselectivity of hydrogenation is also a function of the pressure of hydrogen, the ratio of cis isomer increasing with increasing pressure.

The Horiuti–Polanyi mechanism accommodates the results of Table 3.2. In step (d) (Section III,B,1), which involves the "half-hydrogenated state," hydrogen is adsorbed with retention of configuration, and the process should therefore be aided by an increase in hydrogen pressure.

The Horiuti–Polanyi mechanism also explains the apparent trans addition of hydrogen. In order for *trans*-dimethylcycloalkanes to be formed, 1,2-dimethylcycloalkenes must be transformed to the 2,3 isomers, which could then yield both cis and trans saturated products via cis addition (Scheme 3.1). Isomeric dimethylcyclohexenes indeed are detected when

TABLE 3.2

Stereoselectivity in Hydrogenation of 1,2-Dimethylcycloalkenes[a]

		Cis in product (cis + trans), %		
Hydrocarbon	Catalyst	1 atm[b]	Superatmospheric[b]	
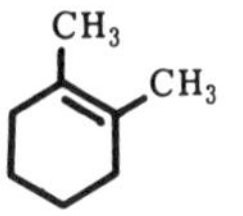	PtO_2	82	96	(300 atm)
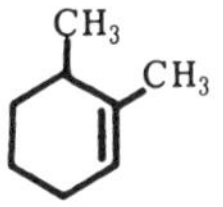	PtO_2	78	71	(150 atm)
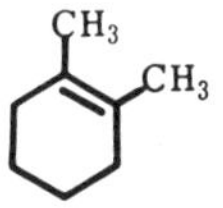	Pd/Al_2O_3	25	—	
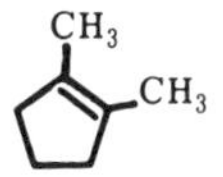	PtO_2	42	67	(230 atm)
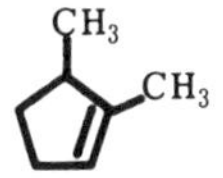	PtO_2	44	37	(290 atm)

[a] From Siegel et al. (1960, 1964).
[b] Pressure of H_2.

hydrogenation is interrupted, especially when palladium is used as catalyst.

c. 9(10)-Octalin. A striking case of apparent trans addition is the formation of about equal parts of *cis*- and *trans*-decalin during the hydrogenation of 9(10)-octalin. Repetition of the hydrogenation with deuterium revealed that only a small amount of *cis*-decalin was formed by the mere addition of two deuterium atoms. Most of the cis and most of the trans compound had at least three deuterium atoms. The major part of the hydrogenation was preceded by isomerization of the 9(10)-octalin to 1(9)-octalin-10-*d* (Scheme 3.2). The slow rate of hydrogenation of 9(10)-octalin makes possible the relatively large rate of isomerization.

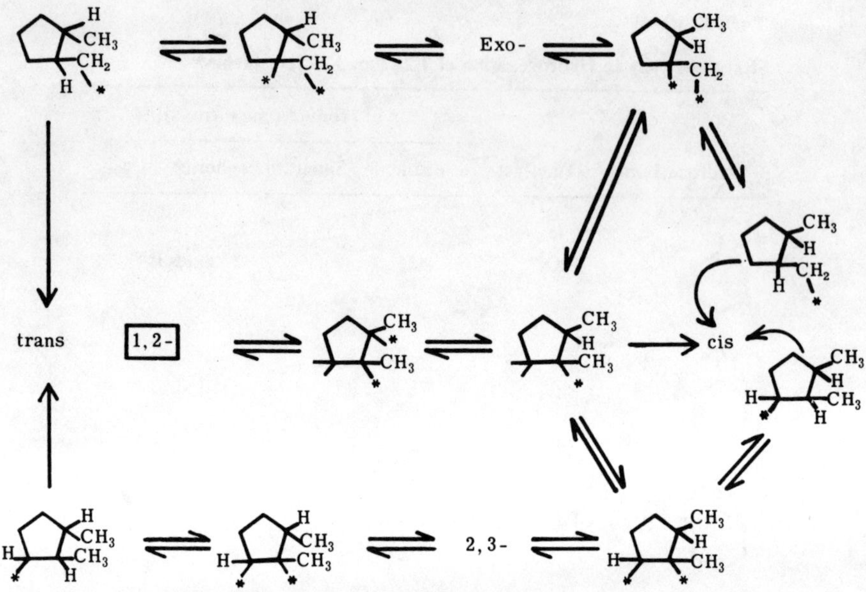

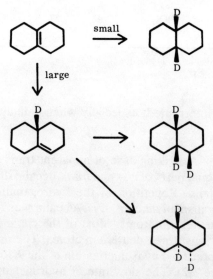

Scheme 3.1. Reduction and isomerization of 1,2-dimethylcyclopentene according to the Horiuti–Polanyi mechanism (from Siegel, 1966, p. 134).

Scheme 3.2. Deuteration of 9(10)-octalin (from Smith and Burwell, 1962).

C. Rates of Hydrogenation of Olefins

1. Individual Olefins

It is difficult to measure the rates of olefin hydrogenation experimentally. These rates are sensitive to operating variables such as temperature, pressure, solvent, type of apparatus, and agitation. Minor traces of impurities associated with the substrate, with the catalyst, or with the storing or handling of material may also influence the rate of hydrogenation. Despite the difficulties, reliable studies on the rate of hydrogenation of a variety of hydrocarbons have been reported (Table 3.3).

2. Competitive Hydrogenation—Binary Systems

A more practical method for determining the relative rates of hydrogenation is based on the measurement of competitive hydrogenation of binary mixtures of hydrocarbons in solution. This method was given a quantitative basis through the derivation of simple kinetic equations based on several postulates (Maurel, 1967).

The first postulate is that both olefins A and B obey Langmuir adsorption isotherms and that both compete for the same active centers on the catalyst. The velocity or rate of hydrogenation v_A of olefin A can be represented by

$$v_A = dn_A/dt = K_A m \theta_A f(pH_2) \tag{1}$$

where n_A is the number of moles of A, K_A is the rate constant of A, m is the weight of catalyst, θ is the fractional coverage of surface by A, and pH_2 is the hydrogen pressure. By assuming that the hydrogenation of B follows the same equation, and by dividing the two, we obtain Eq. (2).

$$v_A/v_B = K_A \theta_A / K_B \theta_B \tag{2}$$

By the use of Langmuir's adsorption isotherm,

$$\theta_A = \frac{\lambda_A n_A}{1 + \Sigma_i \lambda_i n_i} \tag{3}$$

where λ_A is adsorption of olefin A, Eq. (4) can be derived as follows:

$$\frac{v_A}{v_B} = \frac{dn_A}{dn_B} = \frac{K_A \lambda_A n_A}{K_B \lambda_B n_B} \tag{4}$$

After integration, Eq. (5) results:

$$\frac{1}{K_A \lambda_A} \log \frac{n_{0A}}{n_A} = \frac{1}{K_B \lambda_B} \log \frac{n_{0B}}{n_B} \tag{5}$$

TABLE 3.3

Hydrogenation of Cycloalkenes[a,b]

Cycloalkene		Rate
Structure	Name	(mol/gm-atom $\times$ min $\times$ atm)
	Bicyclo[2.2.1]cycloheptene	223 ± 10
	Bicyclo[2.2.2]cyclooctene	174 ± 5
	3-Methylcyclopentene	153 ± 9
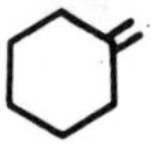	Methylenecyclohexane	131 ± 6
	Cyclopentene	121 ± 2
	Cyclohexene	113 ± 4
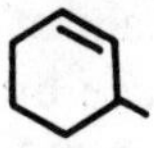	3-Methylcyclohexene	103 ± 2
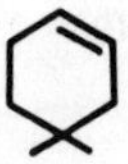	4,4-Dimethylcyclohexene	100 ± 1
	4-Methylcyclohexene	94 ± 4
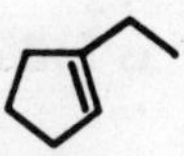	1-Ethylcyclopentene	88 ± 4
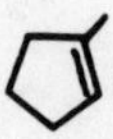	1-Methylcyclopentene	85 ± 1
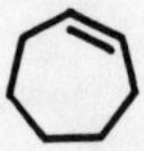	Cycloheptene	78 ± 3

(Continued)

TABLE 3.3   (*Continued*)

Cycloalkene		Rate
Structure	Name	(mol/gm-atom $\times$ min $\times$ atm)
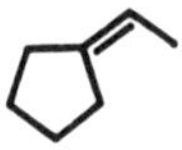	Ethylidenecyclopentane	59 ± 1^c
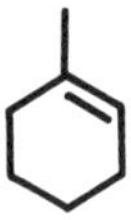	1-Methylcyclohexene	57 ± 6
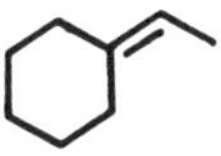	Ethylidenecyclohexane	40 ± 1
	1-Methylcycloheptene	$34 + 1$
	1-Ethylcyclohexene	19 ± 1
	Cyclooctene	10 ± 1

[a] From Hussey *et al.* (1968).
[b] Catalyst, 0.52% Pt on Al_2O_3; 25°C; 1 atm; cyclohexane solvent.
[c] Significant isomerization to 1-ethylcyclopentene.

Here n_{0A} and n_{0B} represent moles of A and B at zero time. A straight line passing through the origin can be drawn by plotting log (n_{0A}/n_A) as a function of log (n_{0B}/n_B). From the slope of the line it is thus possible to determine the relative reactivities of olefins A and B [Eq. (6)].

$$R_{AB} = K_A\lambda_A/K_B\lambda_B \qquad (6)$$

The value of R_{AB} can be established with good precision because it depends, not on the amount of catalyst and the concentration of hydrogen, but on the nature of the catalyst and on the olefins used in the hydrogenation.

The relative reactivities of olefins can be obtained with high precision when the binary mixture is composed of olefins of similar reactivities, namely, when R_{AB} is greater than 0.1 and smaller than 10. However, the relative rates of hydrogenation of olefins having a wide range of reactivities can be established by taking advantage of the relation expressed by Eq. (7).

$$R_{AB} = R_{AC} \times R_{CB} \tag{7}$$

where AC and CB are mixtures of olefins.

The difference in relative rates of hydrogenation of two olefins is a function of the speed of hydrogenation and the degree of adsorption of the hydrocarbons on the surface of the catalyst. In order to ascertain the relative contribution of the two factors, the rate of hydrogenation of olefins A and B was determined under the same conditions [Eqs. (8) and (9), respectively].

$$v_A = k_A m \frac{\lambda_A n_A}{1 + \lambda_A n_A} fp(H_2) \tag{8}$$

$$v_B = k_B m \frac{\lambda_B n_B}{1 + \lambda_B n_B} fp(H_2) \tag{9}$$

TABLE 3.4

Relative Rate of Hydrogenation of Alkenes and Cycloalkenes[a,b]

Hydrocarbon	Relative rate $\times 10^3$	k^c	λ^d
1-Hexene	740	110	6600
1-Octene	560	80	6900
4-Methyl-1-pentene	470	125	3800
2-Methyl-1-pentene	13	120	110
2,3-Dimethyl-1-butene	9	160	55
3-Ethyl-2-pentene	0.34	65	5.2
2,3-Dimethyl-2-butene	0.018	75	0.25
Cyclopentene	130	100	1300
Cyclohexene	10	100	100
4-Methylcyclohexene	8	85	95
1-Methylcyclopentene	3.6	95	40
1-Methylcyclohexene	0.1	40	2.5
1,2-Dimethylcyclohexene	0.006	20	0.03
Camphene	1.3	120	11
α-Pinene	0.087	30	2.9

[a] From Maurel and Tellier (1968).
[b] Catalyst, 7% Pt on SiO_2; 20°C; 1 atm H_2 pressure.
[c] Relative reactivity coefficient vs cyclohexene = 100.
[d] Relative adsorptivity coefficient vs cyclohexene = 100.

Since the reaction is zero order and $\lambda_i n_i \gg 1$, it can be represented by Eq. (10),

$$v_A/v_B = k_A/k_B \tag{10}$$

and, by substitution, Eq. (11) is obtained.

$$\lambda_A/\lambda_B = R_{AB}k_B/k_A \tag{11}$$

With cyclohexene as the reference compound, the relative reactivity k and adsorptivity λ for a variety of olefins are given in Table 3.4. The data indicate that the relative rate of hydrogenation decreases with substitution at the double bond and that steric interference with adsorption is the main deterrent to hydrogenation.

D. Selective Hydrogenation

The essence of selective hydrogenation of unsaturated hydrocarbons lies in the possibility of preferentially adding hydrogen to one unsaturated bond rather than to another, although both are capable of being hydrogenated. Selective hydrogenation can be achieved by a judicious choice of catalysts or experimental conditions, such as temperature, pressure, control of the amount of hydrogen adsorbed, or by partial deactivation of the catalyst.

The selectivity of hydrogenation can often be predicted from the rate of hydrogenation of related compounds or functional groups. In the case of limonene, the olefin bond in the isopropenyl group can be hydrogenated under mild conditions with a variety of catalysts without the endocyclic double bond being affected:

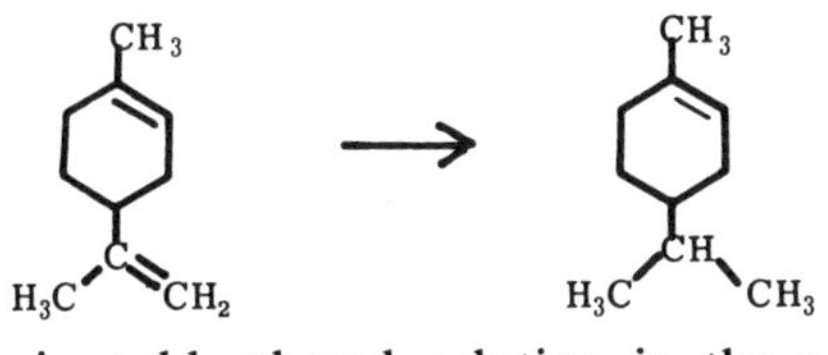

Cyclopentadiene in cold ethanol solution in the presence of Raney nickel is hydrogenated to cyclopentene. There is a sharp drop in hydrogen absorption at the end of the first step. This selective hydrogenation offers a method for preparing cyclopentene:

Certain C_8 to C_{12} conjugated alkadienynes can be hydrogenated at 25°C and under atmospheric pressure in four stages to give trienes, dienes, monoenes, and alkanes. Platinum catalyst hydrogenates the unsaturated hydrocarbons directly to alkanes, but with Raney nickel it is possible to stop the reaction at the various stages:

$$C_2H_5-CH=C(C_3H_7)-C\equiv C-C(C_3H_7)=CH-C_2H_5$$

$$\longrightarrow C_2H_5-CH=C(C_3H_7)-CH=CH-C(C_3H_7)=CH-C_2H_5$$

$$\longrightarrow C_2H_5-CH_2-C(C_3H_7)=CH-CH=C(C_3H_7)-CH_2-C_2H_5$$

$$\longrightarrow C_2H_5-CH_2-CH(C_3H_7)-CH=CH-CH(C_3H_7)-CH_2-C_2H_5$$

$$\longrightarrow C_2H_5-CH_2-CH(C_3H_7)-CH_2-CH_2-CH(C_3H_7)-CH_2-C_2H_5$$

With olefins hindered by branching, it is possible even in the presence of platinum catalysts to interrupt the hydrogenation at the monoolefin stage:

$$CH_3-C(CH_3)(i\text{-}C_3H_7)=C-C\equiv C-C=C(CH_3)(i\text{-}C_3H_7)-CH_3 \longrightarrow CH_3-CH(CH_3)(i\text{-}C_3H_7)-CH-CH=CH-CH(CH_3)(i\text{-}C_3H_7)-CH_3$$

IV. HYDROGENATION OF ALKYNES (ACETYLENES)

Like the simple carbon–carbon double bond, the triple bond in alkynes is hydrogenated with great facility in the presence of a wide variety of catalysts. Platinum, palladium, and nickel are generally preferred. As with olefins the presence of bulky groups, such as *tert*-butyl and phenyl, hinders hydrogenation somewhat.

In general, the hydrogenation of alkynes can be made to yield alkenes or alkanes depending on the conditions and catalyst used. Low temperatures and pressures are recommended if the reaction is to be terminated at the olefin state. With disubstituted acetylenes a break in hydrogenation generally occurs in the hydrogenation curve after the adsorption of 1 mol of hydrogen.

The geometry of the olefins that are obtained from disubstituted acetylenes provides further evidence for the cis addition of hydrogen during catalytic hydrogenation. Thus, cis olefins are usually formed predominantly. A few examples have been reported wherein appreciable quantities of trans olefins are formed during the catalytic hydrogenation of acetylenes. It is probable, however, that cis products are formed primarily and these undergo isomerization in the presence of the catalyst. However, with highly strained acetylenes, such as di-*tert*-butylacetylene, *trans*-1,2-di-*tert*-butylethylene is the primary product of hydrogenation (Kung *et al.*, 1976).

At 14°C, 2-butyne and deuterium react on a 0.03% Pd/Al_2O_3 catalyst to form *cis*-2-butene-2,3-d_2 almost exclusively as long as at least 1% of unreacted butyne remains. The proportion of by-products increases with temperature and reaches about 6% at 58°C.

V. HYDROGENATION OF AROMATIC HYDROCARBONS

A. Mononuclear Aromatic Hydrocarbons

Owing to their resonance stabilization, aromatic benzenoid rings are more difficult to hydrogenate than isolated olefinic double bonds. Nickel catalysts are perhaps the most satisfactory catalysts for temperatures ranging from 100° to 150°C and hydrogen pressures of about 100 atm. At 30°-50°C and 4-5 atm, benzene can be hydrogenated in the presence of platinum, palladium, and rhodium, particularly when a small amount of acetic acid is added to the reaction mixture.

Although the hydrogenation of benzene probably proceeds stepwise by way of one of the cyclohexadienes and cyclohexene, such intermediate products have not been isolated under the usual hydrogenation conditions. The reason for this is apparent from the rate constants for the hydrogenation of these substances in acetic acid solution in the presence of PtO_2 at 30°C (Table 3.5). Small amounts of cyclohexenes have been detected, however, in the liquid phase of hydrogenation of *o*-xylene. With alkyl-substituted benzene containing *tert*-butyl groups as in 1,3- or 1,4-di-*tert*-butylbenzene or 1,3,5-tri-*tert*-butylbenzene, the amount of alkylcyclohexenes can be substantial (Siegel *et al.*, 1979).

Sulfur-containing compounds, such as thiophene, inhibit the rate of hydrogenation of aromatic hydrocarbons by nickel.

TABLE 3.5

Rate of Hydrogenation of Benzene and Related Compounds[a]

Hydrocarbon	Rate, $k \times 10^3$ (liters/gm $\times$ min)
Benzene	290
1,4-Cyclohexadiene	1330
1,3-Cyclohexadiene	1900
Cyclohexene	2350

[a] From Smith and Meriwether (1949).

1. Alkylbenzenes

Alkylbenzenes are hydrogenated at somewhat slower rates than benzene itself. The hydrogenation rate is first order with respect to hydrogen pressure, zero order with respect to the concentration of the arene, and directly proportional to the amount of catalyst used, provided that this amount is sufficiently small. If too much catalyst is used, the rate of hydrogenation depends on the rate of agitation of the reactants. It appears that the rate-determining step involves the surface reaction of hydrogen with chemisorbed arene.

The rate of hydrogenation of alkylbenzene decreases somewhat with the length of unbranched alkyl group for the first two carbon atoms (Table 3.6); a further increase in length has little effect. The results with branched-chain alkyl groups indicate that branching slows down the rate of hydrogenation when such branching occurs in the position α or β to the aromatic nucleus.

2. Polymethylbenzenes

The number and position of the methyl groups are important in determining the rate of hydrogenation of polymethylbenzenes. In the presence

TABLE 3.6

Rate of Hydrogenation of Monoalkylbenzenes[a,b]

Aromatic C_6H_5R, R =	Standard rate, $k_{30°}$(liters/gm × min) × 10^3	Relative rate
H	288	100
CH_3	178	62
C_2H_5	130	45
C_3H_7	117	41
C_4H_9	108	38
C_5H_{11}	117	40
C_6H_{13}	111	38
C_9H_{19}	112	39
$i\text{-}C_3H_7$	96	—
$i\text{-}C_4H_9$	67	—
$sec\text{-}C_4H_9$	84	—
$t\text{-}C_4H_9$	74	—
$(CH_3)_2CHCH_2CH_2$	121	—

[a] From Smith (1957, p. 194).

[b] For each experiment 0.2 gm PtO_2, 50 ml acetic acid, and 4 atm initial hydrogen pressure were used, and the hydrogenation was carried out at room temperature.

of platinum oxide in acetic acid solution at 30°C, 1,3,5-trimethylbenzene is hydrogenated at a rate almost twice as great as that of its 1,2,4 isomer and four times as great as that of 1,2,3-trimethylbenzene (Table 3.7). The relative rate of reduction could be attributed to the steric interference of the methyl groups, which would prevent either the attack of hydrogen on the benzene ring or the adsorption of the benzene ring on the surface of the catalyst. The latter type of hindrance would then be reflected in the relative adsorption of the methylbenzenes on the catalyst surface. As the steric hindrance were increased, the ease of adsorption would decrease.

From the standard rate constants of reduction of various methylbenzenes given in Table 3.7 and from the competitive reduction of binary mixtures of methylbenzenes, the relative adsorption of methylbenzenes was determined. The results indicate that trimethylbenzenes are adsorbed more readily on the active surface of the catalyst than are tetramethylbenzenes, and the latter are adsorbed more readily than are pentamethylbenzenes. Benzene is more readily adsorbed than toluene, and toluene more so than any of the xylenes.

The relative ease of adsorption of the xylenes was found to decrease in the order ortho, meta, para, which is the opposite of the relative reduction rates of the individual xylenes. This indicates that steric hindrance to adsorption is not the only factor that affects the rate of hydrogenation of methylbenzenes.

TABLE 3.7

Rate of Hydrogenation of Polymethylbenzenes at 30°C in the Presence of Platinum Oxide[a]

Aromatic hydrocarbon	Standard rate, k(liters/gm × min)	Relative rate of hydrogenation	Relative adsorption
Benzene	0.205	100	100
Toluene	0.147	72	80
o-Xylene	0.091	44	33.5
m-Xylene	0.126	61	22.1
p-Xylene	0.158	77	16.6
1,2,3-Trimethylbenzene	0.0446	22	8.58
1,2,4-Trimethylbenzene	0.089	43	6.65
1,3,5-Trimethylbenzene	0.149	73	2.60
1,2,3,4-Tetramethylbenzene	0.024	12	0.95
1,2,3,5-Tetramethylbenzene	0.052	25	1.13
1,2,4,5-Tetramethylbenzene	0.051	25	1.94
Pentamethylbenzene	0.015	7.3	0.17
Hexamethylbenzene	0.0029	1.4	0.24

[a] Rader and Smith (1962).

TABLE 3.8

Selective Hydrogenation of Biphenyl[a,b]

Catalyst	mg	Solvent	Rate (ml H_2 min)	Biphenyl (mol %)	Cyclohexylbenzene (mol%)	Bicyclohexyl (mol%)
5% Pd/C	600	Acetic acid	71	10	87	3
		Cyclohexane	52	3	97	0
5% Pt/C	600	Acetic acid[c]	—	—	—	—
		Cyclohexane	210	12	56	32
5% Rh/C	600	Acetic acid	496	15	58	27
		Cyclohexane	203	21	43	36
5% Ru/C	600	Acetic acid[d]	97	36	31	33
		Cyclohexane	78	35	34	31
5% Rh/Al$_2$O$_3$	600	Acetic acid	229	29	61	10
15% Ir/C	300	Cyclohexane	220	40	37	23

[a] Rylander (1967, p. 319).
[b] All experiments carried out at 100°C and 65 atm with 5–10 gm substrate in 25 ml solvent.
[c] Poisoned.
[d] Plus 10 ml water in the ruthenium reductions to increase the rate.

3. Biphenyl and Diphenylmethane

The hydrogenation of aromatic hydrocarbons containing two or more benzene rings can proceed in a stepwise manner in which individual rings are saturated in some definite sequence. There is also a marked influence of catalyst on the selectivity of hydrogenation of biphenyl (Table 3.8) and diphenylmethane (Table 3.9). Palladium is the most selective catalyst for the conversion of biphenyl to cyclohexylbenzene. The other catalysts afford products with the simultaneous presence of the original hydrocarbon and bicyclohexyl. Selectivity depends only to a small degree on the solvent.

Diphenylmethane is hydrogenated with less selectivity than is biphenyl; all partial hydrogenations yield mixtures of benzylcyclohexane and dicyclohexylmethane.

B. Fused-Ring Aromatic Hydrocarbons

Fused-ring aromatic hydrocarbons can undergo selective hydrogenation, one ring or even one multiple bond at a time. Hydrogenation may lead to isomeric and stereoisomeric mixtures of hydrocarbons.

1. Naphthalene

The hydrogenation of naphthalene to cis- and trans-decahydronaphthalene (decalin) proceeds by way of tetrahydronaphthalene (tetralin) as an intermediate:

cis   trans

Small amounts of octahydronaphthalenes (octalins) are also produced.

Hydrogenation over nickel catalysts may stop at tetralin, depending on the pressure, temperature, and catalyst activity. Other catalysts that stop hydrogenation at tetralin include "copper chromite," molybdenum trioxide, and sulfides of molybdenum and tungsten. Of the noble metals platinum, palladium, rhodium, and iridium, only palladium causes reduction to stop exclusively at tetralin (Table 3.10).

The two rings of naphthalene are not likely to be hydrogenated with equal ease. The resonance energy of naphthalene, 61 kcal/mol, is less than twice that of benzene (2 × 36 kcal/mol). Hydrogenation of the first ring involves the loss of 25 kcal/mole. The difference in resonance energy is not the only factor that governs the ease of hydrogenation; competition for sites also should be considered.

The decalins produced in the hydrogenation consist of a mixture of cis

TABLE 3.9

Hydrogenation of Diphenylmethane[a,b]

Catalyst	mg	Temp (°C)	Rate (ml H$_2$/min)	Diphenylmethane (mol %)	Benzylcyclohexane (mol %)	Dicyclohexylmethane (mol%)	H$_2$ (mol)
5% Pd/C	600	119	16	26	66	8	2.46
5% Pt/C	600	102	115	18	62	20	3.06
5% Rh/C	600	87	425	18	39	43	3.75
5% Rh/C[c]	300	35	90	54	34	12	1.74
PtO$_2$	200	32	450	34	47	19	2.55

[a] Rylander (1967, p. 319).
[b] Experiments carried out at 65 atm with 10 ml substrate in 25 ml acetic acid.
[c] With no solvent and 25 ml substrate.

TABLE 3.10

Selectivity of Transition Metals for the Hydrogenation of Naphthalene to Tetralin[a,b]

Catalyst	Temp (°C)	Conversion (%)	Hydrogenated products (%)		
			Tetralin	Octalins	Decalins
0.6% Pd/Al₂O₃	100	99.9	99.7	—	0.2
5% Pd/C	100	93.3	98.7	0.2	1.1
0.6% Pt/Al₂O₃	200	89.0	96.3	0.2	3.5
5% Ir/C	80	77.7	83.9	0.5	15.6
5% Ru/C	80	86.5	82.8	2.3	14.9
0.6% Ru/Al₂O₃	30	82.9	82.0	1.9	16.1
0.6% Rh/Al₂O₃	25	89.8	96.2	0.9	4

[a] From Weitkamp (1968, p. 22).
[b] Initial pressure of hydrogen 65 atm.

and trans isomer. The proportion of *cis*-decalin obtained from the hydrogenation of tetralin and octalins, intermediates in the hydrogenation of naphthalene, depends on the catalyst employed (Table 3.11).

Good selectivities for *cis*-decalin were obtained with iridium, rubidium, and rhodium catalysts. Only the palladium catalyst gave preponderantly the thermodynamically more stable *trans*-decalin. 9(10)-Octalin formed less *trans*-decalin than did 1(9)-octalin. Saturation seems to be a two-center process, and any 9(10)-octalin produced in the reaction must isomerize before any *trans*-decalin is formed (see Section III,B,4,b).

TABLE 3.11

Proportion of *cis*-Decalin from the Saturation of Tetralin and Octalins[a]

Catalyst	cis-Decalin, % in decalin fraction from:		
	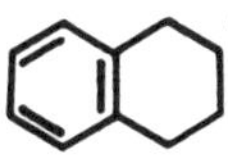 [b]	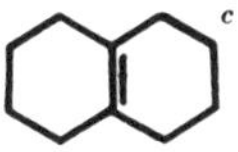 [c]	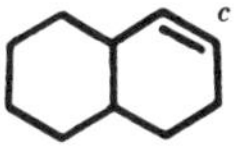 [c]
Pd	47	15	18
Pt	84	62	38
Ir	92.5	98	74
Ru	94.5	95	74
Rh	89	85	59

[a] From Weitkamp (1968, p. 25).
[b] Catalyst, 5% metal on charcoal; 25°C; 65 atm H₂.
[c] Catalyst, 5% metal on charcoal; 25°C; 22–25 atm H₂.

Alkylnaphthalenes. The hydrogenation of alkylnaphthalenes proceeds in two steps, namely, selective formation of tetrahydro derivatives and subsequent hydrogenation to decahydro products.

Qualitative data for the hydrogenation of monoalkylnaphthalenes on various catalysts under a variety of conditions demonstrate that usually the unsubstituted ring is hydrogenated preferentially. Kinetic data for the hydrogenation of monoalkylnaphthalenes are presented in Table 3.12. The overall hydrogenation constant k, the values of k_1 and k_2, and the selectivity $S = k_1/k_2$ are given. The values of S show that in 2-alkylnaphtha-

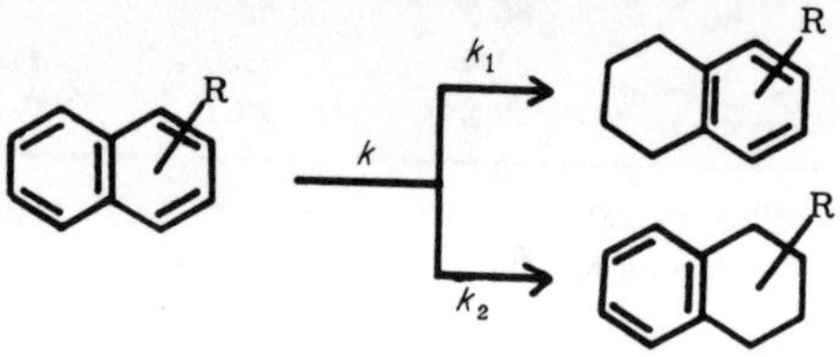

lenes the unsubstituted ring is hydrogenated preferentially. 1-Methylnaphthalene and 2-methylnaphthalene behave similarly. However, in going from 2-ethyl- to 2-isopropyl- and 2-*tert*-butylnaphthalene there is a decrease in rate k and an increase in selectivity S. The hydrogenation rate increases with bulkiness of the 1-alkyl substituent.

TABLE 3.12

Hydrogenation of Monoalkylnaphthalenes over Palladium[a,b]

Alkyl substituent	k	k_1	k_2	S
None	8.6	4.3	4.3	1.00
1-Methyl	4.9	3.3	1.7	1.95
1-Ethyl	6.8	3.8	3.1	1.24
1-Isopropyl	6.2	2.0	4.1	0.48
1-*t*-Butyl	22	0.54	21	0.024
1-Cyclohexyl	2.9	1.2	1.7	0.68
2-Methyl	5.5	3.4	2.1	1.65
2-Ethyl	5.3	3.5	1.8	1.89
2-Isopropyl	5.0	3.1	1.8	1.71
2-*t*-Butyl	2.5	1.5	1.0	1.53
2-Cyclohexyl	2.7	5	—	—

[a] From Nieuwstad *et al.* (1973).

[b] Temp, 80°C; atmospheric H_2 pressure; solvent AcOH; Pd, 10% on carbon; rate constants in mol/sec × gm catalyst × 10^6.

The rates of hydrogenation of dialkylnaphthalenes over palladium are summarized in Table 3.13. The bulkiness effect on the rate of hydrogenation is even more pronounced for 1,4-dialkylnaphthalenes, and it is also observed for 1,5- and 1,8-dialkylnaphthalenes.

The observed hydrogenation phenomena were explained by a partial release of the peri strain of the bulky substituents in the transition state of the rate-controlling step, resulting in an increased hydrogenation rate for the strained molecules (Fig. 3.1). In the hydrogenation of 1-*tert*-butylnaphthalenes the first hydrogen atom is added from below in such a way that the peri strain is released in the transition state. In 1,4-dialkylnaphthalenes the peri strain acts twice, and thus there is a greater difference in the rate of hydrogenation between 1,4-dimethyl- and 1,4-di-*tert*-butylnaphthalene.

2. Anthracene and Phenanthrene

These two hydrocarbons undergo stepwise reduction. The addition of two atoms of hydrogen to form, respectively, 9,10-dihydroanthracene and

TABLE 3.13

Hydrogenation of Dialkylnaphthalenes over Palladium[a,b]

Substituent	k	k_1	k_2	S
None	8.6	4.3	4.3	1.00
2,6-Dimethyl	4.7			
2,6-Diisopropyl	6.0			
2,6-Di-*t*-butyl	10.2			
2,6-Dicyclohexyl	2.0			
2,7-Di-*t*-butyl	5.4			
1,4-Dimethyl	7.6	4.1	3.4	1.21
1,4-Diethyl	7.1	2.1	5.0[c]	0.42
1,4-Diisopropyl	17.0	1.4	15.6[d]	0.093
1,4-Di-*t*-butyl	56	<0.2	56[c]	<0.006
1,5-Dimethyl	8.2			
1,5-Diisopropyl	17.5			
1,5-Di-*t*-butyl	48			
1-8-Dimethyl	59			
1,8-Diethyl	126			
1,8-Diisopropyl	353			

[a] From Nieuwstad *et al.* (1973).
[b] Rate constants in mol/sec × gm catalyst × 10^6.
[c] 47:53 mixture of cis and trans isomer.
[d] 61:36 mixture of cis and trans isomer.
[e] >99% cis isomer formed.

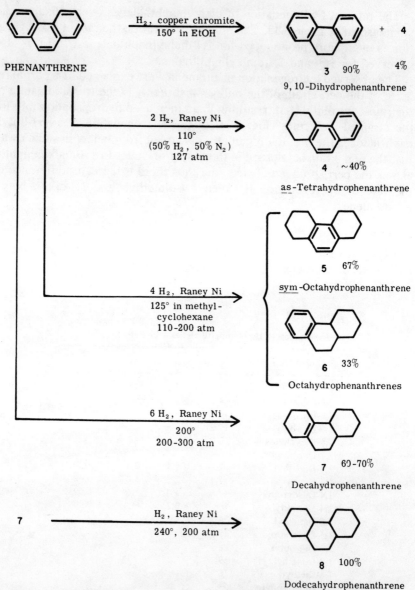

PHENANTHRENE

H$_2$, copper chromite
150° in EtOH

3   90%   4%

+ **4**

9,10-Dihydrophenanthrene

2 H$_2$, Raney Ni
110°
(50% H$_2$, 50% N$_2$)
127 atm

4   ~40%

a̲s̲-Tetrahydrophenanthrene

5   67%

s̲y̲m̲-Octahydrophenanthrene

4 H$_2$, Raney Ni
125° in methyl-
cyclohexane
110-200 atm

6   33%

Octahydrophenanthrenes

6 H$_2$, Raney Ni
200°
200-300 atm

7   60-70%

Decahydrophenanthrene

7

H$_2$, Raney Ni
240°, 200 atm

8   100%

Dodecahydrophenanthrene

Scheme 3.3. Stepwise hydrogenation of phenanthrene (from Durland and Adkins, 1938).

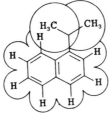

Fig. 3.1. 1-*tert*-Butylnaphthalene with van der Waals spheres (from Nieuwstad *et al.*, 1973).

9,10-dihydrophenanthrene proceeds with relative ease and can be catalyzed by copper chromite.

The relative rates of hydrogenation of anthracene to the stepwise hydrogenated compounds in the presence of tungsten sulfide (WS_2) catalyst, at 400°C and 150 atm, using the rate for benzene to cyclohexane as 100 are as follows:

$$\text{[anthracene]} \xrightarrow[\text{(326)}]{6187} \text{[compound]} \xrightarrow[\text{(147)}]{449} \text{[compound]}$$

$$\text{[compound]} \xleftarrow[\text{(4)}]{291} \text{[compound]}$$

The data in parentheses are relative rates for Ni/Al_2O_3 catalyst at 120°–202°C and 30–50 atm (from Lozovoy and Senyovin, 1958).

Experimental conditions and catalysts used for the stepwise hydrogenation of phenanthrene are presented in Scheme 3.3. The data indicate that the absorption up to 4 mol of hydrogen proceeds readily, whereas the reduction to deca- and dodecahydronaphthalene requires elevated temperature and pressure.

REFERENCES

* Augustine, R. L. (1965). "Catalytic Hydrogenation." Dekker, New York.
* Bond, G. C., and Wells, P. B. (1964). The mechanism of the hydrogenation of unsaturated hydrocarbons on transition metal catalysts. *Adv. Catal.* **15**, 92–2221.
Brown, C. A. (1970). *J. Org. Chem.* **35**, 1900.
Durland, J. R., and Adkins, H. (1938). *J. Am. Chem. Soc.* **60**, 1501.
Frampton, V. I., Edwards, J. D., Jr., and Henze, H. R. (1951). *J. Am. Chem. Soc.* **73**, 4432.
Hussey, A. S., Keulks, G. W., Nowack, P., and Baker, R. H. (1968). *J. Org. Chem.* **33**, 610.

* Kung, H. H., Pellet, R. J., and Burwell, R. L., Jr., (1976). Structure sensitivity in the hydrogenation of hindered hydrocarbons. *J. Am. Chem. Soc.* **98**, 5603–5611.

* Lozovoy, A. V., and Senyovin, S. A. (1958). In "Catalysis" (P. H. Emmett, ed.), Vol. 6, p. 228. Reinhold, New York.

Maurel, R. (1967). In "La Catalyse en Laboratoire et dans l'Industrie" (B. Claudel, ed.), pp. 203–207. Masson, Paris.

Maurel, R., and Tellier, J. (1968). *Bull. Soc. Chim. Fr.* p. 4650.

Nieuwstad, T. J., Klapwijk, P., and Van Bekkum, H. (1973). *J. Catal.* **29**, 404.

Rader, C. P., and Smith, H. A. (1962). *J. Am. Chem. Soc.* **84**, 1443.

Rylander, P. N. (1967). "Catalytic Hydrogenation over Platinum Metals." Academic Press, New York.

* Siegel, S. (1966). Stereochemistry and mechanism of hydrogenation of unsaturated hydrocarbons. *Adv. Catal.* **16**, 123–177.

Siegel, S., and Smith, G. V. (1960). *J. Am. Chem. Soc.* **82**, 6082, 6087.

Siegel, S., and Dmuchovsky, B. (1964). *J. Am. Chem. Soc.* **86**, 2192.

Siegel, S., Outlaw, J., Jr., and Garti, N. (1979). *J. Catal.* **58**, 380.

Smith, G. V., and Burwell, R. L., Jr. (1962). *J. Am. Chem. Soc.* **84**, 925.

Smith, H. A. (1957). The catalytic hydrogenation of aromatic compounds. In "Catalysis" (P. H. Emmett, ed.), Vol. 5, pp. 174–256. Reinhold, New York.

Smith, H. A., and Meriwether, H. T. (1949). *J. Am. Chem. Soc.* **71**, 413.

V. Wessely, F., and Welleba, H. (1941). *Ber. Dtsch. Chem. Ges.* **74**, 777.

Weitkamp, A. W. (1968). Stereochemistry and mechanism of hydrogenation of naphthalenes on transition metal catalysts and conformational analysis of the products. *Adv. Catal.* **18**, 1–110.

4

Dehydrogenation and Cyclodehydrogenation (Aromatization)

I. INTRODUCTION

The dehydrogenation of alkanes to alkenes and alkadienes is thermodynamically unfavorable at moderate temperatures. When the equilibrium is favorable for dehydrogenation below pyrolytic temperatures, the use of appropriate catalysts is greatly beneficial.

Ethane dehydrogenates thermally to ethylene, and therefore it is not suitable for catalytic dehydrogenation. Propene is a product of thermal of catalytic cracking of hydrocarbon and is produced more cheaply than it could be by catalytic dehydrogenation of propane. The dehydrogenation of butane and that of isopentane to the corresponding alkenes and alkadienes are processes of industrial importance. Straight-chain C_{12} to C_{16} alkanes are dehydrogenated to alkenes, which are employed in the production of alkylbenzenes that are used in the manufacture of biodegradable household detergents. Ethylbenzene is dehydrogenated catalytically to styrene, which is used in the manufacture of rubber and plastics. The dehydrogenation of heptanes and octanes leads to the formation of toluene and xylenes. Although aromatic hydrocarbons are obtained in the petroleum industry mainly from the hydroforming reaction, the mechanism of aromatization of alkanes over chromia catalysts is discussed in some detail because it will lead to a better understanding of the complex chemistry of the dehydrogenation of alkanes.

II. ALKANES

The dehydrogenation equilibria for C_3 to C_5 alkanes are given in Table 4.1. Dehydrogenation temperatures of above 500°C are required to achieve commercially practical conversion of alkanes to alkenes. The

185

TABLE 4.1

Dehydrogenation Equilibria for C_3 to C_6 Alkanes[a,b]

Conversion (mol %)	C_3H_6	2-C_4H_8		Mixed	iso-C_4H_8	Pentenes mixed	Hexenes mixed
		cis	trans				
5	410	415	405	375	360	360	335
10	456	470	450	420	405	410	380
20	510	525	505	470	435	440	425
30	540	555	540	505	490	465	445
40	570	590	572	530	515	490	470
50	595	625	598	555	540	520	495
60	625	660	630	580	560	545	520
70	660	700	640	610	590	575	550

[a] From Kearby (1955a, p. 456).
[b] Temperatures (°C) required for given conversions (paraffin → olefin + H_2).

conversion of butane to butenes and butadiene can be greatly increased, however, by carrying out the dehydrogenation reactions at subatmospheric pressures (Table 4.2) or by diluting the hydrocarbon vapors with inert gases, such as nitrogen or helium. Metal catalysts such as nickel, platinum, and iron are less suitable for the dehydrogenation of lower alkanes since at the temperatures required for dehydrogenation they produce a secondary cracking reaction.

Oxide catalysts, principally chromium oxides especially on supports such as alumina, are effective catalysts in dehydrogenation. They cause little cleavage of carbon–carbon bonds and are thermally stable.

A. Butane

Depending on the processing conditions, the dehydrogenation of butane may lead to the production of butenes or to the ultimate production of butadiene. In both cases the preferred catalyst is chromia (12–20%) on γ- or η-alumina.

1. To Butenes

The typical operating conditions for the production of butenes are as follows: temperature, 550°–600°C, pressure, 0.3–3 atm, LHSV, 1–3; and processing period before the catalyst is regenerated, 10–100 min. Conversion per pass is 30–40%, and the ultimate yield of butenes is 70–80%. Around 5% of butane is converted to coke, which is deposited on the catalyst and burned.

TABLE 4.2

Equilibrium between Butenes and 1,3-Butadiene[a]

Type of conversion	Conversion of n-butane at 1.0 atm (mol %)						
	550°C	600°C	650°C	700°C	750°C	800°C	850°C
To 1,3-butadiene	—	6.0	14	27.5	45	61.5	74
To total n-butenes	—	62.5	69	64.5	54	37.5	23
1-Butene	—	22.5	27	26	22.5	17	10
cis-2-Butene	—	16	17	16	13	9	4.5
trans-2-Butene	—	24	25	23	18	12.5	6.0

	Conversion of n-butane at 0.167 atm (mol %)						
	550°C	600°C	650°C	700°C	750°C	800°C	850°C
To 1,3-butadiene	11.5	27.5	48.5	69	82	—	—
To total n-butenes	68.5	64.5	48.5	31.5	17.5	—	—
1-Butene	23.5	23	19	13	7.5	—	—
cis-2-Butene	18	16.5	12	7.5	4	—	—
trans-2-Butene	27	25	17.5	11	6	—	—

[a] From Kearby (1955a, p. 456).

2. To Butadiene

The preferred processing conditions for the dehydrogenation of butane to butadiene are as follows: temperature, 550°–650°C; pressure, 0.2 atm; LHSV, 1–3; and processing period, 7–15 min. By separating the hydrogen and recycling the primary mixture of butane and butenes, a mixture of butanes, butenes, and butadiene is obtained. The butadiene is separated, and the unconverted butane and butenes with added butane are recycled. The overall yield of butadiene is about 65%. The total production of 1,3-butadiene, rubber grade, in 1978 in the United States was 1.60×10^9 kg.

B. Isopentane

Isopentane dehydrogenates over chromia–alumina catalyst to give a mixture of methylbutenes. The technology of conversion of isopentane to the olefins and ultimately to isoprene is similar to that described for the conversion of butane to butadiene. At 70% conversion of methylbutenes, the selectivity to isoprene is 67%.

C. Aromatization

1. Introduction

Alkanes and alkenes containing six carbon atoms or more undergo dehydrogenation and cyclization to aromatic hydrocarbons. The aromatization of paraffins is thermodynamically unfavorable at moderate temperature. On the other hand, owing to the high stability of the aromatic ring, this unfavorable equilibrium can be displaced almost quantitatively in favor of aromatic hydrocarbons at temperatures above 300°C, provided that a suitable catalyst for ring closure is employed.

Two types of catalyst can be used for the aromatization reaction: reduced metals and metal oxides. The reduced metal catalysts are those of Group VIII in the periodic table, preferably nickel or platinum deposited on supports such as alumina, silica, or activated carbon. The metal oxides used as catalysts are mainly chromia, molybdena, or vanadia; the first two appear to be preferred. The oxides as such undergo rapid deactivation during aromatization; however, the oxides deposited on alumina by impregnation are more resistant to rapid deactivation.

In the laboratory the aromatization is usually carried out by passing the hydrocarbon vapors at or near atmospheric pressure over the catalyst. The most frequently used experimental conditions with oxide catalysts include temperatures of 450°–540°C and contact times of a few seconds to 1 min.

The activity of the metal oxides on alumina declines in the course of several hours due to the deposit of carbonaceous material on their surface. However, these catalysts can be restored to their original activity by burning off the deposit.

2. Chromia–Alumina Catalysts

Most of the fundamental information regarding the aromatization of alkanes was obtained using chromia–alumina catalysts. These catalysts are active at 475°–525°C at atmospheric pressure, and their useful lifetime with periodic reactivation is over 1000 h.

The source and method of preparation of the aluminas have an important effect on the aromatization reaction. Aluminas prepared either from aluminum isopropoxide by means of hydrolysis or from aluminum nitrate by precipitation with ammonium carbonate and calcined at 600°C have intrinsic acidity and must be avoided in the preparation of Cr_2O_3/Al_2O_3 catalyst since they may cause cationic skeletal rearrangements and cleavage reactions to occur during aromatization. In order to avoid these cationic reactions it is imperative in the preparation of the catalysts to use aluminas in which the intrinsic acid sites are neutralized by the incorporation of about 0.5% of alkali metal cations (see Chapter 1, Section II,C,1).

It was found that 20% Cr_2O_3/Al_2O_3 catalysts prepared from the acidic alumina A and from the "nonacidic" alumina B gave quite different product distributions (Pines and Chen, 1960; Pines and Goetschel, 1965a):

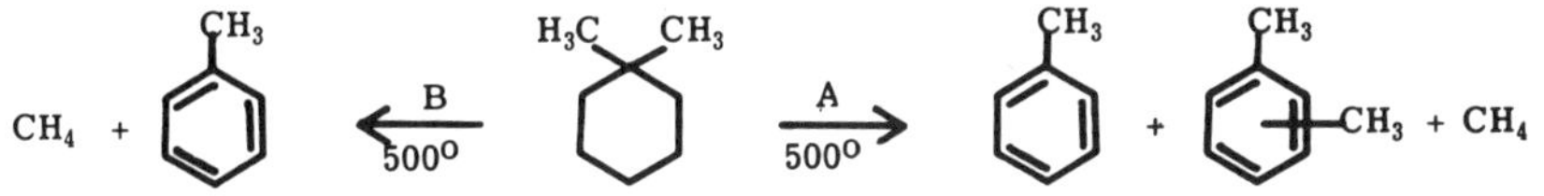

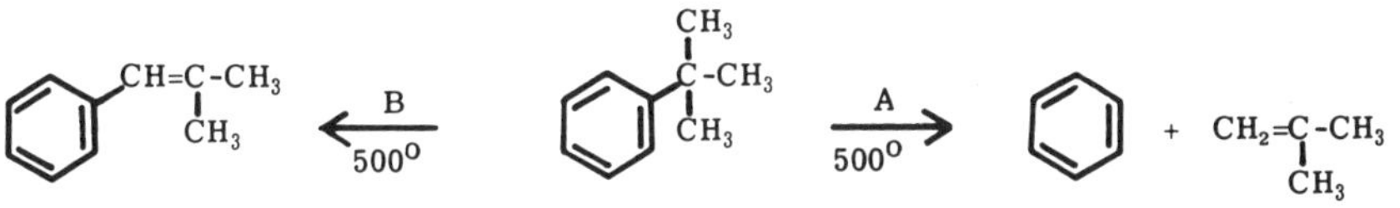

Most of the data presented here are based on results obtained with a "nonacidic" alumina–chromia catalyst. Alkanes, both branched and straight chain, having six carbon atoms or more can undergo aromatization.

a. Heptane. When heptane is passed over a chromia–alumina catalyst, the product consists mainly of toluene along with a small amount of heptenes. At very short contact times the concentration of heptenes rises more rapidly than the concentration of toluene. However, with increasing contact times the concentration of olefins reaches a maximum and then begins to fall, whereas that of toluene continues to rise.

The slow step in the conversion of heptane to toluene is the initial dehydrogenation of heptane to heptenes. The sequence of steps from 1-heptene to toluene is more rapid than the rehydrogenation of 1-heptene to heptane, with the dehydrogenation of methylcyclohexane being the most rapid step of all.

The dehydrogenation to olefins is much less sensitive to poisoning than is the aromatization step. This suggests that the catalytic sites for the dehydrogenation of alkanes to alkenes may be different from those for the dehydrocyclization of alkenes to aromatic hydrocarbons.

The direct 1,6 carbon–carbon closure, as originally suggested for the aromatization of heptane to toluene, is not the only path by which cyclization may occur. Using [1-^{14}C]heptane it was observed that the methyl-labeled toluene, depending on the duration of the run at which the sample

TABLE 4.3

Aromatization of Octane over Chromia–Alumina B Catalyst[a,b]

Fraction	Feed (ml)	Composition (wt%)			Composition of C_8 aromatic hydrocarbons (%)			
		Octane + others[c]	Toluene	C_8 aromatic hydrocarbons	Ethylbenzene	o-Xylene	m- + p-xylene	m-/p-xylene
1	2	28.1	4.0	67.9	11	44	45	1.5
2	2	41.0	3.4	55.6	13	38	49	—
3	2	44.0	3.4	52.6	14	40	46	1.9
4	8	—	—	—	—	—	—	—
5	2	54.1	3.5	42.4	22	47	31	—

[a] From Pines and Chen (1961).
[b] The experiments were carried out at 500°C, LHSV 0.65. The catalyst amounted to 15 ml or 11.0 gm.
[c] Less than 7% of compounds other than octane were present.

was withdrawn, contained from 18 to 32% $^{14}CH_3$, much less than the predicted 50% radioactivity.

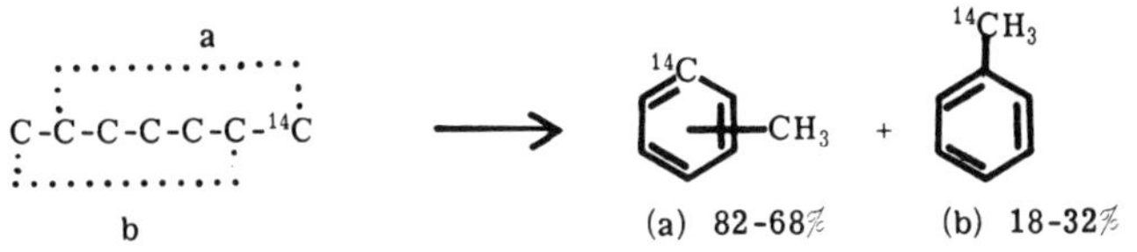

(a) 82-68% (b) 18-32%

b. Octane. On aromatization, octane affords not only ethylbenzene and o-xylene, products of direct 1,6 carbon–carbon ring closure, but also m- and p-xylene, in yields of 31–49% (Table 4.3). Xylenes under similar conditions do not undergo isomerization.

c. Dimethylhexanes. A variety of dimethylhexanes was submitted to aromatization over the "nonacidic" chromia–alumina B (Table 4.4). If only 1,6 carbon–carbon ring closure were operating in the aromatization reaction, each dimethylhexane would form only one aromatic compound; namely 2,2- and 3,3-dimethylhexane would yield only toluene, 2,3- and 3,4-dimethylhexane would form o-xylene, and 2,4- and 2,5-dimethylhexane would produce only m- and p-xylene, respectively.

Toluene was not, however, the only aromatic compound formed from 2,2- and 3,3-dimethylhexane; each of the dimethylhexanes produced all three xylenes and ethylbenzene, and the composition of these aromatic compounds depended on the structure of the starting dimethylhexanes.

TABLE 4.4

Aromatization of Dimethylhexanes over Chromia–Alumina B Catalyst[a,b]

	Dimethylhexane					
	2,2-	3,3-	2,3-	3,4-	2,4-	2,5-
Conversion (%)	19.9	22.5	7.3	15.3	12.7	31.7
Composition of aromatic compounds (mol %)						
Benzene	7.9	12.8	25.7	16.8	12.1	—
Toluene	57.4	53.8	15.6	18.9	32.2	16.8
o-Xylene	8.5	2.1	36.3	22.4	1.2	1.9
m-Xylene	11.6	25.1	5.0	2.1	48.1	7.5
p-Xylene	8.8	4.1	1.7	27.9	4.1	73.8
Ethylbenzene	5.9	2.0	15.6	11.9	2.3	Trace

[a] From Pines et al. (1965).
[b] See footnote b, Table 4.3.

d. Trimethylpentanes. On passage over "nonacidic" chromia–alumina catalyst, trimethylpentanes form xylenes (Table 4.5). Obviously, before the aromatization a skeletal rearrangement must occur to extend the chain length. The composition of the xylenes that are formed depends on the structure of the trimethylpentane used in the reaction.

2,2,4-Trimethylpentane affords only p-xylene, whereas 2,2,3-trimethylpentane produces principally m-xylene and smaller amounts of o- and p-xylene. The composition of xylenes changes with the length of the run. At the beginning of the run m-xylene is the principal xylene produced, and with time o-xylene becomes the main component of the xylenes.

e. Mechanism. On the basis of extensive research carried out on a variety of both nonlabeled and ^{14}C-labeled hydrocarbons, the mechanism of aromatization of normal and branched heptanes and octanes, including trimethylpentanes, is now seemingly well understood.

i. Trimethylpentanes. Kinetic studies have revealed that the aromatization of alkanes is preceded by dehydrogenation to alkenes. It has also been shown that alkanes such as neopentane or tetramethylbutane, which cannot undergo dehydrogenation, also do not undergo skeletal isomerization. The experimental data have demonstrated that skeletal isomerization is not of a cationic type, and the results can be explained by a free-radical mechanism accompanied by vinyl migration.

The mechanism of aromatization of trimethylpentanes can be illustrated by the conversion of 2,2,4-trimethylpentane to p-xylene (Scheme 4.1). Chromia acts as a catalyst for the dehydrogenation (steps *1, 2, 9,* and *11*) and for the double-bond isomerization (steps *3* and *9*). The homolytic

TABLE 4.5

Aromatization of Trimethylpentanes over "Nonacidic" Chromia–Alumina[a,b]

	Trimethylpentane		
	2,2,4-	2,2,3-	2,3,4-
Conversion (%)	36–26	40–29	36–26
Composition of xylenes (%)			
o-Xylene	—	10–7	22–54
m-Xylene	—	77–84	56–20
p-Xylene	100	13–9	22–26

[a] From Pines and Csicsery (1962).
[b] Temp, 523°C, contact time, 3 sec.

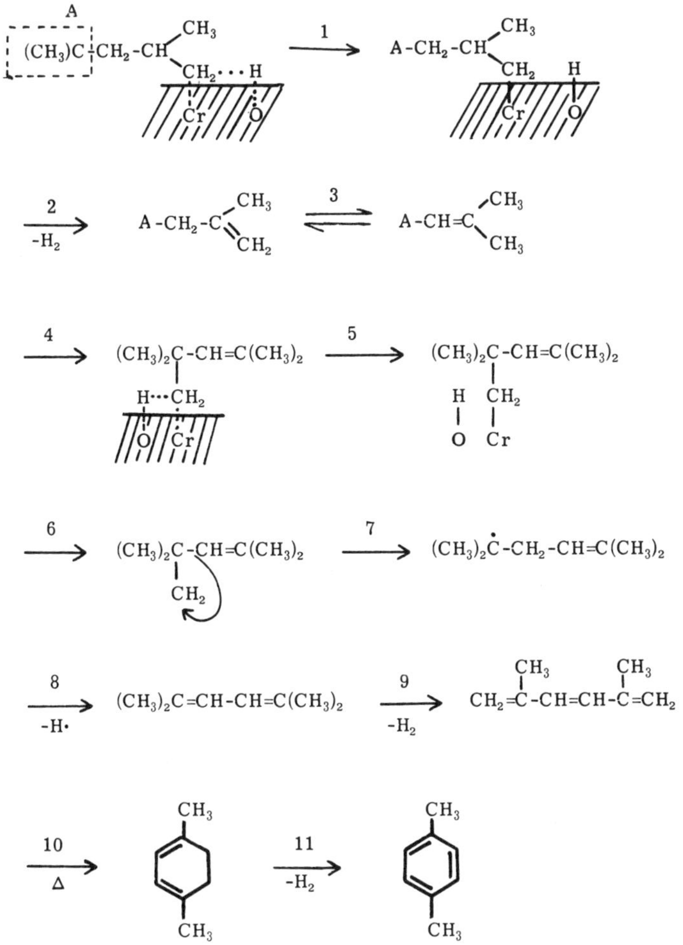

Scheme 4.1. Mechanism of aromatization of 2,2,4-trimethylpentane (from Pines, 1972, p. 24).

rupture of the organometallic bond in step *6* is facilitated by the overlap of electrons between the generated radical and the double bond and is followed in step *7* by vinyl migration, similar to a reported free-radical reaction induced by iodine. The cyclization (step *10*) proceeds thermally and does not require a catalyst.

The mechanism proposed above can also be used to explain the results of aromatization of 2,2-dimethyl[3-*methyl*-^{14}C]pentane in which 71% of the radioactivity in the formed *m*-xylene was in the methyl group and the remainder in the ring (Scheme 4.2).

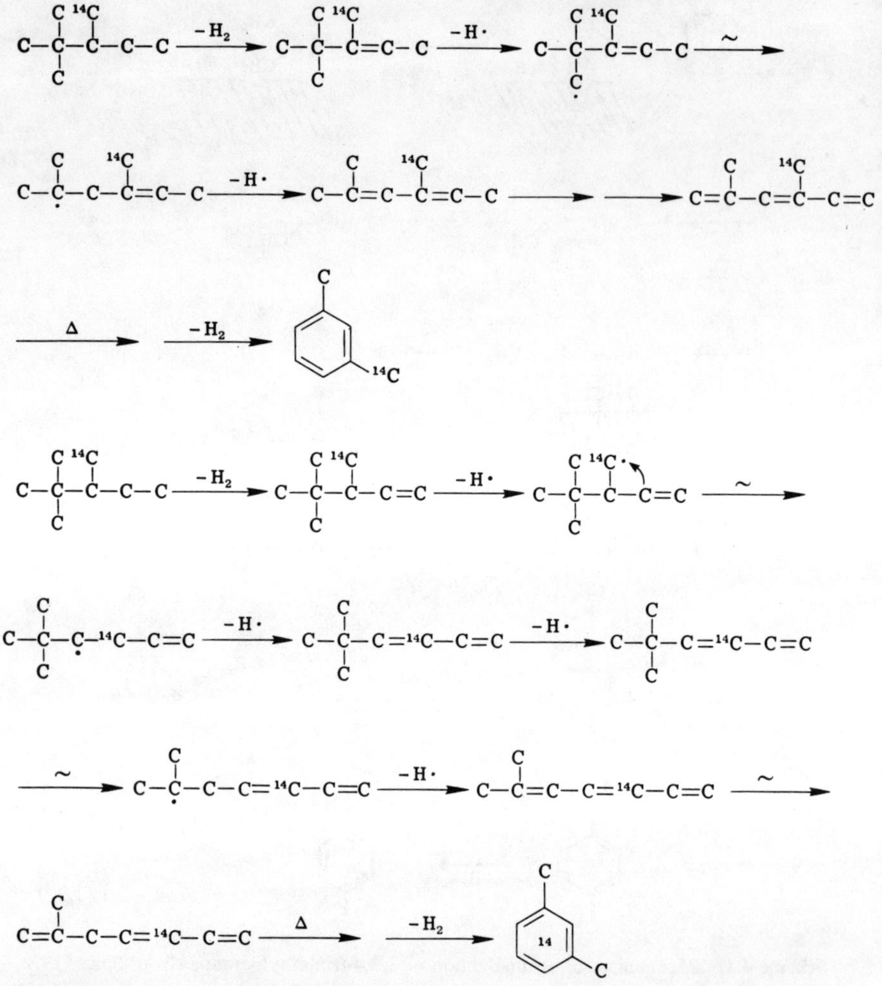

$$C-\underset{\underset{C}{|}}{\overset{\overset{C}{|}}{C}}-\overset{^{14}C}{C}-C-C \xrightarrow{-H_2} C-\underset{\underset{C}{|}}{\overset{\overset{C}{|}}{C}}-\overset{^{14}C}{C}=C-C \xrightarrow{-H\cdot} C-\underset{\underset{\cdot C}{|}}{\overset{\overset{C}{|}}{C}}-\overset{^{14}C}{C}=C-C \sim\rightarrow$$

$$C-\underset{\cdot}{\overset{\overset{C}{|}}{C}}-C-\overset{^{14}C}{C}=C-C \xrightarrow{-H\cdot} C-\overset{\overset{C}{|}}{C}=C-\overset{^{14}C}{C}=C-C \longrightarrow \longrightarrow C=\overset{\overset{C}{|}}{C}-C=\overset{^{14}C}{C}-C=C$$

$$\xrightarrow{\Delta} \xrightarrow{-H_2} \text{[benzene ring with } C \text{ and } ^{14}C \text{ substituents]}$$

$$C-\underset{\underset{C}{|}}{\overset{\overset{C}{|}}{C}}-\overset{^{14}C}{C}-C-C \xrightarrow{-H_2} C-\underset{\underset{C}{|}}{\overset{\overset{C}{|}}{C}}-\overset{^{14}C}{C}-C=C \xrightarrow{-H\cdot} C-\underset{\underset{C}{|}}{\overset{\overset{C}{|}}{C}}-\overset{^{14}C\cdot}{C}-C=C \sim\rightarrow$$

$$C-\underset{\underset{C}{|}}{\overset{\overset{C}{|}}{\cdot C}}-C-^{14}C-C=C \xrightarrow{-H\cdot} C-\underset{\overset{C}{|}}{\overset{\overset{C}{|}}{C}}-C=^{14}C-C=C \xrightarrow{-H\cdot} C-\underset{\overset{C}{|}}{\overset{\overset{C}{|}}{C}}-C=^{14}C-C=C$$

$$\sim\rightarrow C-\underset{\cdot}{\overset{\overset{C}{|}}{C}}-C-C=^{14}C-C=C \xrightarrow{-H\cdot} C-\overset{\overset{C}{|}}{C}=C-C=^{14}C-C=C \sim\rightarrow$$

$$C=\overset{\overset{C}{|}}{C}-C-C=^{14}C-C=C \xrightarrow{\Delta} \xrightarrow{-H_2} \text{[benzene ring with } C \text{ and } C \text{ substituents, labeled } 14\text{]}$$

Scheme 4.2. Aromatization of 2,2-dimethyl[3-*methyl*-^{14}C]pentane (from Pines, 1972).

The aromatization of 2,4-dimethyl[3-*methyl*-^{14}C]pentane produces, in addition to *o*- and *p*-xylene, *m*-xylene, having about 94% of radioactivity in the ring. The formation of the meta isomer can be explained by a double vinyl migration (Scheme 4.3).

ii. Dimethylhexanes. The formation of aromatic hydrocarbons that cannot occur through a direct 1,6 carbon–carbon closure of dimethylhexanes can also be explained by a mechanism similar to that presented for trimethylpentanes. An outline of the suggested steps involved in the con-

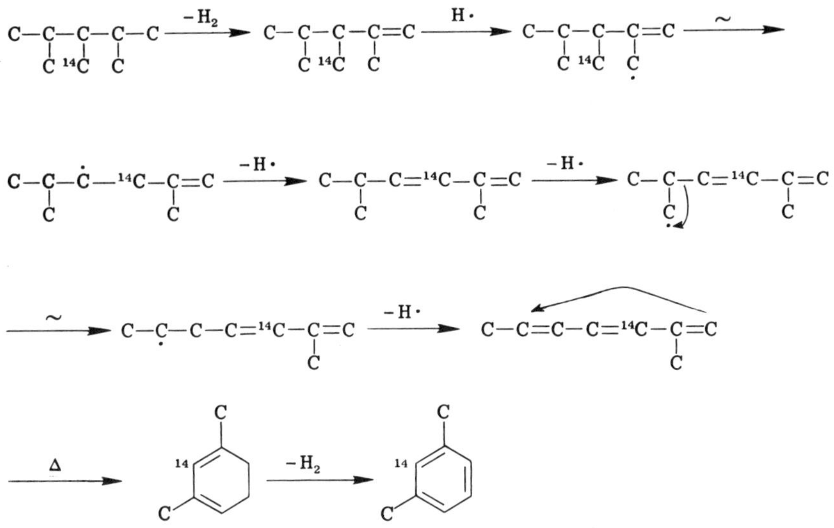

$$
\text{C-C-C-C-C} \quad \xrightarrow{-H_2} \quad \text{C-C-C-C=C} \quad \xrightarrow{H\cdot} \quad \text{C-C-C-C=C} \quad \xrightarrow{\sim}
$$

$$
\xrightarrow{-H\cdot} \quad \text{C-C-C=}{}^{14}\text{C-C=C} \quad \xrightarrow{-H\cdot} \quad \text{C-C-C=}{}^{14}\text{C-C=C}
$$

$$
\xrightarrow{\sim} \quad \text{C-C-C-C=}{}^{14}\text{C-C=C} \quad \xrightarrow{-H\cdot} \quad \text{C-C=C-C=}{}^{14}\text{C-C=C}
$$

Scheme 4.3. Aromatization of 2,4-dimethyl[3-*methyl*-^{14}C]pentane.

version of 3,4-dimethylhexane to *p*-xylene and ethylbenzene is given in Scheme 4.4.

iii. Unbranched alkanes. On aromatization over "nonacidic" chromia –alumina, [1-^{14}C]heptane formed toluene with more than the expected 50% of ^{14}C in the ring. Obviously, therefore, cyclization involved not only a direct 1,6 ring closure, but also other reactions. The excess of radioactivity in the ring compared with that in the methyl group of the toluene could be explained by insertion of the labeled carbon from the terminal position to the 2 position of the original hydrocarbon chain, e.g.,

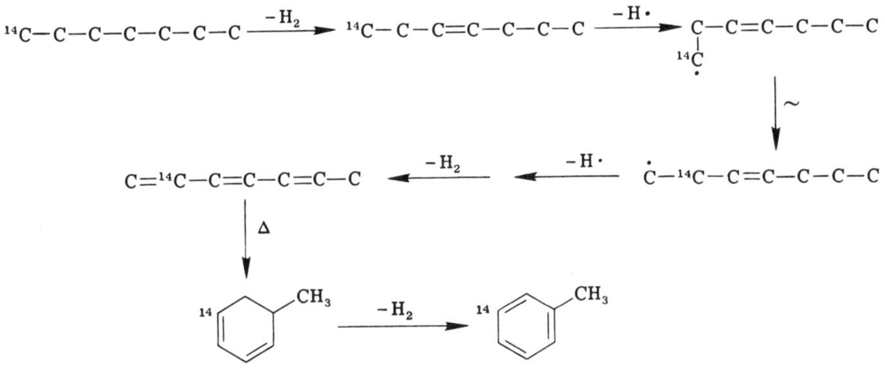

$$
{}^{14}\text{C-C-C-C-C-C-C} \quad \xrightarrow{-H_2} \quad {}^{14}\text{C-C-C=C-C-C-C} \quad \xrightarrow{-H\cdot} \quad \text{C-C=C-C-C-C}
$$

$$
\text{C=}{}^{14}\text{C-C=C-C=C-C} \quad \xleftarrow{-H_2} \quad \xleftarrow{-H\cdot} \quad \text{C-}{}^{14}\text{C-C=C-C-C-C}
$$

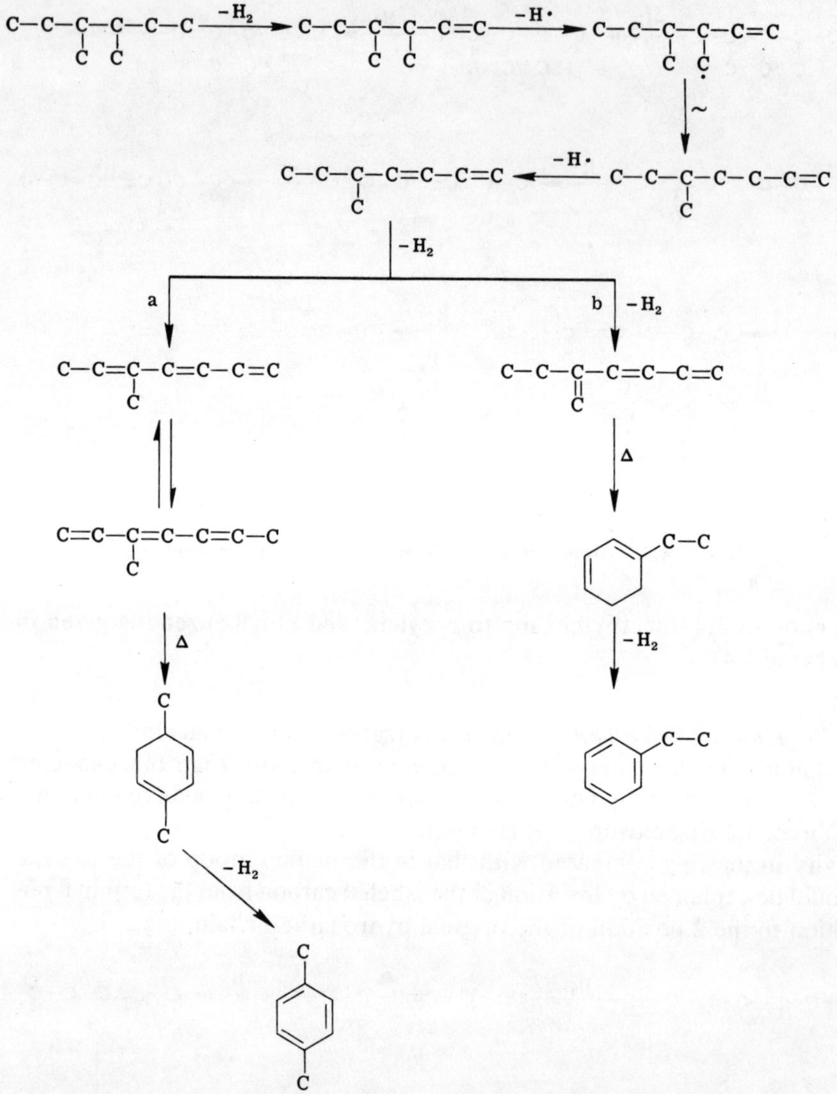

$$C-C-\underset{\underset{C}{|}}{C}-\underset{\underset{C}{|}}{C}-C-C \xrightarrow{-H_2} C-C-\underset{\underset{C}{|}}{C}-\underset{\underset{C}{|}}{C}-C=C \xrightarrow{-H\cdot} C-C-\underset{\underset{C}{|}}{C}-\underset{\underset{\cdot C}{|}}{C}-C=C$$

$$\Big\downarrow \sim$$

$$C-C-\underset{\underset{C}{|}}{C}-C=C-C=C \xleftarrow{-H\cdot} C-C-\underset{\underset{C}{|}}{C}-C-C-C=C$$

$$\Big\downarrow -H_2$$

a   b $-H_2$

$$C-C=\underset{\underset{C}{|}}{C}-C=C-C=C \qquad\qquad C-C-\underset{\overset{\|}{C}}{C}-C=C-C=C$$

$$\Big\updownarrow \qquad\qquad\qquad\qquad\qquad\qquad \Big\downarrow \Delta$$

$$C=C-\underset{\underset{C}{|}}{C}=C-C=C-C$$

image: cyclohexadiene ring with C—C substituent

$$\Big\downarrow \Delta \qquad\qquad\qquad\qquad\qquad\qquad \Big\downarrow -H_2$$

image: dimethylcyclohexadiene ring (1,4)

image: benzene ring with C—C substituent (ethylbenzene)

$$\Big\searrow -H_2$$

image: p-xylene ring

Scheme 4.4. Aromatization of 3,4-dimethylhexane to p-xylene and ethylbenzene.

This mechanism, which involves the generation of a radical followed by vinyl migration, is similar to the mechanism proposed for the aromatization of trimethylpentanes (Schemes 4.1–4.3).

Scrambling of ^{14}C in alkanes was also observed in [1-^{14}C]butene over chromia–alumina catalyst, and the results can be interpreted by a similar mechanism:

$$C-C-C=^{14}C \xrightarrow{\sim} C=C-C-^{14}C \xrightarrow{-H\cdot} C=C-C$$
$$| \atop ^{14}\overset{|}{C}\cdot$$

$$| \sim$$

$$C=^{14}C-C=C \xleftarrow{-H\cdot} \cdot C-^{14}C-C=C$$

3. Metal Catalysts

Transition metals of Group VIIIB, except for iron and osmium, catalyze the aromatization of alkanes. Of these, platinum deposited on carbon, platinum on alumina, and platinum on silica–alumina are the most widely studied catalysts. Other metals that cause aromatization to occur are rhenium, copper, and cobalt.

Platinum, palladium, and iridium may also catalyze C_5 cyclization of alkanes. The selectivity toward C_5 cyclization depends on the catalyst used and the experimental conditions. The following selectivities were obtained by passing hexane over platinum and platinum catalysts at 520°C in a pulse system using hydrogen carrier (Fadeev *et al.*, 1969):

	Selectivity[a]	
Catalyst	C_5 cyclic	Benzene
Pt/Al_2O_3	0.21	0.22
Pd/Al_2O_3	0.07	0.19

[a] Total conversion = 1.00.

At temperatures below 300°C and in the presence of Pt/Al_2O_3, cyclic C_5 products predominate over cyclic C_6 products. From 2- and 3-methylpentane, C_5 ring products appear almost to the exclusion of C_6 ring hydrocarbons or benzene when passed over either 0.2 or 10% Pt/Al_2O_3 catalysts at temperatures of 270°–280°C. At higher temperatures, however, the yield of benzene tends to increase.

It has not yet been resolved whether the preferred route to aromatization of methylpentanes is their prior conversion to a C_6 linear hydrocarbon species adsorbed on the surface or whether a closure to a C_5 ring is followed by ring enlargement. Once the six-carbon ring is formed on the metal, dehydrogenation to aromatic hydrocarbons takes place rapidly.

The selectivity for dehydrocyclization and aromatization can be changed substantially when the size of Pt particles on carriers is varied. Small particles strongly favor the dehydrocyclization of heptane to aromatic hydrocarbons. Low coordination and isolated platinum atoms are

TABLE 4.6

Reaction of Heptane over 2% (Pt + Re)/Al_2O_3[a,b]

Re (%)	Relative conversion to gaseous hydrocarbons	Selectivity (%)			
		Hydrogenolysis	Isomerization	To C_5 ring closure	Aromatization
0	100	28	16	24	32
40	35	31	15	23	31
80	49	60	8	5	27
100	42	87	4	2	7

[a] From Tournayan et al. (1978).
[b] At 400°C, 1 atm H_2; pulse reactor.

the most effective for aromatization, but it is difficult to prepare a platinum catalyst with a large and controlled proportion of isolated Pt atoms. By alloying, however, a situation related to the extreme dilution of Pt on carriers can be simulated.

Alloys of platinum and rhenium on γ-Al_2O_3 modify the conversion and selectivity of heptane toward dehydrocyclization. In the absence of rhenium the ratio of aromatization to cyclic C_5 ring hydrocarbons is 1.3 : 1. In the presence of 80% rhenium in platinum the ratio increases to 5.4 : 1 (Table 4.6).

III. CYCLOALKANES

A. Six-Membered Ring System

Hydrocarbons containing a six-carbon ring configuration with at least one hydrogen on each carbon atom undergo dehydrogenation to arenes. Both metals and metal oxides catalyze this reaction.

Platinum and palladium deposited on a suitable carrier, such as alumina, silica, or carbon, afford high yields and selectivity. The dehydrogenation proceeds in the vapor phase at 250°–310°C with LHSV of 0.3–0.5 without the formation of cycloalkenes or cycloalkadienes as isolable intermediates:

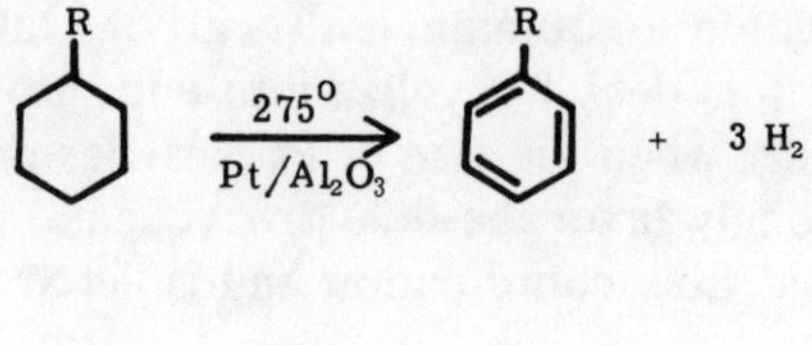

Nickel deposited on alumina requires higher temperatures, 350°–400°C, and gives lower yields due to its hydrogenolysis activity, which includes partial cleavage of the alkyl side chain.

Oxide catalysts such as chromia, molybdena, and vanadia as such or preferably deposited on alumina can also be used for dehydrogenation of cyclohexanes. Of the oxides, chromia–alumina is the preferred catalyst. The temperature required for dehydrogenation over chromia-containing catalysts is much higher than that over reduced metal catalysts. The dehydrogenation of cyclohexane at 400°–500°C over chromia catalyst gives, in addition to 90% benzene, some cyclohexene.

The dehydrogenation of cyclohexane can be explained by the reversal of the steps of benzene hydrogenation and may proceed by the α, β, or π

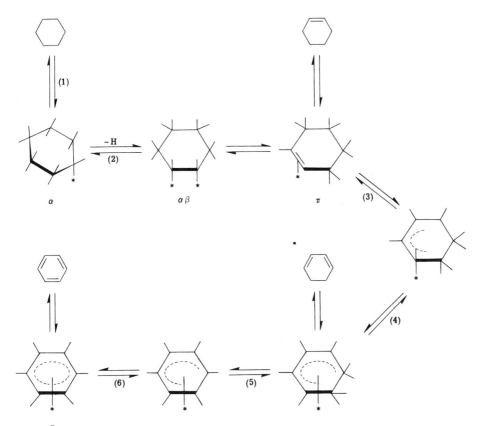

Fig. 4.1. Mechanism of dehydrogenation of cyclohexane (* represents catalytic site) (from Germain, 1969).

adsorbed species followed by a stepwise removal of hydrogen atoms and the extension of the π system (Fig. 4.1). All the steps must be fast because only traces of cyclohexene and cyclohexadienes were observed in the gas phase, the rate-determining steps being (1) and (2).

A variety of cyclic hydrocarbons containing a cyclohexane group undergo dehydrogenation over platinum deposited on charcoal or alumina by straight dehydrogenation or accompanied by a ring-opening or a ring-forming reaction, as in the case of fluorene and phenanthrene (Table 4.7). The ring forming occurs at about 350°C.

Spirocyclohexanes containing a geminal carbon atom undergo dehydrogenation at higher temperatures than do the corresponding alkylcyclohexanes. With Pt/Al_2O_3 and with Cr_2O_3/Al_2O_3 temperatures above 320° and 500°C, respectively, are required. In the presence of chromia on "nonacidic" alumina, demethanation of 1,1-dimethylcyclohexane occurs and toluene is formed exclusively. With platinum deposited on a "nonacidic" alumina, the aromatization is accompanied by methyl migration and results in the formation of xylenes. The ratio of xylenes to toluene increases with the relative intrinsic acidity of the alumina. The reaction can be interpreted as follows:

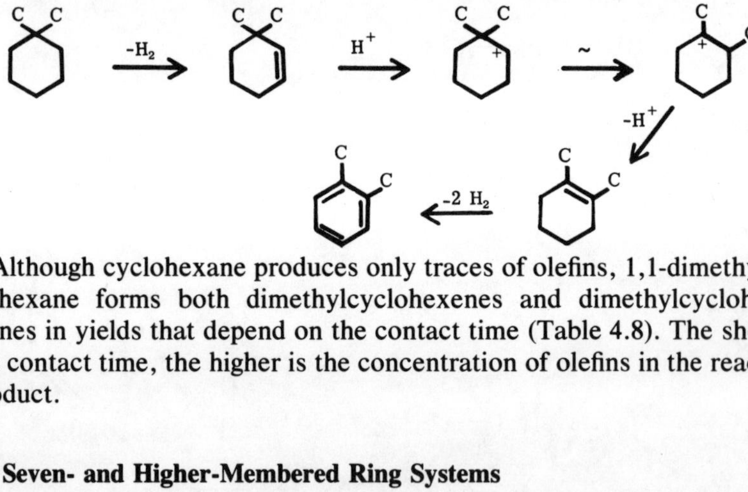

Although cyclohexane produces only traces of olefins, 1,1-dimethylcyclohexane forms both dimethylcyclohexenes and dimethylcyclohexadienes in yields that depend on the contact time (Table 4.8). The shorter the contact time, the higher is the concentration of olefins in the reaction product.

B. Seven- and Higher-Membered Ring Systems

Cycloheptane and larger ring compounds undergo aromatization at higher temperatures than do C_6 ring cyclanes. In the presence of platinum catalysts a temperature of at least 310°C, and in the presence of chromia a temperature of 500°C, is required. On dehydrogenation in the presence of Pd/C catalyst, C_9 to C_{18} cyclanes generate new rings and yield polycyclic aromatic hydrocarbons.

TABLE 4.7

Examples of Catalytic Dehydrogenation of Cyclic Hydrocarbons in the Presence of Pt/Carriers at 275°–350°C

Cyclic hydrocarbon	Product
 Menthane	 p-Cymene
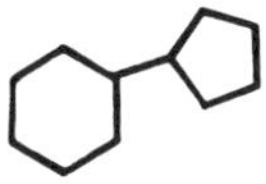 Cyclohexylcyclopentane	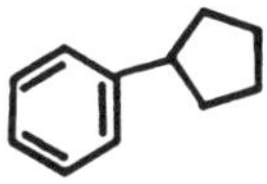 Cyclopentylbenzene
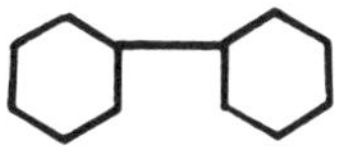 Dicyclohexyl	 Biphenyl
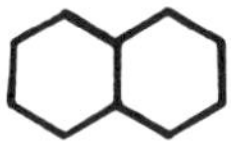 Decalin	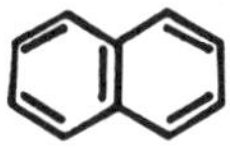 Naphthalene
 Pinane	 p-Cymene (main product)
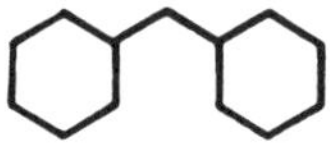 Dicyclohexylmethane	 Fluorene
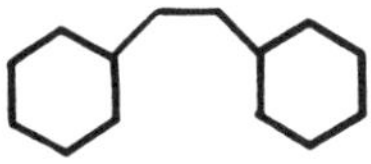 1,2-Dicyclohexylethane	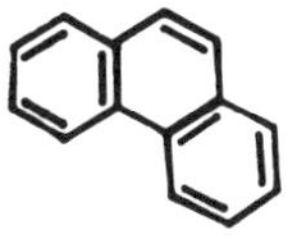 Phenanthrene
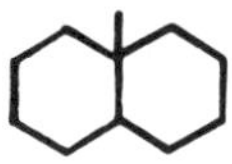 9-Methyldecalin	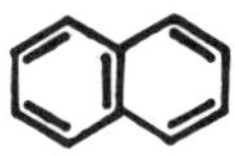 Naphthalene
 Spirotetramethylenecyclohexane	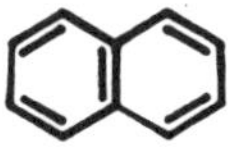 Naphthalene

TABLE 4.8

Dehydrogenation of 1,1-Dimethylcyclohexane over Pt/Al_2O_3[a,b]

		Composition of products (%)				
LHSV	Conversion (%)		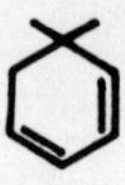	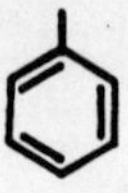	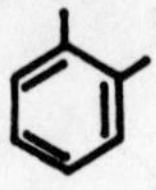	m- and p-xylene
0.5	33	7	—	52	34	6
1.0	15	12	Trace	38	46	3
2.0	9	20	6	29	40	4
4.0	4	48	7	21	23	

[a] From Pines and Greenlee (1961).
[b] Catalyst, 1% Pt on "nonacidic" alumina; temp, 350°C.

1. Cycloheptanes

Cycloheptane on dehydrogenation yields toluene.

a. Platinum Catalyst. Methylcycloheptane formed the following compounds over a "nonacidic" 1% Pt/Al_2O_3 at 350°C, at HLSV 0.5, and at 36% conversion (Pines and Greenlee, 1961):

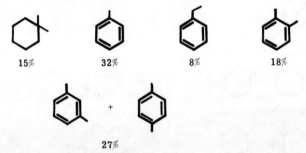

15%   32%   8%   18%

27%

Toluene was most likely derived from the dehydrogenation of 1,1-dimethylcyclohexane accompanied by demethanation. In accordance with the proposal of N. D. Zelinskii (cited in Germain, 1969, p. 108), the ring contraction occurs via a bicycloalkane intermediate (Scheme 4.5).

It is unlikely that the ring contraction occurs through a cationic mechanism since this would lead to the eventual formation of ethylbenzene rather than toluene as the principal aromatic hydrocarbon. The migration of a methyl group to the primary cation to form the most stable tertiary cation, ethylcyclohexyl cation, would be the preferred reaction (Scheme 4.6).

The generation of xylenes and of ethylbenzene over platinum catalysts

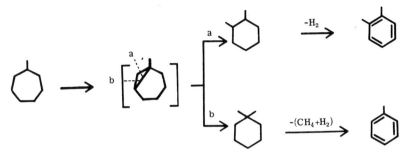

Scheme 4.5. Aromatization of methylcycloheptane via bicycloalkane.

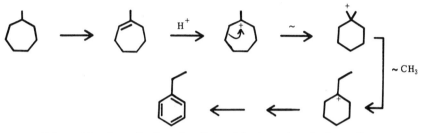

Scheme 4.6. Aromatization of methylcycloheptane via cationic reaction.

could likewise be explained by an intermediate formation of 2,4-, 3,5-, and 3,6-bicycloalkanes.

b. Chromia–Alumina Catalyst. The aromatization of [¹⁴C]methylcy-cloheptane over chromia–alumina catalyst produced toluene, ethylbenzene, and xylenes, in which the radioactivity was located both in the side chain and in the aromatic ring (Table 4.9). The distribution of radioactivity in the aromatic hydrocarbon indicates that the mechanism of aromatization of methylcycloheptane over chromia–alumina catalyst is not the same as that proposed for platinum–alumina catalyst. The almost equal distribution of radioactivity between the ethyl side chain and the benzene ring is an indication that cleavage of the cycloheptane ring occurred to generate a transient C_8 straight-chain species.

TABLE 4.9

Aromatization of [^{14}C]Methylcycloheptane over "Nonacidic"
Chromia–Alumina Catalyst^{a,b}

		Radioactivity distribution (%)	
		Side chain	Ring
Conversion (%)	36.9		
Composition of aromatic compounds (%)			
Benzene	1.7		
Toluene	16.3	87.4	13.6
Ethylbenzene	10.5	α, 3.5; β 44.3	50.5
o-Xylene	30.8	96.6	4.5
m-Xylene	25.9	72.0	27.5
p-Xylene	14.1	96.0	5.6

^{a} From Goetschel and Pines (1965).
^{b} Temp, 533°C; contact time, 3 sec.

The almost identical distribution of radioactivity in o-xylene and p-xylene suggests a common intermediate in the formation of the two xylenes.

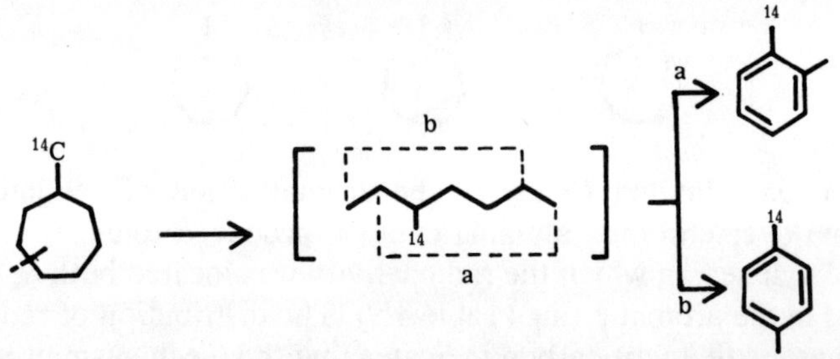

The presence in m-xylene of ^{14}C, both in the methyl group and in the ring, suggests that the transient open-chain C_8 species is a triene and that the 1,6 carbon–carbon ring closure proceeded by a thermal reaction:

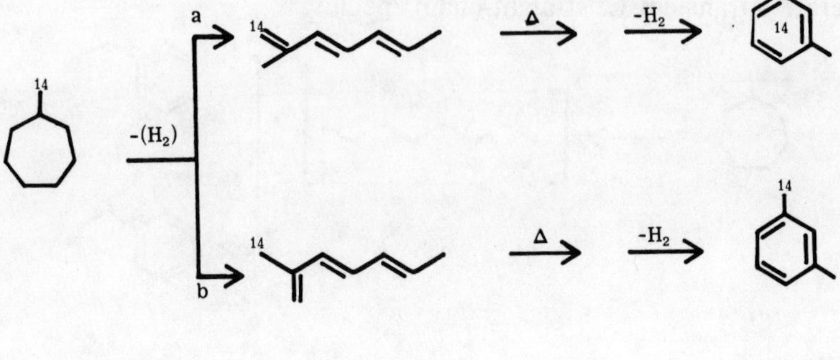

2. Cyclooctane

Over a platinum catalyst, cyclooctane forms pentalane as the main product and *o*-xylene and ethylbenzene as minor products. Over a "non-

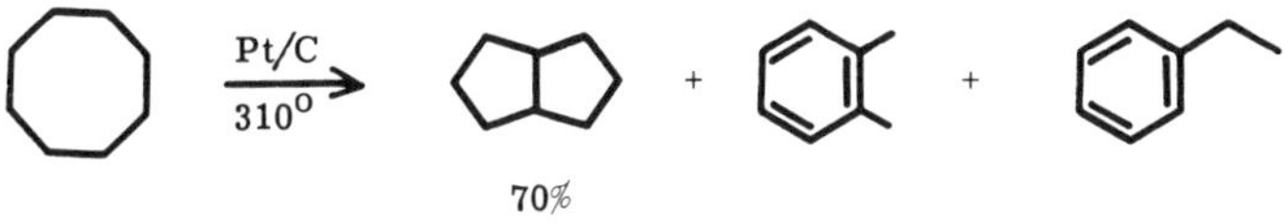

acidic" chromia–alumina at 500°C, cyclooctane undergoes isomerization to methylcycloheptane and aromatization to toluene, xylene, and ethylbenzene (Table 4.10). Methylcycloheptane is the principal product of reaction over the "acidic" chromia–alumina A. The formation of methylcycloheptane can be best explained by a cationic mechanism as described in Chapter 1, Section III.

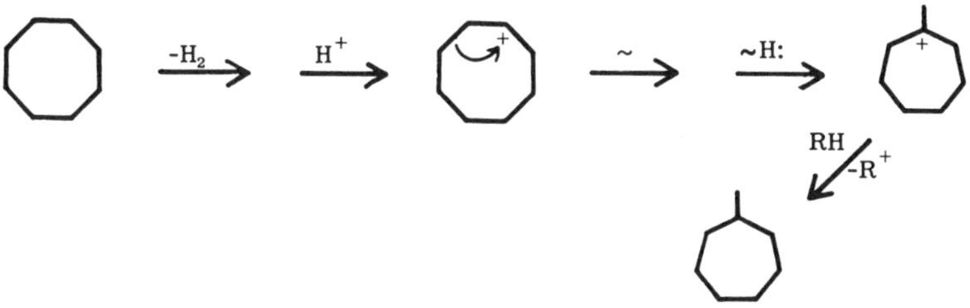

3. Cyclononane and Larger Ring Compounds

The C_9 to C_{18} cycloalkanes form polycyclic aromatic hydrocarbons when heated at 400°C in the presence of a Pd/C catalyst (Table 4.11).

C. Cyclopentane Ring System

Cyclopentane ring hydrocarbons do not dehydrogenate under the conditions in which cyclohexanes react smoothly to form aromatic compounds. Apparently adsorbed cyclopentadienyl species do not desorb easily, and therefore they could be reponsible for the deactivation of the catalyst surface. It is also possible that once a desorbed cyclopentadiene species is desorbed it might undergo condensation and dehydrogenation, which would lead to the formation of carbonaceous material that would be adsorbed and deactivate the catalyst.

Alkylcyclopentanes, however, may undergo a facile dehydroisomeriza-

TABLE 4.10

Reaction of Cyclooctane over Chromia–Alumina Catalysts[a,b]

Cut[c]	Catalyst[d]	Conversion (%)	Products		Composition of aromatic hydrocarbons (%)				
			MCHP[e] (%)	Aromatics (%)	Toluene	Ethyl-benzene	o-Xylene	m- + p-xylene	m-/p-xylene
1	A	79.2	32.2	67.8	1.8	29.2	36.5	32.4	
2	A	57.3	39.1	60.9	4.3	32.2	31.3	32.2	1.5
5	A	45.1	38.8	61.2	4.7	32.6	29.7	33.0	1.5
1	B	45.1	8.2	91.8	8.2	19.8	31.1	40.8	
3	B	17.8	17.4	82.6	16.9	26.5	27.2	29.4	1.4

[a] From Pines and Chen (1961).
[b] Temp, 500°C; LHSV 0.5.
[c] Cuts were collected after the following milliliters of cyclooctane passed over the catalyst: cut 1, 0–2; cut 3, 4–6; cut 5, 8–10.
[d] 17% Cr_2O_3/Al_2O_3: A, alumina prepared from $Al(OC_3H_7i)_3$; B, alumina prepared from potassium aluminate.
[e] Methylcycloheptane.

TABLE 4.11

Aromatization of Large-Ring Cycloalkanes[a,b]

Starting hydrocarbon[c]	Product
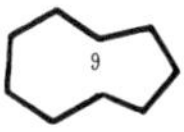 Cyclononane	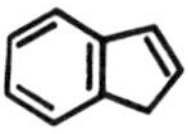 Indene
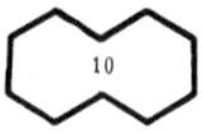 Nonane	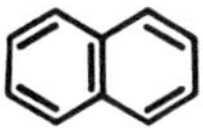 Naphthalene
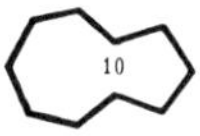 Cyclodecane	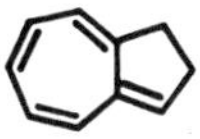
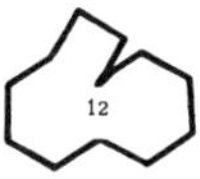 Cyclododecane	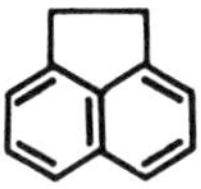 Acenaphthene
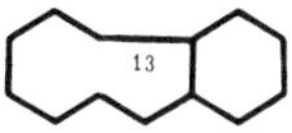 Cyclotridecane	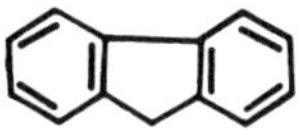 Fluorene
 Cyclotetradecane	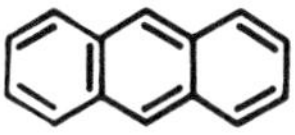 Anthracene
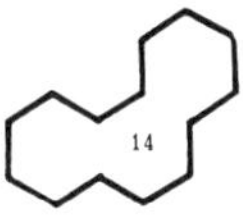	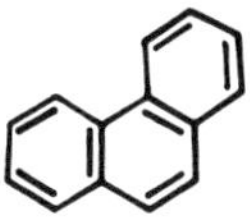 Phenanthrene
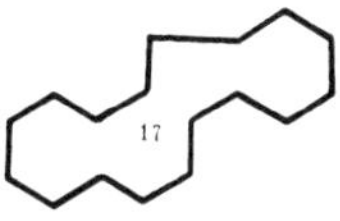 Cycloheptadecane	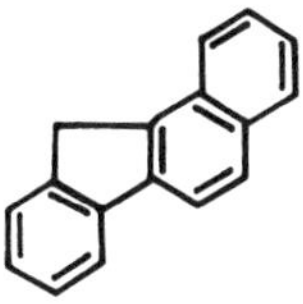 11H-benzo[a]fluorene, 1,2-benzfluorene
 Cyclooctadecane	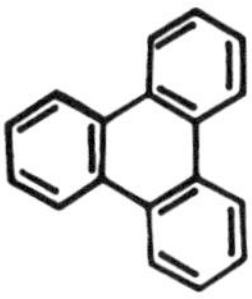 Triphenylene

[a] From Prelog *et al.* (1955).
[b] Catalyst, 5% Pd/C; temp, 400°C; duration, 30 min.
[c] Number within ring indicates number of carbons comprising ring.

207

tion to aromatic hydrocarbons if a dual-function catalyst is used. Platinum deposited on an acidic support such as alumina containing small amounts of halides or silica (re-forming catalyst) is especially effective for this

$$\text{(structure)} \longrightarrow \text{(structure)} \quad + \quad 3\,H_2$$

reaction, e.g., the change of ethylcyclopentane to toluene. A more detailed discussion of the subject is presented in Chapter 1, Section VIII.

IV. ALKYLBENZENES

A. Ethylbenzene

The dehydrogenation of ethylbenzene to styrene is important industrially. In 1978 there were 3.08 billion kilograms of styrene produced in the United States alone. Ethylbenzene is relatively easy to dehydrogenate thermally and over a variety of catalysts. However, in only a limited number of catalytic systems are high selectivity and long catalyst life without regeneration achieved.

Ethylbenzene is effectively dehydrogenated in the presence of chromia –alumina or even alumina, when no steam diluent is used. Steam deactivates these catalysts. Steam dilution, however, has several advantages for commercial operation. An equilibrium is reached in this reaction, and a diluent shifts it toward higher styrene concentration in the product. Commercially a nonregenerative process using steam was developed with a catalyst containing 7% K_2CO_3 and about 5% CrO_3 on iron oxide. The dehydrogenation is carried out at 590°–650°C, and the molar ratio of steam to ethylbenzene is about 12:1. Steam as an oxidant equilibrates iron oxide catalysts to an oxidation state that proves to be highly selective to styrene. In addition, steam reduces the partial pressure of the feed and serves to remove carbonaceous residue from the catalyst in the form of CO and CO_2, and this allows continuous operation for 1 or 2 years with a single catalyst charge. Of all the catalysts tested the alkali-promoted iron catalyst is the most efficient for the commercial production of styrene (Table 4.12).

The activity of iron oxide reaches a plateau with about one equivalent monolayer of potassium ions, which indicates that the potassium promoter operates specifically on the surface of the iron oxide. Potassium-promoted iron oxide catalysts are so active that intraparticle diffusion may limit the rate of dehydrogenation. For high selectivity it is important to have a catalyst of low specific surface area. Selectivity is maximized by

TABLE 4.12

Relative Activities of Unsupported Metal
Oxides in the Presence of Steam[a,b]

Metal oxide	Relative rate, mol styrene (sec)(m²) × 10⁸
Al + K	~0
Cu	6
Cr	9
Fe	6–18
Fe + K	300–500
Mo	16
Zn	20
Zn + K	40
Co	30
Co + K	42
Mn	50
Mn + K	160

[a] From Lee (1973).
[b] Temp, 600°C; steam/ethylbenzene = 13/1 mol.

calcining the catalyst at 900°–950°C in air to reduce the surface area to about 2 m²/gm.

Commercial catalysts usually have a minor amount of chromium oxide Cr_2O_3, which acts as a structural stabilizer. There is a rapid loss of activity of the catalyst without chromium, and this is attributed to sintering. Aside from structural stabilization, minor amounts of chromium have little or no effect on overall catalytic properties of the catalyst, but the catalyst is very sensitive to chlorides. Organic chlorides in as low a concentration as 3–5 ppm in ethylbenzene cause a drastic drop in catalytic activity.

In commercial operation over 90% selectivity to styrene is achieved with conversion of over 50%. The side products are benzene and toluene.

B. Skeletal Isomerization Accompanying Dehydrogenation: Chromia–Alumina Catalysts

The dehydrogenation of short-chain alkylbenzenes to corresponding alkenylbenzenes can be accomplished readily in the presence of chromia-alumina catalysts. The study of these reactions has theoretical significance because it demonstrates the unifying mechanism of the dehydrogenation of alkylbenzenes and the aromatization of alkanes (Section II,C,2,c).

The dehydrogenation of alkylbenzenes including ^{14}C-labeled ethylbenzene over chromia–alumina catalysts is accompanied by skeletal isomerization caused by the participation in the reaction of free radicals generated *in situ* by the intervention of the catalyst. The catalysts used in this study consisted of "nonacidic" chromia–alumina containing 14.8 wt % of Cr_2O_3. The experiments were carried out in a flow system at 485°–520°C and at LHSV 0.8–1.0.

1. [β-^{14}C]Ethylbenzene

On passing over chromia–alumina catalyst, [β-^{14}C]ethylbenzene produces styrene and a rearranged ethylbenzene (Pines and Abramovici, 1969).

$$
\begin{array}{ccc}
 & \xrightarrow{15\%} & C_6H_5CH{=}^{14}CH_2 \quad + \quad C_6H_5{}^{14}CH{=}CH_2 \\
 & & 86.7\% \qquad\qquad\quad 3.3\% \\
C_6H_5CH_2{}^{14}CH_3 & & \\
 & \xrightarrow{83\%} & C_6H_5CH_2{}^{14}CH_3 \quad + \quad C_6H_5{}^{14}CH_2CH_3 \\
 & & 83.4\% \qquad\qquad\quad 3.6\%
\end{array}
$$

2. [2-^{14}C]Phenylpropane

On passing over chromia–alumina catalyst, the labeled isopropylbenzene [$C_6H_5{}^{14}CH(CH_3)_2$] afforded 38% conversion. The products of reaction consisted of benzene, toluene, and ethylbenzene (16 mol %); the remaining 84 mol % were products of dehydrogenation and rearrangement:

$$
\begin{array}{ccc}
C_6H_5CH_2CH_2CH_3 & C_6H_5CH{=}CHCH_3 & C_6H_5{}^{14}C{\Big\langle}{\begin{array}{c}CH_3\\CH_2\end{array}} \\
25\% & 8\% & 51\%
\end{array}
$$

None of the radioactivity in the skeletally isomerized product resided on the α-carbon atom, which indicates that the skeletal isomerization had occurred through a phenyl migration only.

3. tert-Butylbenzene

The dehydrogenation of *tert*-butylbenzene over "nonacidic" chromia–alumina is of interest because it requires skeletal isomerization before dehydrogenation can occur. When the title hydrocarbon was passed over the catalyst at 481°C and LHSV 0.77, about one-tenth of it changed to form isobutylbenzene, β,β-dimethylstyrene, and isobutenylbenzene, in approximate yields of 30, 50, and 20%, respectively. The remaining 90%

of the starting *tert*-butylbenzene was recovered unchanged. The skeletal rearrangement accompanying the dehydrogenation occurs via a radical mechanism as outlined in Scheme 4.7.

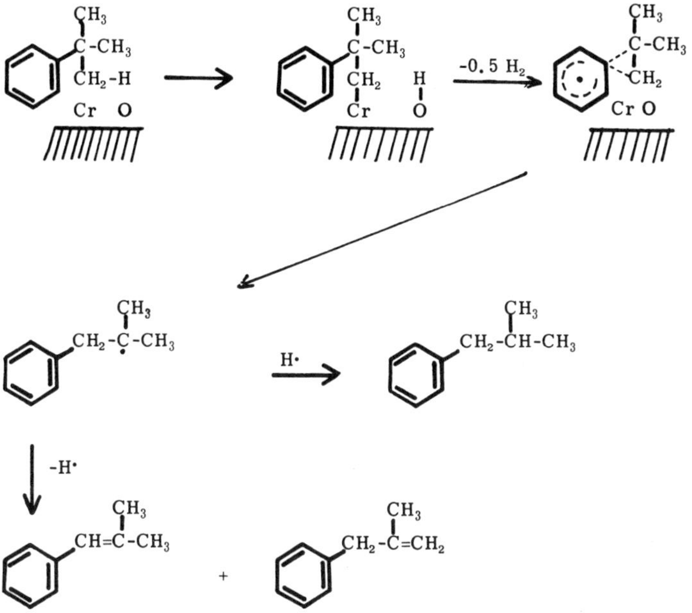

Scheme 4.7. Mechanism of rearrangement and dehydrogenation of *tert*-butylbenzene (from Pines and Goetschel, 1965a).

REFERENCES

Fadeev, V. S., Gostunskaya, I. V., and Kazanskii, B. A. (1969). *Dokl. Akad. Navk SSSR* **189**, 788.
* Germain, J. E. (1969). "Catalytic Conversion of Hydrocarbons," Chap. 3. Academic Press, New York.
Goetschel, C. T., and Pines, H. (1965). *J. Org. Chem.* **30**, 3544.
Kearby, K. K. (1955a). Catalytic dehydrogenation. *In* "Catalysis" (P. H. Emmett, ed.), Vol. 3, pp. 453–491. Reinhold, New York.
* Kearby, K. K. (1955b). Catalytic dehydrogenation. *In* "Chemistry of Petroleum Hydrocarbons" (B. T. Brooks, C. E. Boord, S. S. Kurtz, Jr., and L. Schmerling, eds.), Reinhold, New York.
* Lee, E. H. (1973). Iron oxide catalysts for dehydrogenation of ethylbenzene in the presence of steam. *Catal. Rev.* **8**, 285–305.
* Paal, Z. (1980). Metal-catalyzed cyclization reactions of hydrocarbons. *Adv. Catal.* **29**, 273–334.
Pines, H. (1972). *Intra-Sci. Chem. Rep.* **6**(2), 1–41.
Pines, H., and Abramovici, M. (1969). *J. Org. Chem.* **34**, 70.
Pines, H., and Csicsery, S. M. (1962). *J. Catal.* **1**, 313.

Pines, H., and Chen, C. T. (1960). *J. Am. Chem. Soc.* **82,** 3562.

Pines, H., and Chen, C. T. (1961). *Proc. Int. Congr. Catal., 2nd, Paris, 1960* pp. 367–387. Editions Technip, Paris.

Pines, H., and Goetschel, C. T. (1965a). *J. Am. Chem. Soc.* **87,** 4207.

* Pines, H., and Goetschel, C. T. (1965b). Discussion of the mechanism of the aromatization of alkanes in the presence of chromia–alumina catalysts. *J. Org. Chem.* **30,** 3530–3536.

Pines, H., and Greenlee, T. W. (1961). *J. Org. Chem.* **26,** 1052.

Pines, H., Goetschel, C. T., and Dembinski, J. W. (1965). *J. Org. Chem.* **30,** 3540.

Prelog, V. (1955). *Helv. Chim. Acta* **38,** 434.

Tournayan, L., Bacaud, R., Charcosset, H., and Leclercq, G. (1978). *J. Chem. Res.* (M), 3582.

5

Oxidation

I. INTRODUCTION

The conversion of hydrocarbons to products containing oxygen is of great industrial importance. Many key organic compounds are formed by controlled oxidation of selected hydrocarbons. This chapter is limited principally to a discussion of the methods and mechanisms of catalytic oxidation of hydrocarbons to compounds of major industrial importance.

II. ALKANES: BUTANE (TO ACETIC ACID)

The three major methods for producing acetic acid commercially consist of the oxidation of butane or low-boiling liquid petroleum hydrocarbons with air in the presence of small amounts of cobalt or manganese salts, the carbonylation of methanol, as discussed in Chapter 6, Section III,B, and the oxidation of acetaldehyde, which is produced by the palladium-catalyzed oxidation of ethylene (Wacker process).

In 1973 about 20% of all acetic acid produced in the U.S., or 2.5 × 10⁸ kg, was formed by the oxidation of butane with air, with cobalt(III) acetate as catalyst. Under operating conditions of 60 atm and 180°C the conversion of butane was almost complete. The yield of acetic acid based on carbon efficiency was 57%. By-products consisted of CO and CO_2 (17%) and esters and ketones (22%) (Lowry and Aquilo, 1974).

This oxidation of butane proceeds by a radical chain reaction. The cobalt ion is assumed to participate in the decomposition of *sec*-butyl hydroperoxide, and it could possibly participate in the initial step as well. The main propagation step between the peroxy radical and butane does not seem to involve a metal ion. The aldehyde formed is oxidized in the presence of the catalyst to acetic acid.

The reaction is proposed to proceed according to the following steps:

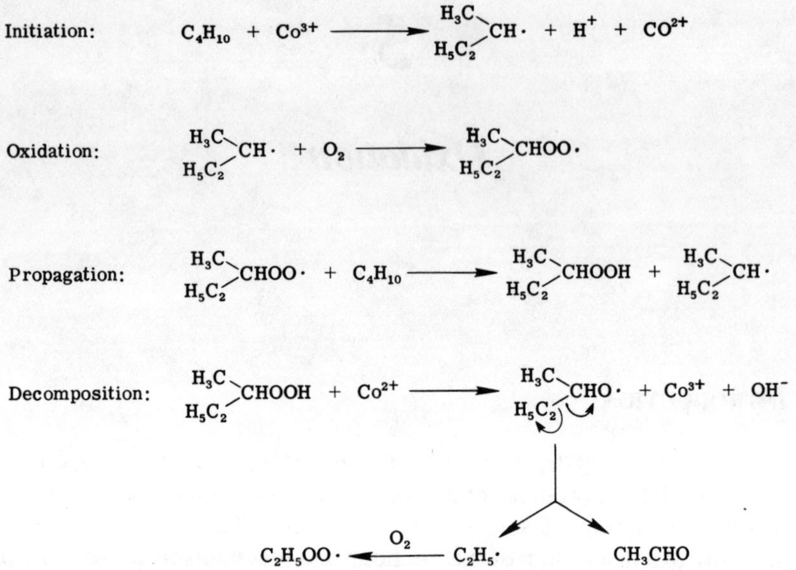

Initiation: C_4H_{10} + Co^{3+} ⟶ $\begin{matrix} H_3C \\ H_5C_2 \end{matrix}$CH· + H^+ + CO^{2+}

Oxidation: $\begin{matrix} H_3C \\ H_5C_2 \end{matrix}$CH· + O_2 ⟶ $\begin{matrix} H_3C \\ H_5C_2 \end{matrix}$CHOO·

Propagation: $\begin{matrix} H_3C \\ H_5C_2 \end{matrix}$CHOO· + C_4H_{10} ⟶ $\begin{matrix} H_3C \\ H_5C_2 \end{matrix}$CHOOH + $\begin{matrix} H_3C \\ H_5C_2 \end{matrix}$CH·

Decomposition: $\begin{matrix} H_3C \\ H_5C_2 \end{matrix}$CHOOH + Co^{2+} ⟶ $\begin{matrix} H_3C \\ H_5C_2 \end{matrix}$CHO· + Co^{3+} + OH^-

$C_2H_5OO·$ ⟵^{O_2}— $C_2H_5·$ CH_3CHO

This method of oxidation is used less frequently with higher boiling alkanes as raw material, because the oxidation product is more complex and is composed of a great variety of compounds which would require expensive separation.

III. CYCLOALKANES

Industrially, the two most important cycloalkanes used for oxidation are cyclohexane and cyclododecane, which produce adipic acid and dodecanedioic acid respectively. The U.S. production of these two acids in 1975 was 5.5×10^7 and 9.1×10^5 kg, respectively. Both acids are used in the manufacture of synthetic fibers.

A. Cyclohexane (to Adipic Acid)

The oxidation of cyclohexane to adipic acid is generally carried out in a two-step process. The first stage, which produces a mixture of cyclohexanone and cyclohexanol, is accomplished by oxidation with air. This mixture is then oxidized to adipic acid either by air directly or by nitric acid, which is an air-regenerable reagent. Both stages require metal ions as catalysts.

The air oxidation of cyclohexane is carried out in the presence of

20 ppm of cobalt naphthenate until 10–12% reaction occurs. The unreacted cyclohexane is distilled off from the higher boiling products, mainly cyclohexanone and cyclohexanol.

With cyclohexane the air oxidation proceeds via a free-radical chain mechanism, and the major initial product in the oxidation is cyclohexyl hydroperoxide. Controlled decomposition of hydroperoxide to cyclohexanone and cyclohexanol is the major catalytic function of the cobalt ions.

In the second stage, the oxidation is usually carried out at about room temperature by mixing cyclohexanol and cyclohexanone with nitric acid plus copper(II) and vanadium(V) salts, which serve as catalysts, in a vigorously stirred reactor. The nitric acid, which is the immediate oxidant, is converted to NO and NO_2. These oxides are reoxidized to nitric acid in a conventional nitric acid facility usually available at the plant site. Air is the ultimate oxidant. Adipic acid, in a yield of over 90%, crystallizes from the reaction mixture. The reaction sequence from cyclohexanol is given in Scheme 5.1.

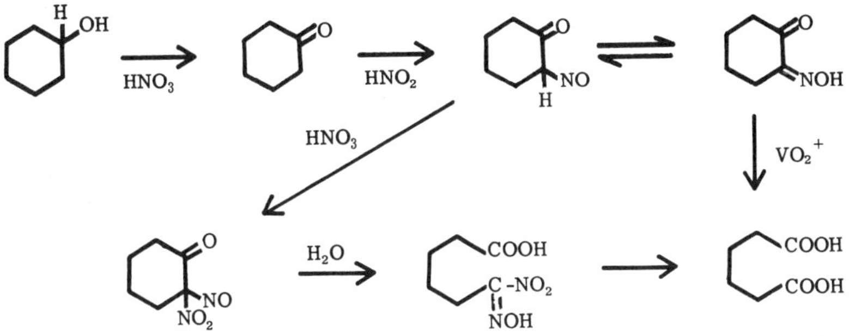

Scheme 5.1. Pathways in the oxidation of cyclohexanol to adipic acid (Van Asselt and Van Krevelen, 1963).

B. Cyclododecane (to 1,12-Dodecanedioic Acid)

Cyclododecane, produced by the hydrogenation of 1,5,9-cyclododecatriene obtained from the trimerization of butadiene (Chapter 6, Section IV,C,2), is oxidized to dodecanedioic acid by a method similar to that described for the oxidation of cyclohexane. Oxidation of cyclododecane by air is conducted in the presence of boric acid in order to stabilize cyclododecyl hydroperoxide, the initial oxidation product. The hydroperoxide is decomposed catalytically in the presence of Co^{2+} to a mixture of cyclododecanol and cyclododecanone, which is then oxidized with nitric acid and vanadium salts, used as the catalyst, to 1,12-dodecanedioic acid.

IV. ALKENES

A. Ethylene

1. Ethylene Oxide

The production of ethylene oxide in the United States in 1977 was 2.1×10^9 kg. Most of it was used in the manufacture of fibers, plastics, and glycol for antifreeze. Practically all of the ethylene oxide is currently produced by the oxidation of ethylene using as catalyst silver oxide mounted on a refractory support.

In practice the reaction is carried out at 230°–300°C and 10–20 atm pressure. The catalyst is contained in tubes bathed in an external organic cooling medium. Ethylene and air are mixed with recycled gas containing mainly nitrogen and unconverted ethylene, and the mixture is passed over the catalyst. The conversion per pass is 30–40% with a selectivity of about 60 mol %. The remaining product is mostly CO_2. The selectivity is improved by the addition of traces of moderators to the feed, such as a few parts per million of an organic chloride, e.g., ethylene dichloride.

In the preparation of a silver-supported catalyst it is important for the carrier to have a low surface area and to be catalytically inert. A fairly large silver area is desirable, 0.05–5 m²/gm Ag, and diffusion effects in catalyst particles are to be avoided. Promoters such as Ca and Ba are usually added to the catalyst, mainly to stabilize the silver and thus prevent it from sintering.

A catalyst useful for laboratory exploration was prepared as follows. Seventy grams of 8-mesh corundum, 22 gm Ag_2O, and 2.25 gm Ba_2O_2 were mixed with 100 ml water. The mixture was stirred while being heated to evaporate the water, dried, and heated in air at 115°C for 10 hr (McBee et al., 1945).

Silver is unique among metals in its capacity to catalyze the oxidation of ethylene to ethylene oxide. The understanding of this reaction is based on kinetics, adsorption studies, the use of isotopes of oxygen, and the use of IR spectroscopy. There are two types of adsorption of oxygen on silver: dissociative and nondissociative. In the nondissociative adsorption the diatomic oxygen species requires one adsorption site and is presumably bonded perpendicular to the surface of the silver. In the case of dissociative adsorption of oxygen four bare neighboring silver atoms are required. Therefore, the effect of chlorine, which is added to the ethylene feed to increase the selectivity of the reaction toward oxide formation, is to destroy the quartets of silver atoms and to prevent the dissociative adsorption of oxygen, which is associated with carbon dioxide formation.

The interaction of ethylene with the nondissociatively adsorbed oxygen leads to the formation of ethylene oxide and to an adsorbed oxygen atom (Kilty *et al.*, (1973)):

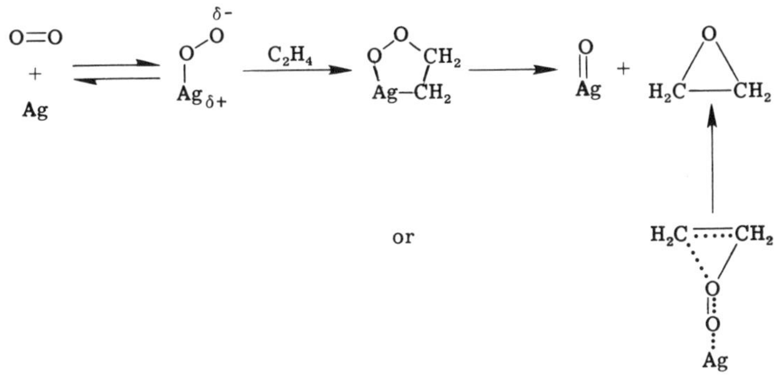

From the above reaction scheme it can be deduced that theoretically a maximum of six-sevenths (86%) of the ethylene can go to ethylene oxide:

$$6C_2H_4 + 6O_2 \longrightarrow 6C_2H_4O + 6O(ads)$$

$$C_2H_4 + 6O(ads) \longrightarrow 2CO_2 + 2H_2O$$

A selectivity greater than 86% has not been observed.

2. Acetaldehyde (Wacker Process)

The "direct" homogeneous oxidation of ethylene to acetaldehyde with air or oxygen in aqueous solution and in the presence of catalytic amounts of palladium(II) chloride and copper(II) chloride is called the Wacker process, which was developed in 1969. This process is based on two known reactions:

$$PdCl_2 \cdot C_2H_4 \text{ (complex)} + H_2O \longrightarrow Pd + 2HCl + CH_3CHO \qquad (1)$$

$$Pd + 2HCl + \tfrac{1}{2}O_2 \longrightarrow PdCl_2 + H_2O \qquad (2)$$

The finely divided palladium produced in Eq. (1) is rendered soluble in aqueous HCl by oxygen. At room temperature reaction (2) is 100 times slower than reaction (1). However, in the presence of catalytic amounts of CuCl$_2$ the reaction is speeded up greatly, and this modification makes the conversion of ethylene to acetaldehyde commercially feasible. The function of CuCl$_2$ is to reoxidize Pd(0) to Pd(II), and the CuCl that is generated is reoxidized to CuCl$_2$ by air.

$$2CuCl_2 + Pd(0) \longrightarrow 2CuCl + PdCl_2$$

$$2CuCl + HCl + O_2 \longrightarrow 2CuCl_2 + H_2O$$

Commercially the Wacker process is carried out in either one or two stages at 120°–130°C and about 10 atm pressure. In the two-stage process ethylene is passed into a hot aqueous solution of copper(II) chloride and palladium(II) dichloride, which oxidizes ethylene to acetaldehyde with reduction of the copper salt. After separation of the acetaldehyde, the catalyst solution containing copper(I) chloride is then transferred to a tube oxidizer in which the cuprous salt is reoxidized to copper(II) chloride with air at 10 atm pressure and recycled. The yield of acetaldehyde amounts to about 90%.

The oxidation of ethylene occurs within the coordination sphere of the palladium catalyst, and the palladium is the oxidant. The generalized scheme for the conversion of ethylene to acetaldehyde is given in Fig. 5.1.

The transformation of the ethylene complex (1) to the hydroxyethyl derivative (2) appears to involve nucleophilic attack of water on the coordinated ethylene. The conversion of the 2-hydroxyethyl group in 2 to acetaldehyde presumably involves hydrogen transfer from C-2 in 2 to palladium to give a vinyl alcohol complex (3). Transfer of the same hydrogen to C-1 yields protonated acetaldehyde.

Part of the acetaldehyde produced commercially is used for the synthesis of crotonaldehyde, an intermediate in organic syntheses. The bulk of the aldehyde, however, is converted by oxidation to acetic acid and acetic anhydride. The oxidation of acetaldehyde to acetic acid is carried out in

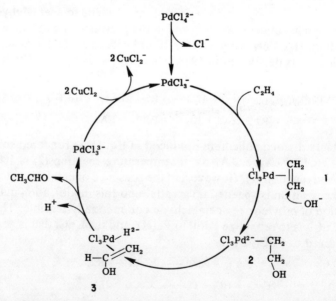

Fig. 5.1. Scheme for the conversion of ethylene to acetaldehyde by $PdCl_4^{2-}$ (from Parshall, 1978, p. 250).

the liquid phase with a transition metal such as Mn^{2+}, Co^{2+}, or Cu^{2+} as catalyst.

In the last few years the Wacker process for the ultimate production of acetic acid has been replaced by the more economical method of carbonylation of methanol, in which carbon monoxide and hydrogen are used as primary raw materials (Chapter 6, Section III,B).

3. Vinyl Acetate

Vinyl acetate is produced commercially from ethylene by a modified Wacker process. The oxidation avoids water solutions and uses palladium(II) acetate in an acetic acid medium containing sodium acetate. As in the Wacker process the reaction was developed into a catalytic process by the use of copper salts to reoxidize the palladium.

$$CH_2=CH_2 + HOAc + \tfrac{1}{2}O_2 \xrightarrow[Cu^{2+}]{Pd^{2+}} CH_2=CHOAc + H_2O$$

Ethylene reacts readily with $Pd(OAc)_2$ at atmospheric pressure and 70°C to form palladium metal and vinyl acetate with a selectivity of 40–50%. The rate of reaction increases considerably in the presence of sodium acetate, and the selectivity to vinyl acetate becomes 80–90%.

The mechanism of the reaction appears to be similar to that of the oxidation of ethylene to acetaldehyde (Fig. 5.1). In acetic acid solution, nucleophilic attack of coordinated acetate is accompanied by a rearrangement to a σ-bonded complex, which is then decomposed to products via β-hydrogen elimination.

$$[Pd(OAc)_3]^- + CH_2=CH_2 \longrightarrow [Pd(OAc)_3CH_2=CH_2]^-$$

$$\searrow -AcO^-$$

$$Pd + HOAc \longleftarrow [HPdOAc] + CH_2=CHOAc \longleftarrow AcOPdCH_2CH_2OAc$$

Vinyl acetate is used widely in industry as a monomer for coating and adhesive manufacture. The U.S. production in 1977 was 7.7×10^8 kg.

B. Propene

1. Propene Oxide (1,2-Propylene Oxide)

Unlike ethylene oxide, 1,2-propylene oxide cannot be produced by the direct oxidation of propene with air. However, propene as well as other alkenes undergo almost quantitative epoxidation when they are reacted with alkyl or aryl hydroperoxides in the presence of catalytic amounts of acetyl acetonates of molybdenum, vanadium, or tungsten. The hydroperoxides are converted to the corresponding alcohols.

$$\text{>C=C<} \quad + \quad \text{>C-O-O-H} \quad \xrightarrow{\text{cat.}} \quad \underset{O}{\text{>C--C<}} \quad + \quad \text{>C-OH}$$

Industrially, the epoxidation reaction is accomplished in two stages: conversion of a suitable hydrocarbon to hydroperoxide by oxidation with air, and interaction of the hydroperoxide with propene in the presence of a soluble molybdenum catalyst, such as acetyl acetonate or naphthenates.

The catalyzed epoxidation of propene with hydroperoxide from ethylbenzene is the basis for a commercial process, the Halcon process (1967), in which propylene oxide and styrene are produced from propene, ethylbenzene, and air via the following sequence of reactions:

$$\underset{\underset{OOH}{|}}{C_6H_5CHCH_3} \quad + \quad CH_3CH{=}CH_2 \xrightarrow{Mo} \underset{\underset{OH}{|}}{C_6H_5CHCH_3} \quad + \quad CH_3HC\overset{O}{\underset{}{\triangle}}CH_2$$

$$-H_2O \Big| \text{ catalyst}$$

$$\downarrow$$

$$C_6H_5CH{=}CH_2$$

Every hydroperoxide tested could be made to epoxidize propene in an efficient manner. Industrially, emphasis is placed on hydroperoxides that are readily obtained by air oxidation of hydrocarbons and on by-products that can be utilized commercially. Ethylbenzene, which has a benzylic hydrogen, and isobutane, which has a tertiary carbon atom, undergo hydroperoxidation readily and are suitable for this process. The by-product *tert*-butyl alcohol can be converted to methyl *tert*-butyl ether, an octane booster in gasoline.

Commercially, ethylbenzene is oxidized in part to α-methylbenzyl hydroperoxide. The mixture is then contacted with propene and the molybdenum catalyst at 110°C for 1 hr under 15–60 atm. About 0.002 mol of catalyst and 2–6 mol of propene are used per mole of hydroperoxide (*Hydrocarbon Process.*, (4) 141, 1967).

The 1977 U.S. production of propylene oxide was reported to be 8.6×10^8 kg. It is used as a monomer in the production of plastics.

Mechanism. The active agents in the epoxidation of propene are complexes of the hydroperoxides with the catalyst in its high oxidation state: Mo(VI), W(VI), Ti(IV), and V(V). The principal function of the catalyst is to attract electrons from the peroxidic oxygens and make these oxygens more susceptible to attack by the electrons of the alkenes. The catalyst thus acts as a Lewis acid. The Lewis acidity of metal complexes generally increases with increasing oxidation state of the metal. The catalytic epoxidation proceeds by a mechanism involving an intact alkyl hydroperoxide and can be presented by the following scheme:

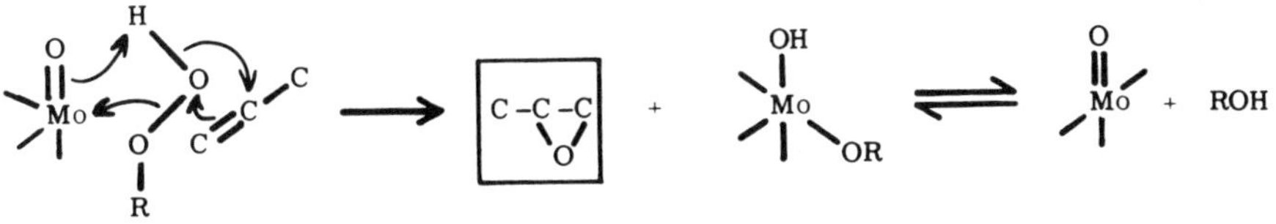

A modified version of this scheme, which is based on a study with isotopic oxygen, suggests that (a) the coordinated peroxide is polarized by a three-centered interaction with an empty coordination site on the metal or (b) the peroxidic bond is polarized by an acidic hydrogen through a five-membered interaction (Chong and Sharpless, 1977). In both mechanisms the epoxide is initially coordinated to the metal.

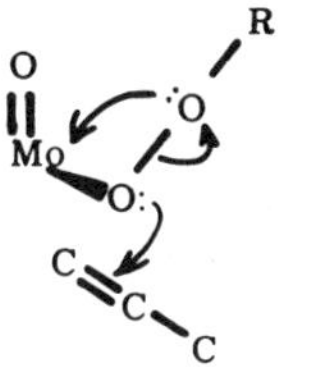

(a) 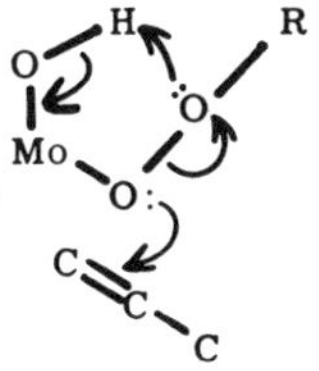   (b)

2. Acrolein and Acrylonitrile

a. Introduction. In 1948 the oxidation of propene to acrolein in a yield of 50% was carried out with cuprous oxide as the catalyst. A higher yield of acrolein was obtained in 1959 with air over bismuth–molybdenum catalyst, referred to as "bismuth molybdate catalyst". At the same time it was observed that a gaseous mixture composed of propene, air, and ammonia when passed over this catalyst formed acrylonitrile. The commercial vapor-phase oxidation and ammoxidation was developed by Standard Oil of Ohio (SOHIO).

Shortly afterward an even more selective catalyst was introduced, a uranium antimonate catalyst, and in 1970 SOHIO developed a multicomponent catalyst. The commercial vapor-phase oxidation and ammoxidation process is used worldwide with a total yearly capacity in 1977 of 2.3×10^9 kg.

b. Acrolein. The oxidation of propene with air over bismuth molybdate catalyst affords good yields of acrolein. When propene, air, and steam in a molar ratio of $1.0:9.3:3.5$ were passed over 200 gm catalyst at 450°C, at atmospheric pressure, and with a contact time of 1.6 sec., 92% of propene was converted to acrolein with 60.5% selectivity. On the basis of converted propene, the remaining product consisted of CO_2 (20.5%) and acrylic acid and other oxygenated products (16%).

The catalyst was prepared by adding to an aqueous silica solution containing 225 gm of SiO_2 a solution of 9.3 ml 85% H_3PO_4, 272 gm molybdic acid (85% MoO_3), 40 ml HNO_3, and 582 gm $Bi(NO_3)_3 \cdot 5 H_2O$ in 400 ml H_2O. The mixture was dried to 538°C for 16 hr and then ground to obtain particles in the range of 0.1–0.4 mm diameter (Veatch *et al.*, 1961).

Using 0.2 gm of bismuth molybdate catalyst, Bi/Mo = 1.0, the selectivity of acrolein produced as a function of propene reacted (in parentheses) was found to be as follows: 90% (10%), 86% (40%), and 73% (80%); the main side reaction was a complete combustion of propene to CO_2. The selectivity is inversely related to the percentage of propene converted.

c. Acrylonitrile. Stoichiometric quantities of propene, ammonia, and air were reacted in a fluidized catalyst reactor at substantially atmospheric pressure and at 400°–510° to produce primarily acrylonitrile. The catalysts used included both the bismuth molybdate type and a silica-supported uranyl antimonate. The latter gave higher yields of acrylonitrile and less acetonitrile than the former. With both catalysts the conversion of per pass was complete.

$$2CH_2{=}CHCH_3 + 2NH_3 + 3O_2 \longrightarrow 2CH_2{=}CHCN + 6H_2O$$

The catalyst currently used consists of a multicomponent oxide containing Bi, Fe, Co, Ni, and Mo, in addition to several other cations. Only a few reports concerning multicomponent catalysts have appeared. Yields in excess of 85 wt % of acrylonitrile per weight of propene feed are achieved. Over 0.10 kg of by-product HCN can be recovered per kilogram of acrylonitrile.

d. Mechanism. Evidence of the mechanism of oxidation of propene over bismuth molybdate has come mainly from kinetic studies and tracer experiments using isotopic carbon and oxygen. When labeled propene $^{14}CH_3CH{=}CH_2$ is oxidized, the ^{14}C label in the acrolein produced is equally divided between the methyl group and the carbonyl group. The experiments thus indicate the initial abstraction of an allylic hydrogen to form a symmetric allylic intermediate. The next step is hydrogen abstraction from either end of the allylic intermediate, followed by incorporation of an oxygen atom.

The oxidation steps of this reaction can be explained by means of a redox mechanism. The catalysts used in the formation of acrolein and acrylonitrile must be capable not only of undergoing reduction, but also of undergoing a subsequent reoxidation reaction whereby the molecular oxygen is incorporated into the lattice of the catalyst as O_2^-. That the oxidizing agent is the lattice oxygen was determined by the use of isotopic oxygen.

$$C_3H_8 + MO \longrightarrow C_3H_4O + H_2O + M$$

$$M + \tfrac{1}{2}O_2 \longrightarrow MO$$

The cations in the bismuth molybdate catalyst serve as sites for propene activation and for oxygen adsorption (catalyst reactivation). Oxygens of molybdenum polyhedra are credited with oxygen insertions, whereas the rate-determining α-hydrogen abstraction is caused by Bi—O polyhedra. In the case of the ammoxidation reaction to form acrylonitrile, NH from NH_3 is inserted from the lattice associated with molybdenum atoms. The mechanism of oxidation and ammoxidation of propene is presented in Scheme 5.2.

In the case of "uranyl antimonate" catalyst it appears that the surface layers contain U^{5+} and Sb^{5+} with intensities corresponding to USb_3O_{10}. It is envisioned that propene chemisorbs on an active site, postulated as an oxygen ion vacancy on Sb^{5+}, and loses an allylic hydrogen to uranyl oxygen.

C. Oxidative Dehydrogenation of Alkenes

When olefins containing at least four carbon atoms in the chain are submitted to oxidation in the presence of a bismuth molybdate catalyst and under conditions specified for the oxidation of propene to acrolein, the main initial products of reaction are conjugated dienes (Table 5.1).

When allowed by the structure of the original olefin, conjugated dienes are formed at a much faster rate than the conjugated unsaturated aldehydes. Thus, the unbranched allylic methyl group of 2-butene or 2-pentene is not oxidized. However, monosubstituted alkenes having a methyl group in the side chain give monounsaturated aldehydes as primary parallel products. It is apparent from Table 5.1 that the initial products may react further if an additional conjugated system can be formed (Fig. 5.2).

Some exceptions to the general rules have been observed. Cyclopentene is completely converted to CO_2, presumably because of the high reactivity of cyclopentadiene. 3,3-Dimethyl-1-butene, which is not expected to react at all under general rules, affords 2,3-dimethylbutadiene. It seems that the catalyst must have intrinsic acidic properties causing methyl migration before an oxidation reaction occurs.

Whenever possible conjugated dienes undergo further dehydrogenation to a triene, as in the case of 1-hexene. The triene is then converted to benzene.

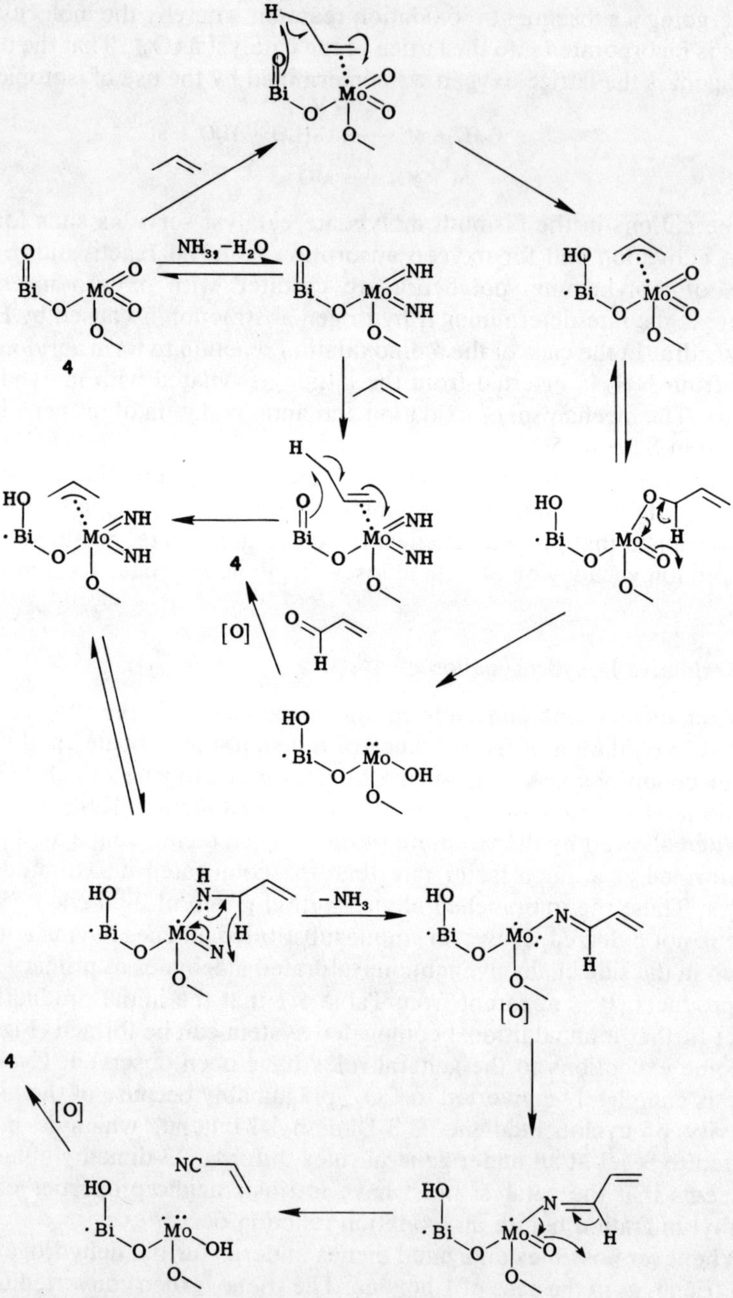

Scheme 5.2. Mechanism for oxidation and ammoxidation of propene over bismuth molybdate catalyst (from Grasselli and Burrington, 1981).

TABLE 5.1

Oxidation and Oxidative Dehydrogenation of
Olefins over Bismuth Molybdate at $460°C^{a,b}$

1-Butene	20	40	80
1,3-Butadiene	95	95	90
2-Butene	20	40	80
1,3-Butadiene	90	90	85
2-Methyl-1-propene	10	40	70
Methacrolein	72	72	72
1-Pentene	10	20	60
1,3-Pentadiene	93	87	38
2-Pentene	5	15	—
1,3-Pentadiene	88	79	—
2-Methyl-2-butene	20	40	70
Isoprene	73	62	44
C_5H_8O	9	9	9
C_5H_6O	4	9	16
Cyclopentene	Only CO_2 + CO		
1-Hexene	20	40	70
Hexadienes	71	59	32
1,3,5-Hexatriene	2	5	7
Benzene	0	7	32
2-Methyl-1-pentene	5	10	20
2-Methylpentadiene	55	49	38
$C_6H_{10}O$	9	9	9
C_6H_8O	5	5	5
3-Methyl-1-pentene	10	50	80
3-Methylpentadiene	90	54	27
3-Methylene-1,4-pentadiene	10	16	12
3,3-Dimethyl-1-butene	2	5	—
2,3-Dimethylbutadiene	60	56	—
$C_6H_{10}O$	3	3	—
C_6H_8O	6	8	—
$C_6H_6O_2$	1	2	—
4,4-Dimethyl-1-pentene	2	5	—
2,3-Dimethylpentadiene	74	53	—

[a] From Adams (1965).

[b] Numbers opposite olefins are percent olefin conversions; numbers below these are percent selectivities to stated products. Other products are mainly CO and CO_2.

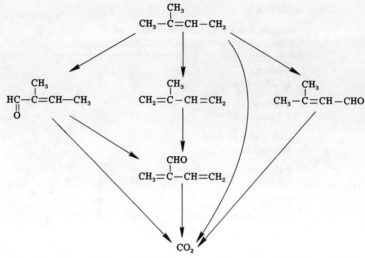

Fig. 5.2. Reactions of 2-methyl-2-butene over bismuth molybdate catalyst (from Adams, 1965).

V. AROMATIC HYDROCARBONS

A. Introduction

Aromatic hydrocarbons are a valuable source of industrially important carboxylic acids. Benzene and naphthalene afford on oxidation maleic and phthalic anhydride, respectively. Toluene and *m*- and *p*-xylene are converted on oxidation to benzoic acid and to the corresponding dicarboxylic acids.

The oxidations of benzene, naphthalene, and *o*-xylene, which lead to the formation of volatile anhydrides, usually are carried out industrially in the vapor phase. The oxidations of *m*- and *p*-xylene, which afford nonvolatile acids that are not easily separated from the catalyst, are performed in the liquid phase. The oxidation of toluene to benzoic acid can be done in either the liquid or the vapor phase. The catalyst used for oxidation contains metals with variable valences, such as V, Mo, Co, and Mn.

B. Vapor-Phase Oxidation

1. Benzene (to Maleic Anhydride)

The commercial production of maleic anhydride by the catalytic oxidation of benzene has been in operation since the late 1930's. Although the composition of the catalyst used in industry is not described, it is assumed to be composed of V_2O_5 and MoO_3 on a nonporous support.

The yield of maleic acid obtained from commercial oxidation is reported to be 75–80 wt % or 60–64 mol % based on the benzene. In addition 5–10 wt % of fumaric acid is produced.

The reaction proceeds through a number of intermediates including phenol and hydroquinone. However, maleic anhydride is the only product that has been obtained in economically attractive yields.

A kinetic study of the oxidation of benzene was made using as catalyst oxides composed of V, Mo, and Al (1.0:0.3:1.0), a temperature of 375°C, and an air to benzene mole ratio of about 150. The principal overall reaction involved in the oxidation of benzene is summarized by the following scheme:

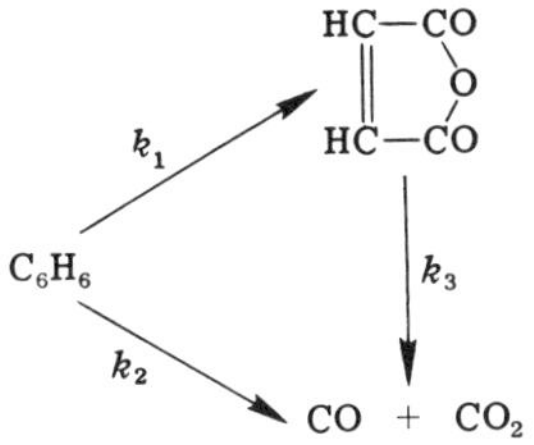

2. Naphthalene or Xylene (to Phthalic Anhydride)

The oxidation of naphthalene and o-xylene to phthalic anhydride is commercially one of the most important of the catalytic vapor-phase oxidation reactions. The major reactions in the oxidation of naphthalene are presented in the following scheme:

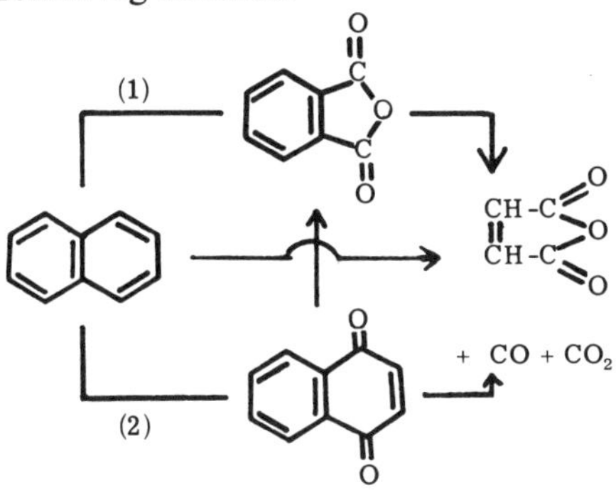

Reactions (1) and (2) are the most important primary reactions. Supported vanadium pentoxide is the predominant catalyst. The oxidation reaction is highly exothermic, and it is important that the heat of reaction be removed. To accomplish this, two types of reactors are used commercially: a fixed bed reactor and a fluidized catalyst bed.

In a fixed bed reactor air at 0.5 atm and 135°–150°C is saturated with naphthalene vapor, the molar ratio of air to hydrocarbon being 20–30:1. The gas mixture is then passed through a reactor containing catalyst pellets, and the heat of reaction is removed by cooling tubes located in the catalyst bed. The temperature of oxidation is maintained at 315°–375°C.

In a fluidized bed reactor the movement of the catalyst tends to spread out the heat of reaction and prevents hot spots from forming on the catalyst. Once the reaction is started some of the heat is absorbed by the cold air introduced into the reactor. The ratio of air to naphthalene for the system is 10–15:1. The residence time of the reactants in the reactor is 10–15 sec.

Phthalic anhydride can be made by a similar vapor-phase oxidation using o-xylene instead of naphthalene as a starting material. The yield of phthalic anhydride from either naphthalene or o-xylene is 93–98 wt %, based on hydrocarbon feed, and the life of the catalyst is from 2 to 4 years. The longer catalyst life is attained with naphthalene. Phthalic anhydride is used in the manufacture of paints, plastics, and coatings.

C. Liquid-Phase Oxidation

Isophthalic and terephthalic acids are not very volatile; under oxidation conditions they remain on the catalyst and thus undergo further oxidation. For that reason the oxidation of m- and p-xylene is carried out commercially by a liquid-phase process. Liquid-phase oxidation is also applied to the conversion of toluene to benzoic acid.

m-Xylene and p-Xylene

The liquid-phase oxidation of m- and p-xylene is carried out in acetic acid solution in the presence of high concentrations of cobalt and/or manganese acetate catalysts together with promoters such as aldehydes, ketones, or bromides, the latter acting as a regenerable source of free radicals. Terephthalic acid is produced by one-step oxidation with high yield and purity.

In a laboratory experiment the following procedure was used. p-Xylene, 75 gm, was dissolved in 150 gm of acetic acid, and 0.75 gm of NH_4Br and 1 gm of cobalt acetate tetrahydrate were added. The mixture was heated to around 200°C at 27 atm pressure and oxidized with air for 2 hr. The yield of terephthalic acid was 75–80 mol % (Towle et al., 1964).

It is preferable to oxidize p-xylene independently of m-xylene because in this way the problem of separation of the two acids is avoided.

Bromide ions were discovered to have a pronounced synergistic effect on cobalt- and manganese-catalyzed oxidation of p-cymene. This finding provided an important breakthrough in the manufacture of terephthalic acid. In the absence of bromide, p-xylene produces p-toluic acid, and further oxidation is slow.

The rate-enhancing effects of bromide ion are ascribed to the formation of bromine atoms through electron transfer of bromide ion by Co(III):

$$\text{Co(III)} + \text{Br}^- \longrightarrow \text{Co(II)} + \text{Br}\cdot$$

$$\text{ArCH}_3 + \text{Br}\cdot \longrightarrow \text{ArCH}_2\cdot + \text{HBr}$$

$$\text{ArCH}_2\cdot + \text{O}_2 \longrightarrow \text{ArCH}_2\text{O}_2\cdot$$

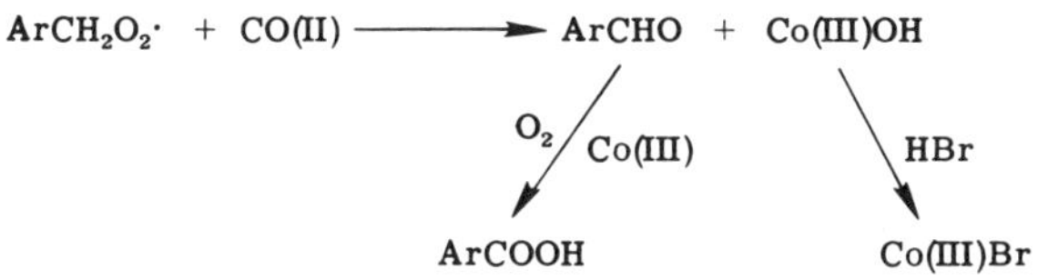

$$\text{ArCH}_2\text{O}_2\cdot + \text{CO(II)} \longrightarrow \text{ArCHO} + \text{Co(III)OH}$$

$$\text{O}_2 \diagdown \text{Co(III)} \qquad \text{HBr}$$

$$\text{ArCOOH} \qquad\qquad \text{Co(III)Br}$$

In the presence of bromide ion there is apparently no direct reaction of Co(III) with the p-xylene.

D. Isopropylbenzene (to Phenol and Acetone)

Isopropylbenzene (cumene) is the key hydrocarbon for the synthesis of phenol and acetone. The reaction proceeds in two steps. The first step is considered to be noncatalytic and involves a free-radical type of oxidation of cumene to cumene hydroperoxide:

$$\text{C}_6\text{H}_5\text{CH}_2\text{CH}_3 + \text{O}_2 \longrightarrow \underset{\underset{\text{OOH}}{|}}{\text{C}_6\text{H}_5\text{CHCH}_3}$$

The second step is acid-catalyzed rearrangement and cleavage of the hydroperoxide to phenol and acetone. Any type of protonic acid can be used, including ion-exchange resins. The rearrangement of the hydroperoxide is analogous to carbocation rearrangements discussed in Chapter 1, except that positive oxygen instead of positive carbon is involved in the intermediate stage:

$$C_6H_5\overset{\underset{\displaystyle CH_3}{|}}{\underset{\underset{\displaystyle CH_3}{|}}{C}}-OOH \; \underset{\longleftarrow}{\overset{H^+}{\rightleftharpoons}} \; \left[C_6H_5\overset{\underset{\displaystyle CH_3}{|}}{\underset{\underset{\displaystyle CH_3}{|}}{C}}-\overset{..}{\underset{..}{O}}-\overset{+}{\underset{}{O}}\overset{H}{\underset{}{}}{-}H \right] \; \overset{-H_2O}{\longrightarrow} \; \left[C_6H_5-\overset{\underset{\displaystyle CH_3}{|}}{\underset{\underset{\displaystyle CH_3}{|}}{C}}-\overset{+}{\underset{..}{O}} \right]$$

$$\left[H-\overset{..}{O}-\overset{\underset{\displaystyle H_3C}{|}}{\underset{\underset{\displaystyle H_3C}{|}}{C}}\overset{H}{\underset{+}{}}-O-C_6H_5 \right] \; \overset{H_2O}{\longleftarrow} \; \left[\overset{\underset{\displaystyle CH_3}{|}}{\underset{\underset{\displaystyle CH_3}{|}}{C}}\overset{+}{\underset{}{-}}\overset{..}{\underset{..}{O}}-C_6H_5 \right]$$

$$\downarrow -H^+$$

$$\boxed{O=C\overset{\diagup CH_3}{\diagdown CH_3}} \quad + \quad \boxed{HOC_6H_5}$$

Acetone   Phenol

After the catalyst is removed the cleavage mixture is fractionated to purify phenol and acetone products. The yield of phenol based on cumene is 97 mol % and that of acetone 94.5%.

REFERENCES

Adams, C. R. (1965). *Proc. Int. Congr. Catal., 3rd, Amsterdam, 1964* **1**, 240. North-Holland Publ., Amsterdam.

Chong, A. O., and Sharpless, K. B. (1977). *J. Org. Chem.* **42**, 1587.

* Grasselli, R. G., Burrington, J. D. (1981). Selective oxidation and ammoxidation of propylene by heterogeneous catalysis. *Adv. Catal.* **30**, (in press).

* Keulks, G. W., Krenzke, L. D., and Notermann, T. M. (1978). Selective oxidation of propylene. *Adv. Catal.* **27**, 183–209.

* Kilty, P. A., Rol, N. C., and Sachtler, Ŵ. M. H. (1973). *Proc. Intl. Congr. Catal, 5th, Miami Beach, Florida, 1972*, **64**, 929–943. North-Holland Publ., Amsterdam.

Lowry, R. P., and Aquilo, A. (1974). *Hydrocarbon Process.* **53**(11), 103.

McBee, E. T., Haas, H. B., and Wiseman, T. A. (1945). *Ind. Eng. Chem.* **37**, 432.

* Parshall, G. W. (1978). Industrial applications of homogeneous catalysis. A review. *J. Mol. Catal.* **4**, 243–270.

* Sheldon, R. A., and Kochi, J. K. (1976). Metal-catalyzed oxidations of organic compounds in the liquid phase: A mechanistic approach. *Adv. Catal.* **25**, 272–413.

Towle, P. H., and Baldwin, R. H. (1964). *Hydrocarbon Process.* **43**(11), 149.

Van Asselt, W. J., and Van Krevelen, D. W. (1963). *Rec. Trav. Chim. Pays-Bas* **82**, 51.

Veatch, F., Callahan, J. L., Milberger, E. C., and Foreman, R. W. (1961). *Actes Int. Congr. Catal., 2nd, Paris, 1960* **2**, 2647. Editions Technip, Paris.

* Voge, H. H., and Adams, C. R. (1967). Catalytic oxidation of olefins. *Adv. Catal.* **17**, 151–221.

6

Homogeneous Catalysis by Transition Metal Organometallic Compounds

I. INTRODUCTION

Homogeneously catalyzed reactions involving transition metal organometallic compounds have relatively recently attracted much attention, and research activity in this field is expanding rapidly. In certain large-scale industrial processes, such as oxidation of ethylene to acetaldehyde, hydroformylation of olefins to aldehydes and ketones, and the reaction of methanol with carbon monoxide to produce acetic acid, soluble transition metal complexes are used as catalysts.

In heterogeneous systems, the catalytic process necessarily takes place on the surface of the catalyst, since only this is exposed to the reactants. Homogeneous catalysts are soluble in the same medium as the reactants. Consequently, all the molecules of the catalyst are available for interaction.

The most distinctive virtue of homogeneous catalysis by transition metal organometallic compounds is the high selectivity obtainable in reactions that can yield several possible products. The ultimate example appears to be the asymmetric hydrogenation of organic compounds leading to the production of optically active organic compounds, as in the synthesis of optically active amino acids in optical yields of over 85%. Such stereospecificity is rarely, if ever, achieved with heterogeneous catalysts. In addition, homogeneous catalysts may have an advantage in highly exothermic reactions because heat exchange with a catalytic solution is more efficient than that with a solid bed of catalysts.

Although homogeneous transition metal complexes are active catalysts for a variety of industrial processes, certain disadvantages are involved in handling the catalysts and in separating them from reactants. For that reason in some instances the catalysts are heterogenized by attachment to various supports. In order to achieve chemical linkage between the catalyst and support, the support must have functionality. An example of heerogenation of a homogeneous catalyst is illustrated in Section VI.

The 16- and 18-Electron Rule

The fact that, with some exceptions, transition metals tend to form complexes in which the effective atomic number corresponds to the higher inert gas has long been recognized. Most of the homogeneous catalyst mechanisms involve a series of oscillations between 16 and 18 valence electrons in the transition metal complex. The number of valence electrons (NVE) consists of the valence electrons of the metal and those electrons donated or shared by the ligands and would be 18 for a noble gas configuration.

The ligands or groups attached to the transition metals can be classified according to the number of electrons that they donate to the metals, as follows:

1. One-electron donors: $\cdot CH_3$, $\cdot H$, $\cdot Cl$

2. Two-electron donors: $:CO$, $:PR_3$, $:NR_3$, $:N\equiv N$, $\overset{\ominus}{\underset{\ominus}{\diagup}}$ (olefins)

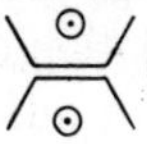

3. Three-electron donors: π-allyl $\overset{\ominus}{\underset{\ominus}{\diagup}}$

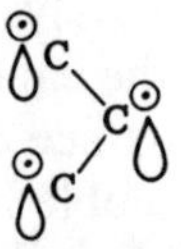

4. Four-electron donors: conjugated dienes, such as butadiene and 1,4-cyclooctadiene
5. Five-electron donors: cyclopentadiene

The compounds $Cr(CO)_6$, $Fe(CO)_5$, $Fe(C_5H_6)_2$, and $Ni[P(C_6H_5)_3]_4$ are all 18-electron complexes, as indicated by the following calculations based on the electronic configuration of the metals:

$$
\begin{array}{lll}
Cr & 3d^5 4s^1 = 6e & \left.\vphantom{\begin{array}{l} a \\ b \end{array}}\right\} 18e \\
(CO)_6 & 6 \times 2 = 12e & \\
Ni & 3d^8 4s^2 = 10e & \left.\vphantom{\begin{array}{l} a \\ b \end{array}}\right\} 18e \\
[P(C_6H_5)_3]_4 & 4 \times 2 = 8e & \\
Fe & 3d^6 4s^2 = 8e & \left.\vphantom{\begin{array}{l} a \\ b \end{array}}\right\} 18e \\
\text{⬠} & 2 \times 5 = 10e &
\end{array}
$$

The compound $Rh[P(C_6H_5)_3]_3Cl$ is a 16-electron complex:

$$\left.\begin{array}{lll} Rh & 4d^8 = 8e \\ [P(C_6H_5)_3]_3 & 3 \times 2 = 6e \\ Cl^- & 1 \times 2 = 2e \end{array}\right\} 16e$$

Reactions that involve the 16- and 18-electron oscillations are divided into five elementary groups. The classification is based on the changes in the number of metal valence electrons, formal oxidation state, and coordination number which accompany each reaction (Table 6.1). The term "Lewis base ligand" denotes ligands such as CO that are not normally considered bases, in addition to more common bases such as phosphines, since both ligands contribute two electrons to the metal valence shell. The term "oxidative addition" in reaction 3 of Table 6.1 describes an overall reaction in which the formal oxidation state of the metal and the coordination number increase by either one or two.

II. HYDROGENATION

The first practical catalyst for the homogeneous hydrogenation of unsaturated hydrocarbons was reported by Wilkinson (1965) (recipient of the Nobel Prize in 1976), who used a rhodium(I) complex, $Rh(PPh_3)_3Cl$, where $Ph = C_6H_5$. The reduction of alkenes, alkynes, and other unsaturated compounds takes place in benzene in the presence of this catalyst at 25°C and 1 atm hydrogen pressure.

Tris(triphenylphosphine)chlororhodium(I) catalyst is prepared by adding a solution of $RhCl_3$ in hot ethanol to an excess of freshly recrystallized triphenylphosphine in hot ethanol. The mixture is then heated under reflux for 30 min. After filtration and drying the product melts at 157°–158°C.

The relative ease of hydrogenation of hydrocarbons in a homogeneous system differs from that in a heterogeneous system in which reduced nickel or platinum is used as the catalyst. Unlike the latter two catalysts, which promote the change of ethylene to ethane, rhodium catalyst in a homogeneous system seems to form a stable complex with the ethylene so that the ethylene does not hydrogenate. The rate of hydrogenation of alkynes and of conjugated dienes is slower than that of the corresponding alkenes (Table 6.2). Substitution around the double bond of alkenes reduces the rate of hydrogenation, e.g., 1-alkenes vs 2-methyl-1-pentene. However, the lengthening of the alkyl chain in 1-alkenes does not seem to affect the rate of hydrogenation. The double bond of terminal olefins, including nonconjugated, nonchelating dienes, is reduced more readily than

TABLE 6.1

Elementary Organometallic Reactions[a]

Reaction	ΔNVE[b]	ΔOS[c]	ΔN[d]	Example	Reverse reaction	ΔNVE[b]	ΔOS[c]	ΔN[d]
1. Lewis acid ligand dissociation	0	0	-1	$CpRh(C_2H_4)_2SO_2 \rightleftharpoons SO_2 + CpRh(C_2H_4)_2$	Lewis acid association	0	0	+1
	0	-2	-1	$HCo(CO)_4 \rightleftharpoons H^+ + CoCO_4^-$		0	+2	+1
2. Lewis base ligand dissociation	-2	0	-1	$NiL_4 \rightleftharpoons NiL_3 + L$	Lewis base association	+2	0	+1
3. Reductive elimination	-2	-2	-2	$H_2IrCl(CO)L_2 \rightleftharpoons H_2 + IrCl(CO)L_2$	Oxidative addition	+2	+2	+2
4. Insertion (or ligand migration)	-2	0	-1	$MeMn(CO)_5 \rightleftharpoons MeCOMn(CO)_4$	Deinsertion	+2	0	+1
5. Oxidative coupling	-2	+2	0	$(C_2F_4)_2Fe(CO)_3 \rightleftharpoons \begin{array}{c} CF_2\!-\!CF_2 \\ \quad\quad\ Fe(CO)_3 \\ CF_2\!-\!CF_2 \end{array}$ 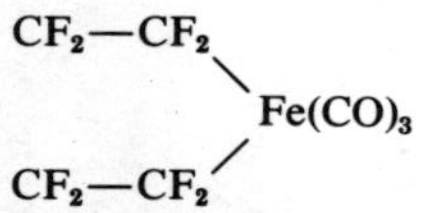	Reductive decoupling	+2	-2	0

[a] From Tolman (1972).
[b] Change in the number of metal valence electrons.
[c] Change in the formal oxidation state of the metal. The usual convention which regards hydrides, alkyls, π-allyls, and π-cyclopentadienyls as uninegative ions is used.
[d] Change in coordination number.

TABLE 6.2

Relative Rates of Hydrogenation Catalyzed by RhCl(PPh$_3$)$_3$[a]

Substrate	Relative rate[b]	Substrate	Relative rate[b]
C$_6$–C$_{12}$ 1-alkenes	1.0	Cycloheptene	0.46
Cyclopentene	1.0	2-Methyl-1-pentene	0.43
C$_6$–C$_8$ 1-alkynes	0.85	Cyclooctene	0.42–0.27
2,4,4-Trimethyl-1-pentene	0.70	2-Octene	0.25
Cyclohexene	0.71–0.55	Phenylacetylene	0.15
1,3-Cyclooctadiene	0.60–0.50	1,3-Pentadiene	~0.12
1,3-Octadiene	0.60	Isoprene	0.045
1,5-Hexadiene	~0.55	1,5-Cyclooctadiene	Very slow
		1,3-Cyclopentadiene	Very slow
		Norbornadiene	Very slow

[a] In benzene at 22°C, 10^{-2} M catalyst, 1.0 M substrate, 1 atm H$_2$; $t_{0.5}$(octene), 10 min.

[b] Calculated from initial rates or half-lives of the substrates (from Candin and Oldham, 1968).

that of internal olefins. The rate of hydrogenation of cyclic olefins decreases with larger ring size.

Kinetic data for the hydrogenation of cycloalkenes (Table 6.3) reveal that benzene–ethanol is a more effective solvent than benzene alone. The rate of hydrogenation of cyclohexene is 80–100 times faster than that of 1-methylcyclohexene, and the rate of hydrogenation of 1,4- and 2,3-dimethylcyclohexenes is less than half that of 1-methylcyclohexene, while 1,2- and 1,3-dimethylcyclohexenes do not undergo hydrogenation under these conditions. The slow hydrogenation of dimethylcyclohexenes is attributed to the lesser stability of their π complexes and to steric interference. The rate of hydrogenation seems to depend on a delicate balance between the stability of the π complex and the ease of insertion of hydrogen. The addition of hydrogen to olefins or acetylene is stereospecifically cis.

A. Mechanism

The most generally accepted mechanism for homogeneous hydrogenation is based on a combination of chemical and kinetic evidence. The catalyst RhCl(PPh$_3$)$_3$ in solvent (S) presumably forms a solvated RhCl(PPh$_3$)$_2$ in kinetically significant amounts, and this then adds hydrogen [Eq. (1)].

$$\text{RhCl(PPh}_3)_3 \underset{\text{PPh}_3}{\overset{S}{\rightleftharpoons}} \text{RhCl(S)(PPh}_3)_2 \xrightarrow[-S]{\text{H}_2} \text{RhClH}_2\text{(PPh}_3)_2 \qquad (1)$$

TABLE 6.3

Kinetic Data for the Hydrogenation of Cycloalkenes Catalyzed by RhCl(PPh$_3$)$_3$[a,b]

Substrate	[Substrate] (M)	k^c (min^{-1})	k^d (min^{-1})	% cis[c]
Cyclohexene	1.10	17	13.7	—
1-Methylcyclohexene	0.94	0.49	0.13	—
1,4-Dimethylcyclohexene	0.82	0.16	0.09	50
1-Methyl-4-isopropylcyclohexene	0.67	0.25	0.11	30
4-Methylmethylenecyclohexane	0.81	7.2	—	67
1,3-Dimethylcyclohexene	0.82	0.0	—	—
2,3-Dimethylcyclohexene	0.74	0.20	—	50
1,2-Dimethylcyclohexene	>0.60	0.0	—	—
2,4-Dimethylcyclohexene	0.41	0.16	—	48
Norbornene	1.22	—	4.4	—

[a] From Hussey and Takeuchi (1969, 1970).
[b] In benzene–ethanol (3:1) and benzene at 25°C; 2.4 × 10^{-3} M catalyst, 1 atm H$_2$.
Initial rate = k [catalyst].
[c] Benzene–ethanol (3:1).
[d] Benzene.

In weakly basic solvents, such as benzene, coordination of the olefins to the coordinatively unsaturated complex occurs [Eq. (2)]. The final step in the reaction involves the insertion of hydrogen by a rapid two-step process; the first step involves the generation of an alkyl attached to the complex, and this is then followed by a reductive elimination step resulting in the formation of alkane and completion of the cycle [Eq. (3)].

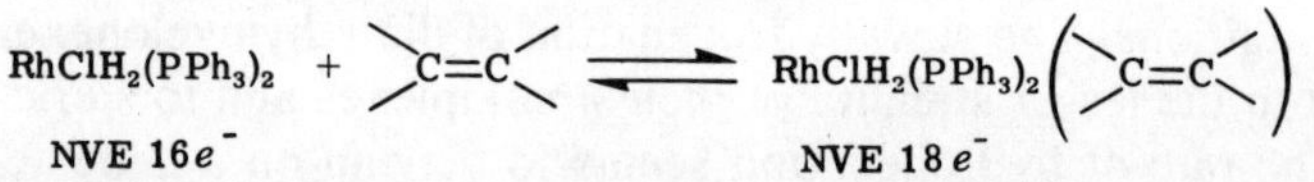

$$RhClH_2(PPh_3)_2 \;+\; {>}C{=}C{<} \;\rightleftharpoons\; RhClH_2(PPh_3)_2\left({>}C{=}C{<}\right) \qquad (2)$$

NVE 16e^- NVE 18e^-

(3)

NVE 18e NVE 16e NVE 16e

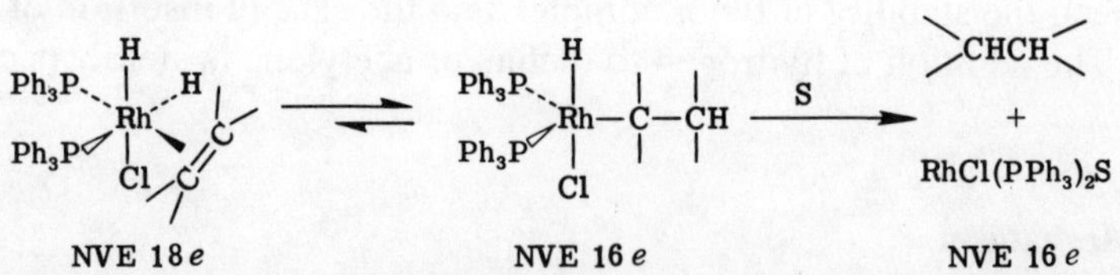

During the catalytic process the NVE oscillates between 16 and 18. A more detailed study of the mechanism using ^{2}H (deuterium) as a hydrogenating agent showed that its insertion to form an attached alkyl group may be a reversible reaction.

B. Asymmetric Hydrogenation

One of the most important commercial applications of homogeneous catalysis is the synthesis of optically active compounds that can be used therapeutically; an example of this is the synthesis of L-dopa, which is applied to the treatment of Parkinson's disease.

Most of the asymmetric hydrogenation reactions are achieved by using as catalysts rhodium–phosphine complexes of the type $RhClL_n$, where L is an optically active tertiary phosphine such as (R)-$(-)$-methylphenylpropylphosphine. The catalysts are prepared *in situ* from π-dichlorobis(1,5-hexadiene)dirhodium(I), $[RhCl(1,5-C_6H_{10})]_2$, and the $(-)$-phosphine.

Hydrogenation of α-substituted styrenes, C_6H_5—CA=CH_2, with the chiral phosphine as catalyst, would be expected to result in the formation of both stereoisomers x and y. Both were formed, but x predominated

$$H_5C_6\!-\!\overset{\displaystyle A}{\underset{\displaystyle H}{C}}\!\!-\!CH_3 \qquad\qquad H_5C_6\!-\!\overset{\displaystyle A}{\underset{\displaystyle CH_3}{C}}\!\!-\!H$$

$$x \qquad\qquad\qquad y$$

over y to the extent of 7–8% if A was C_2H_5, and 3–4% if A was OCH_3. In Cahn–Ingold–Prelog terminology, x is the S form for A = C_2H_5, whereas it is the R form for A = OCH_3. The mode of formation that gives more x than y obviously is not altered by changing A from ethyl to methoxyl.

The hydrogenation of a number of other α-substituted styrenes, $C_6H_5C(A)$=CH_2 (A = C_2H_5, n-C_3H_7, i-C_3H_7, OC_2H_5, $CH_2C_6H_5$, or α-naphthyl), is accomplished using rhodium complexes with chiral phosphines. The catalyst with chiral phosphine, $CH_3PC_6H_5R'$ (R' = n-C_3H_7, i-C_3H_7, n-C_4H_9, or *tert*-C_4H_9), produced hydrogenation products with optical purities varying between 2 and 19%.

The chirality of hydrogenation products increases when α,β-unsaturated acids instead of hydrocarbons are used as substrates and when the ratio of the chiral phosphine to rhodium is increased from 2 to 8; the enantiomeric excess in the product is 28%:

$$H_2C\!=\!C\!\!\begin{array}{c}C_6H_5\\COOH\end{array}\xrightarrow[\text{phosphine}]{\text{excess chiral}} CH_3C\!\!\begin{array}{c}H\quad C_6H_5\\COOH\end{array}$$

The effect of chiral phosphine ligands in the catalyst on the optical purity of products derived from the hydrogenation of unsaturated amino acids is given in Table 6.4.

Other types of rhodium complexes used for asymmetric hydrogenation contain a chiral p-menth-3-yl group (A) or *threo*-erythritol (B) attached to diphenylphosphine. In Tables 6.5 and 6.6 these are referred to as A and B,

TABLE 6.4

Reduction of Some α-Amidoacrylic Acids (Mainly Dopa Precursors) with Rhodium(I)–Tertiary Phosphine Catalysts[a,b]

Chiral phosphine PRR¹R²			Phosphine (% ee)[a]	Substrate R³CH=C(NHCOR⁴)CO₂H		Product optical purity (% ee)[a]
R[c]	R¹	R²		R³	R⁴	
o-Anisyl	Me	Ph	95	3-OMe-4-(OH)phenyl	Ph	58[e]
Me	Ph	n-Pr	90	3-OMe-4-(OH)phenyl	Ph	28[f]
Me	Ph	i-Pr	90	3-OMe-4-(OH)phenyl	Ph	28[f]
m-Anisyl	Me	Ph	80	3-OMe-4-(OH)phenyl	Ph	1[f]
o-Anisyl	C₆H₁₁ (cyclohexyl)	Me	95	3-OMe-4-(OH)phenyl	Ph	87[f]
C₆H₁₁	Me	Ph	75	3-OMe-4-(OH)phenyl	Ph	32[f]
o-Anisyl	Ph	i-Pr	80	3-OMe-4-(OH)phenyl	Ph	1[f]
o-Anisyl	Me	Ph	95	3-OMe-4-(OAc)phenyl	Ph	55[e]
o-Anisyl	Me	n-Pr	95	3-OMe-4-(OAc)phenyl	Ph	20[f]
o-Anisyl	C₆H₁₁	Me	95	3-OMe-4-(OAc)phenyl	Me	85[g]
o-Anisyl	C₆H₁₁	Me	95	Ph	Me	85[g]
o-Anisyl	C₆H₁₁	Me	95	Ph	Ph	85[g]
o-Anisyl	C₆H₁₁	Me	95	H	Me	60[g]

[a] From Knowles et al. (1972).

[b] L-Dopa = (dihydroxyphenyl)alanine:

$$3,4\text{-(HO)}_2C_6H_3CH_2\underset{\overset{|}{H}}{\overset{\overset{NH_2}{|}}{C}}\text{---COOH}$$

[c] For brevity, anisyl denotes methoxyphenyl.

[a] % ee denotes percent enantiomeric excess.

[e] These reductions were run in a stirred autoclave in methanol at 55 psi (absolute) of H₂, 50°C, with 1 equivalent of NaOH. The mole ratio of catalyst to substrate was 1:3000.

[f] These reductions were run in a Parr shaker in methanol at 55 psi (absolute) at 25°C.

[g] These reductions were run in a Parr shaker in 95% ethanol at 10 psi (absolute).

TABLE 6.5

Optical Purity of Olefinic Compounds Hydrogenated in the Presence of Rhodium Catalyst Containing A Ligands[a]

Substrate	Product	% ee
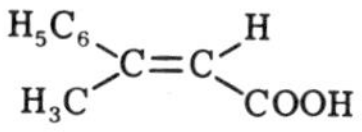 H_5C_6 $C=C$ H / H_3C $COOH$	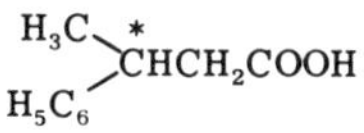 H_3C $\overset{*}{C}HCH_2COOH$ / H_5C_6	61
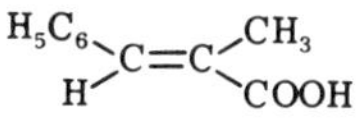 H_5C_6 $C=C$ CH_3 / H $COOH$	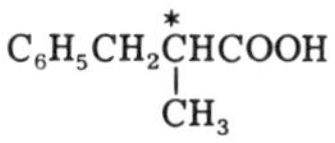 $C_6H_5CH_2\overset{*}{C}HCOOH$ / CH_3	52
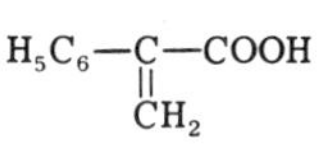 $H_5C_6-C-COOH$ / CH_2	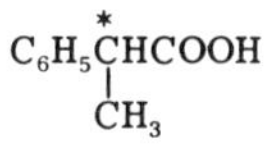 $C_6H_5\overset{*}{C}HCOOH$ / CH_3	28
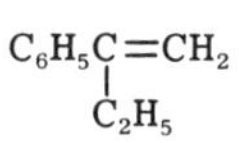 $C_6H_5C=CH_2$ / C_2H_5	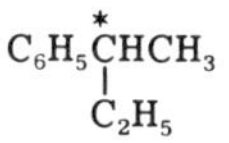 $C_6H_5\overset{*}{C}HCH_3$ / C_2H_5	9

[a] The hydrogenation was carried out at 60°C and 20 atm H_2.

TABLE 6.6

Hydrogenation of Substituted N-Vinylamides in the Presence of Catalyst Containing B Ligands

Substrate		Product		
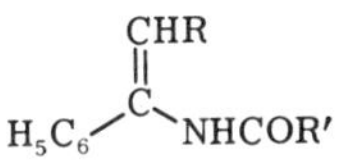 CHR $\parallel$ C / H_5C_6 $NHCOR'$		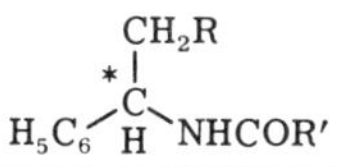 CH_2R $\overset{*}{\mid}$ C / H_5C_6 H $NHCOR'$		
R	R'	R	R'	% ee
H	CH_3	H	CH_3	45
CH_3	CH_3	CH_3	CH_3	83
CH_3	C_6H_5	CH_3	C_6H_5	>73 ± 3
CH_3	$CH(CH_3)_2$	CH_3	$CH(CH_3)_2$	85 ± 4

[a] From Kagan et al. (1975).

respectively. These chiral ligands are prepared from commercially avail-

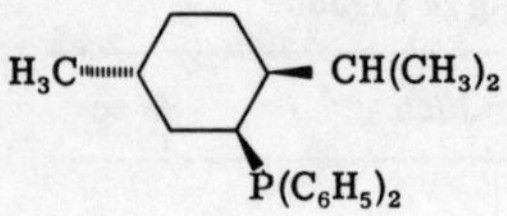

A

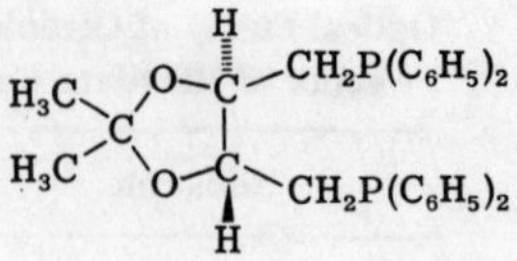

B

able neomenthol and tartaric acid, respectively. The high stereoselectivity observed with B catalyst is attributed to the appreciable conformational rigidity due to the trans-fused dioxolane ring.

III. CARBONYLATION

This section includes a discussion of the two major carbonylation reactions of commercial importance. The first is hydroformylation; this term was coined by Homer Adkins in 1948 to describe the reaction of an olefin with carbon monoxide. This reaction is also called the oxo reaction, which was named by its discoverer, Otto Roelen, in 1938. The second reaction is the carbonylation of methanol to produce acetic acid.

A. Hydroformylation (Oxo Reaction)

The reaction of an olefin with carbon monoxide and hydrogen is one of the oldest and most commonly used carbonylation reactions. Its main application is the carbonylation of propene to butyraldehyde (butanal), which can be hydrogenated to 1-butanol or converted to 2-ethylhexanol

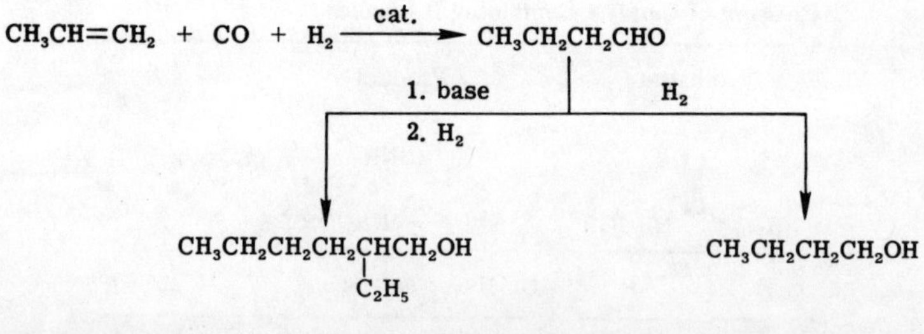

for the synthesis of ester-type plasticizers. The total production of butyraldehyde in the United States in 1975 was 6.5×10^8 kg.

The other major application of hydroformylation is the synthesis of alcohols from higher olefins for use in the production of detergents.

The predominant commercial catalyst consists of cobalt carbonyl, which is generated *in situ* from cobalt salts and synthesis gas, CO and H_2.

In this respect the process has changed little from that described by Roelen (1938). However, advances in catalyst technology derived from studies of the mechanism of reaction have occurred in the last 15 years. Another catalyst that has attracted attention recently is rhodium carbonyl.

1. Cobalt Carbonyl Catalysts

The oxo synthesis is a homogeneous reaction in which soluble cobalt carbonyls are the catalysts. Its efficiency depends on the form of cobalt used. Although cobalt carbonyl can be formed from powdered cobalt and carbon monoxide, it is preferable to prepare the catalyst *in situ* from cobalt salts, such as cobalt acetate, and in the presence of hydrogen in order to reduce Co^{2+} to $Co(0)$ in cobalt carbonyl.

$$2Co(OAc)_2 + 8CO + 2H_2 \longrightarrow Co_2(CO)_8 + 4HOAc$$

a. Procedure and Variables. The hydroformylation reaction takes place under pressure at temperatures varying from 115° to 190°C, depending on the activity of the catalyst and the type of olefins used. The reaction is carried out in the laboratory as illustrated by the following example. Pentenes, 25 gm, in 60 ml of ether or benzene and 0.6 gm of reduced cobalt on kieselguhr or cobalt acetate are placed in a stirred autoclave. The autoclave is heated at 125°–150°C with an initial pressure of 160 atm of carbon monoxide, and then hydrogen is added to 240 atm. The duration of heating is 3 hr. At 50 atm only about 10% conversion is obtained in 1 hr, whereas at 150 atm the conversion is 70–80%.

The temperature of the reaction influences the nature of the products. For a maximal yield of aldehydes, the reaction is performed at as low a temperature as possible but compatible with a reasonable rate. This temperature lies between 110° and 140°C. At higher temperatures the aldehydes are reduced to alcohols.

Variation in the pressure of synthesis gas ($1CO/1H_2$) from 75 to 150 atm has no effect on the rate of hydroformylation. The minimal pressure of operation is dependent on the dissociation pressure of dicobalt octacarbonyl, and the concentration of the latter depends on the carbon monoxide pressure.

The successful operation of the oxo reaction is less dependent on hydrogen pressure than on the partial pressure of carbon monoxide. The hydrogen is required for converting dicobalt octacarbonyl to cobalt hydrocarbonyl, a yellow liquid which is the catalyst for the reaction.

$$[Co(CO)_4]_2 + H_2 \rightleftharpoons 2HCo(CO)_4 \rightleftharpoons 2HCo(CO)_3 + 2CO$$

b. Effect of Olefin Structure. There is an appreciable difference in the rate at which olefins of various structures undergo hydroformylation

(Table 6.7). Straight-chain terminal olefins react most rapidly. There is a small drop in rate with an increase in chain length, but the effect becomes smaller with an increase in the number of carbons. The rate of reaction of straight-chain internal olefins is about one-third that of the corresponding terminal olefins. The position of the double bond, as long as it is internal, has little or no effect on the rate. Branching at one of the carbon atoms of

TABLE 6.7

Rates of Hydroformylation of Olefins at 110°C^{a,b}

Type of olefins	Specific reaction rate, $10^3\ k(\text{min}^{-1})$
Straight-chain terminal olefins	
1-Pentene	68.3
1-Hexene	66.2
1-Heptene	66.8
1-Octene	65.6
1-Decene	64.4
1-Tetradecene	63.0
Straight-chain internal olefins	
2-Pentene	21.3
2-Hexene	18.1
2-Heptene	19.3
3-Heptene	20.0
2-Octene	18.8
Branched terminal olefins	
4-Methyl-1-pentene	64.3
2-Methyl-1-pentene	7.82
2,4,4-Trimethyl-1-pentene	4.79
2,3,3-Trimethyl-1-butene	4.26
Branched internal olefins	
4-Methyl-2-pentene	16.2
2-Methyl-2-pentene	4.87
2,4,4-Trimethyl-2-pentene	2.29
2,3-Dimethyl-2-butene	1.35
2,6-Dimethyl-3-heptene	6.22
Cyclic olefins	
Cyclopentene	22.4
Cyclohexene	5.82
Cycloheptene	25.7
Cyclooctene	10.8
4-Methylcyclohexene	4.87

^{a} From Wender et al., 1956.
^{b} Conditions: 0.50 mol olefin, 65 ml methylcyclohexane as solvent, 2.8 gm (8.2 × 10^{-3} mol) dicobalt octacarbonyl, and an initial pressure at room temperature of 233 atm of 1:1 synthesis gas.

the double bond causes an appreciable decrease in the rate of hydroformylation, as in the case of 1-hexene vs 2-methyl-1-pentene, 2-hexene vs 2-methyl-2-pentene, and 4-methyl-2-pentene vs 2-methyl-2-pentene. Further branching along the chain results in a further decrease in rate. The effect of the structure of the olefins on the rate of hydroformylation may be interpreted in terms of steric effects in complex formation between the catalyst and the olefins.

On hydroformylation, straight-chain terminal olefins afford aldehydes consisting of about 60% straight-chain and about 40% branched-type isomers. The distribution of aldehydes from 1-pentene and 2-pentene obtained at 120°–160°C is about the same: 50–55% of 1-hexanal, 35–40% of 2-methyl-1-pentanal, and 10% of 2-ethyl-1-butanal.

Olefins possessing the structure $R—C(CH_3)=CH_2$ add the formyl group predominantly to the terminal carbon atom.

$$CH_3\underset{H_3C}{C}=CH_2 \xrightarrow[\text{(cat.) }120°C]{CO + H_2} CH_3\underset{CH_3}{C}HCH_2CHO + (CH_3)_3CCHO$$

$$96\% \qquad\qquad 4\%$$

c. Mechanism. The mechanism of the hydroformylation reaction is consistent with the 16- to 18-electron rule and in agreement with experimental data. The active catalyst for hydroformylation was inferred from kinetic data.

The key step in the reaction is the insertion of CO into the cobalt–alkyl complex. The insertion reaction is one of the basic reactions of all the processes involving transition metal complexes as catalysts. On the basis of detailed experiments with $(CH_3)Mn(CO)_5$ it is thought that most if not all CO insertions proceed by migration of the alkyl group onto the carbon of the coordinated CO. The mechanism of hydroformylation is presented in Scheme 6.1.

The formation of 2-methylalkanols in the hydroformylation of 1-alkenes can be explained by the addition of the hydrogen of the hydridocobalt tricarbonyl to the terminal carbon of the 1-alkenes. The reduction of aldehydes by hydrogen, especially when the hydroformylation is carried out at higher temperatures, was explained as occurring according to Scheme 6.2.

2. Phosphine-Modified Cobalt Catalysts

The direction of addition of cobalt hydrocarbonyl complexes to terminal alkenes can be altered by introducing tertiary phosphine groups as ligands. The catalysts thus produced are active at low pressure, 6–18 atm, and exhibit a high preference for reactions at the terminal carbon position of 1-alkenes. The products of the reaction are, however, alcohols rather

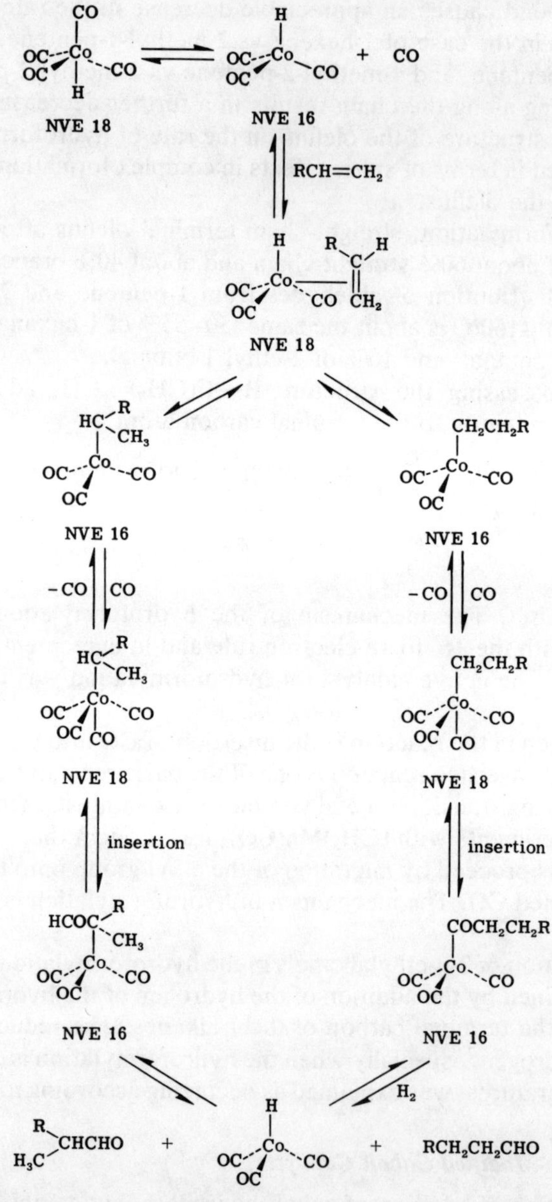

$$
\begin{array}{c}
\text{CO} \\
\text{OC--Co--CO} \\
\text{OC} \quad \text{H}
\end{array}
\text{NVE 18}
\quad\rightleftharpoons\quad
\begin{array}{c}
\text{H} \\
\text{OC--Co--CO} \\
\text{OC}
\end{array}
\text{NVE 16}
\;+\; \text{CO}
$$

RCH=CH₂

$$
\begin{array}{c}
\text{H} \qquad \text{R} \quad \text{H} \\
\text{OC--Co} \quad \text{C} \\
\text{OC} \quad \text{CO} \quad \text{CH}_2
\end{array}
\text{NVE 18}
$$

HC⟨R⟩CH₃   CH₂CH₂R

$$
\begin{array}{c}
\text{HC} \overset{\text{R}}{\underset{\text{CH}_3}{}} \\
\text{OC--Co--CO} \\
\text{OC}
\end{array}
\text{NVE 16}
\qquad\qquad
\begin{array}{c}
\text{CH}_2\text{CH}_2\text{R} \\
\text{OC--Co--CO} \\
\text{OC}
\end{array}
\text{NVE 16}
$$

−CO ‖ CO   −CO ‖ CO

$$
\begin{array}{c}
\text{HC} \overset{\text{R}}{\underset{\text{CH}_3}{}} \\
\text{OC--Co--CO} \\
\text{OC} \quad \text{CO}
\end{array}
\text{NVE 18}
\qquad\qquad
\begin{array}{c}
\text{CH}_2\text{CH}_2\text{R} \\
\text{OC--Co--CO} \\
\text{OC} \quad \text{CO}
\end{array}
\text{NVE 18}
$$

insertion   insertion

$$
\begin{array}{c}
\text{HCOC} \overset{\text{R}}{\underset{\text{CH}_3}{}} \\
\text{OC--Co--CO} \\
\text{OC}
\end{array}
\text{NVE 16}
\qquad\qquad
\begin{array}{c}
\text{COCH}_2\text{CH}_2\text{R} \\
\text{OC--Co--CO} \\
\text{OC}
\end{array}
\text{NVE 16}
$$

$$
\underset{\text{H}_3\text{C}}{\overset{\text{R}}{}}\text{CHCHO}
\;+\;
\begin{array}{c}
\text{H} \\
\text{OC--Co--CO} \\
\text{OC}
\end{array}
\;+\; \text{RCH}_2\text{CH}_2\text{CHO}
$$

H₂

Scheme 6.1. Mechanism of hydroformylation in the presence of cobalt carbonyls [from Heck and Breslow (1961), as modified by Moffat (1970)].

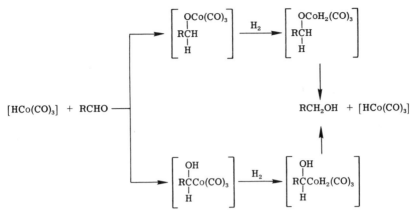

Scheme 6.2. Reduction of aldehydes to alcohols (Heck, 1974).

than aldehydes. The linearity of the alcohols produced depends on the type of phosphines used as ligands (Table 6.8). Bulky ligands of low basicity favor branched products.

The strong increase in the hydrogenation properties of phosphine-modified cobalt catalysts to convert aldehydes to alcohols can be attributed to the decreased acidity and increased hydridic character of Co—H. Phosphine-modified cobalt catalysts are used commercially for the conversion of higher molecular weight 1-alkenes to the corresponding alcohols.

TABLE 6.8

Effect of Phosphines on the Selectivity of Hydroformylation of 1-Pentene[a,b]

	Products formed (%)	
Phosphine PR_3, R =	Straight chain	Branched chain
C_2H_5	80.9	19.1
$n\text{-}C_4H_9$	84.1	15.9
Phenyl	66.0	34.0

[a] From Slaugh and Mullineaux (1968).
[b] Conditions: 1-pentene, 65 mmol in 20 ml $n\text{-}C_6H_{14}$; catalyst, 4 mmol PR_3 and 2 mmol $Co_2(CO)_8$; $H_2/CO \sim 2$; temp, 195°C; maximum pressure, ~30 atm.

3. Rhodium Catalysts

Rhodium carbonyls are especially suitable catalysts for oxo reactions. They are 10^3–10^4 times more active than the corresponding cobalt carbonyls, and in addition they show high selectivity to aldehyde formation. Hydroformylation in the presence of rhodium catalysts proceeds at moderate temperatures and pressures.

The active rhodium catalysts are $HRh(CO)_2(PR_3)_2$ (R = Ph or butyl), and the mechanism of hydroformylation is similar to that proposed for the cobalt-catalyzed reaction. Unlike phosphine-modified cobalt catalysts, rhodium catalysts do not cause the formation of alcohols.

The experimental conditions for a commercial hydroformylation of propene to butanal are as follows: pressure 35–40 atm; temp, 90°–120°C; CO/H_2 ratio, from 1:2 to 1:4; concentration of catalyst based on Rh, 0.01%. In order to obtain a high ratio of butanal to methylpropanal and to achieve the desired thermal stability of the catalyst complex, a molar ratio of 50 to 100 of the phosphine to catalyst is required.

TABLE 6.9

Effect of Catalytic Precursors of Cobalt and Rhodium on the Selectivity and Reactivity of Allyl- and Propenylbenzenes[a,b]

Substrate C_6H_5R, R =	Catalyst precursor	Selectivity[c]	Aldehyde distribution (%)[d]		
			A	B	C
—CH_2—CH=CH_2	$Co_2(CO)_8$	78.3	17.9	12.6	69.7
	Rh	97.4	13.7	37.6	48.7
—CH=$CHCH_3$ (trans)	$Co_2(CO)_8$	52.0	68.8	4.6	26.6
	Rh	94.3	72.0	26.0	2.0
	Rh—$(PPh_3)_3$	92.2	87.1	12.9	0
—CH=$CHCH_3$ (cis)	$Co_2(CO)_8$	41.5	68.2	6.0	25.8
	Rh	98.6	56.6	41.6	1.8
	Rh—$(PPh_3)_3$	99.5	64.5	34.3	1.1

[a] From Lai and Lucciani (1974).

[b] The experiments were carried out in a 125-ml autoclave under 160 atm initial total pressure ($1H_2$/$1CO$) as measured at room temperature. The catalyst precursors were the following: 1×10^{-2} mol $Co_2(CO)_8$; 1×10^{-3} atm Rh and 2×10^{-2} mol PPh_3 if required per mole of olefins.

[c] Moles of aldehydes per 100 mol of reacted olefin.

[d] The aldehydes were the following:

$$\underset{\textbf{A}}{\underset{\overset{|}{C_2H_5}}{C_6H_5CHCHO}} \qquad \underset{\textbf{B}}{\underset{\overset{|}{CH_3}}{C_6H_5CH_2CHCHO}} \qquad \underset{\textbf{C}}{C_6H_5(CH_2)_3CHO}$$

The use of excess ligand, such as triphenylphosphine, inhibits the rate of hydroformylation and has a profound effect on the products. It decreases the competing hydrogenation and isomerization of the alkenes and increases the ratio of straight-chain to branched aldehydes.

A comparison of the effect of catalyst precursors of $Co_2(CO)_8$, Rh, and $Rh(PPh_3)_3$ on the hydroformylation of allyl- and propenylbenzenes is given in Table 6.9. Unlike rhodium, cobalt catalysts cause extensive hydrogenation of propenylbenzenes to propylbenzene and apparent isomerization of the double bond to produce 26% of 4-phenylbutyraldehyde (C).

B. Carbonylation of Methanol

The low-pressure carbonylation of methanol to yield acetic acid, the Monsanto process, was first described in 1968, and in 1978 the worldwide licensed capacity of this process was 1.4×10^9 kg of acetic acid per year. The process is based on a catalyst precursor solution of rhodium component [e.g., $RhCl_3 \cdot 3H_2O$, $Rh_2O_3 \cdot 5H_2O$, or $RhCl(CO)(PPh_3)_2$, $5 \times 10^{-3}\ M$] and a halogen promoter (e.g., aqueous HI, MeI, or I_2, $0.05\ M$) in a solvent such as benzene, water, acetic acid, or methanol. The carbonylation of methanol to acetic acid occurs at 175°C and about 25 atm. The total pressure of CO is 16 atm. The selectivity to acetic acid is reported to be

$$CH_3OH + HI \rightleftharpoons CH_3I + H_2O$$

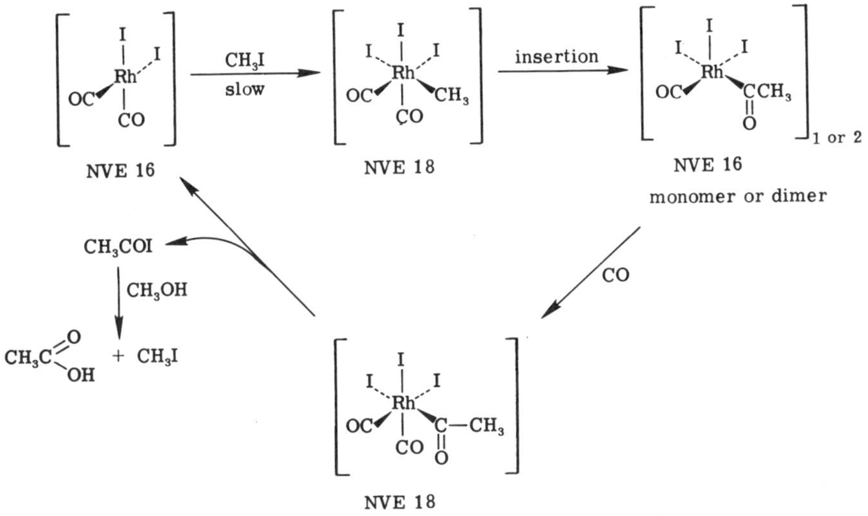

Scheme 6.3. Mechanism of carbonylation of methanol (from Forster, 1976).

>99%. The catalyst may be used continuously by controlled addition of reactants and removal of products (Paulik and Roth, 1968).

The mechanism of the carbonylation of methanol resembles in many respects the hydroformylation mechanism. The elucidation of this mechanism, which is based on the spectroscopic identification of the key intermediates, is presented in Scheme 6.3 (p. 247).

IV. OLIGOMERIZATION OF ALKENES AND DIENES

A. Introduction

The conversion of alkenes such as ethylene and propene to high-molecular-weight polymers has received a large amount of attention. Such polymers are of great industrial importance. A vast number of catalysts have been used for this reaction including transition metal compounds in conjunction with trialkylaluminum. This type of polymerization leading to macromolecules is not be discussed here since it lies outside of the scope of this book.

The oligomerization of alkenes to dimers and trimers has been studied extensively using transition metal compounds containing rhodium, nickel, cobalt, and other metals. These oligomers are usually formed by a metal hydride adding reversibly to an olefin, and the metal alkyl thus formed may add to another olefin via an olefin π complex. π-Complexing ability is a factor in determining the tendency of an olefin to undergo oligomerization. Increasing the alkyl substituents in alkenes decreases the stability of the π complex formed and consequently decreases the ability of the olefins to oligomerize. The rate of oligomerization decreases in the order $CH_2{=}CH_2 > CH_3CH{=}CH_2 > $ cycloolefin $> CH_3CH{=}CHCH_3$. The rate of dimerization of ethylene using a nickel complex catalyst is about 40 times faster than that of propene.

Catalyst adducts that undergo metal hydride elimination more slowly than they add to another molecule of olefin lead to the formation of high-molecular-weight polymers. The two characteristics of the catalyst cited above have to be properly balanced in order for oligomerization reactions to form dimers and trimers selectively. This can be achieved by modifying the transition metal compounds by means of the ligands attached to them, as shown in the examples below.

B. Alkenes

1. Dimerization of Ethylene

The conversion of ethylene to butenes is difficult to achieve by either cationic or anionic polymerization. The dimerization, however, proceeds satisfactorily in the presence of transition metal complexes, such as those of rhodium, ruthenium, nickel, and cobalt.

The rhodium-catalyzed dimerization of ethylene is carried out conveniently in an autoclave provided with agitation at 40°C in ethanol containing 0.1 mol % of rhodium trichloride based on ethylene charged, which is supplied from a pressure vessel. The conversion of ethylene appears to be limited by the size of the reaction vessel. The yield of butenes is greater than 99%, as determined on the basis of consumed ethylene. 1-Butene is the primary product of reaction; in time and especially at higher temperatures it is isomerized to 2-butenes.

The conversion of ethylene to butenes with rhodium catalyst is typical of the many alkene dimerization reactions and is given in Scheme 6.4. The

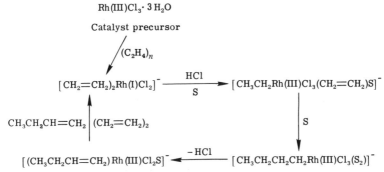

Scheme 6.4. Mechanism for the dimerization of ethylene [S = solvent (ethanol)] from Cramer, 1965).

four-step cycle involves protonation of a bis(ethylene)rhodium(I) complex to an ethylrhodium(III) compound, the rate-determining rearrangement of this compound to a butylrhodium(III) complex, collapse of the latter to 1-butene and rhodium(I), and coordination of the rhodium(I) with ethylene to continue the cycle.

2. Dimerization of Propene

Complex catalysts based on alkylaluminum halides and nickel compounds are very active toward the dimerization of propene under mild reaction conditions. A catalyst prepared from 1 mmol nickel acetylacetonate, 45 mmol $C_2H_5AlCl_2$, and 16 mmol $P(C_6H_5)_3$ gives 2.6×10^5 mole propene dimers, 90% yield, in 2 hr at 60°C and 10 atm in toluene solvent. The dimers consist of the following:

$$
\underset{37\%}{C-\overset{\overset{\displaystyle C}{|}}{C}-C=C-C}
\qquad
\underset{26.5\%}{C-\overset{\overset{\displaystyle C}{|}}{C}=C-C-C}
\qquad
\underset{17.4\%}{C=\overset{\overset{\displaystyle C}{|}}{C}-C-C-C}
$$

$$\underset{\text{1.6\%}}{C-\overset{\overset{\displaystyle C}{|}}{C}-C-C=C} \qquad \underset{\text{6\%}}{C=\overset{\overset{\displaystyle C}{|}}{C}-\overset{\overset{\displaystyle C}{|}}{C}-C} \qquad \underset{\text{11.2\%}}{\text{2- and 3-hexenes}}$$

The composition of the products can be varied widely by changing ligands, especially phosphines (Table 6.10).

The oligomerization of alkenes in the presence of transition metal catalysts involves the insertion of a hydride into the alkene to form an alkyl metal compound. The mechanism of the generation of a nickel hydride species in the nickel catalyst complex precursor is not completely clear. Hydrogen is not evolved in the oligomerization reaction, which indicates that the transfer is accompanied by the formation of a nickel hydride species. Unlike the dimerization of ethylene, the oligomerization of propene is complicated by the asymmetry of the molecule.

The catalytically active species seems to consist of a square planar hybridized nickel atom interacting with a hydride or alkyl group, a phosphine, an electronegative group X, and an alkene. The Lewis acid $R_2Al_2Cl_3$, which is bridged to the nickel by a chlorine atom, is assumed to decrease the charge on the nickel. The first steps in the dimerization of propene can be represented as follows:

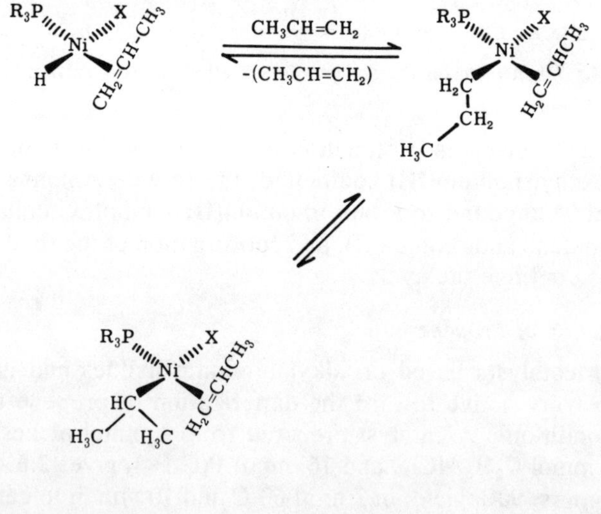

X = $R_2Al_2Cl_3$

The first step is kinetically controlled and affords principally Ni–n-propyl species (Ni → C_1), which rearranges to the thermodynamically more

TABLE 6.10

Influence of Phosphines on the Nickel-Catalyzed Dimerization of Propene[a,b]

Phosphine	Σ n-Hexene	Σ 2-Methylpentene	Σ 2,3-Dimethylbutene	Dimer formation (gm/hr)
—	19.8	76.0	4.2	ca. 300
$P(C_6H_5)_3$	21.6	73.9	4.5	126
$(C_6H_5)_2PCH_2P(C_6H_5)_2$	12.2	83.0	4.7	58
$(C_6H_5)_2PCH_2C_6H_5$	19.2	75.4	5.1	291
$(C_6H_5)_2P(CH_2)_3P(C_6H_5)_2$	20.1	73.3	6.6	7
$P(CH_3)_3$	9.9	80.3	9.8	ca. 400
$(C_6H_5)_2P\text{-iso-}C_3H_7$	14.4	73.0	12.6	388
$P(C_2H_5)_3$	9.2	69.7	21.1	ca. 350
$P(C_4H_9)_3$	7.1	69.6	23.3	87
$P(CH_2C_6H_5)_3$	6.7	63.6	29.2	318
$(cyclo\text{-}C_6H_{11})_2P\text{—}P(cyclo\text{-}C_6H_{11})_2$	4.4	46.5	49.2	132
$P(cyclo\text{-}C_6H_{11})_3$	3.3	37.9	58.8	250
$P(iso\text{-}C_3H_7)_3$	1.8	30.3	67.9	300
$(tert\text{-}C_4H_9)_2PCH_3$	1.2	24.5	74.0	189
$(tert\text{-}C_4H_9)_2PCH_2CH_3$	0.6	22.3	77.0	292
$(iso\text{-}C_3H_7)_2P\text{-}tert\text{-}C_4H_9$	0.1	11.9	87.8	332
$(tert\text{-}C_4H_9)_2P\text{-iso-}C_3H_7$	0.6	70.1	29.1	347

[a] From Jolly and Wilke (1975a, p. 12).

[b] Catalyst: $(\pi\text{-}C_3H_5NiX)_2 - (C_2H_5)_3Al_2Cl_3 - \text{Lig}$; $-20°C$; 1 atm.

stable Ni–isopropyl species (Ni → C_2). The same choice exists in the second step, and this is presented by Scheme 6.5.

The effect of phosphine on the structure of hexenes obtained from the dimerization of propene is largely steric in origin. The direction of addition in the first step is approximately independent of the phosphine, and the value for Ni → C_1/Ni → C_2 is about 20:80. The second step, in contrast, depends on the nature of the phosphine. The bulkier the phosphine group, the more the Ni → C_1 addition occurs.

C. Butadiene

Nickel-containing catalysts can either cyclodimerize or cyclotrimerize 1,3-butadiene, depending on the preparation of the catalyst. A catalyst prepared *in situ* from nickel acetylacetone and an alkylaluminum in the presence of butadiene is effective for the trimerization of butadiene to cyclodecatrienes, of which the trans,trans,trans isomer **1** predominates. This catalyst can be modified by adding a ligand, a phosphine or a phosphite, to cause the cyclodimerization of the butadiene to 1,5-cyclooctadiene (**2**) and to a much smaller extent 4-vinylcyclohexene (**3**) and 1,2-divinylcyclobutane (**4**).

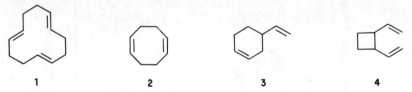

| 1 | 2 | 3 | 4 |

1. Cyclodimerization

The effect of ligands in nickel-containing catalysts on the rate and selectivity of dimerization of butadiene is given in Table 6.11.

Cyclooctadiene is obtained in 96% yield with $P(OC_6H_4\text{-}o\text{-}C_6H_5)_3$ (**5**) as ligand and a reaction temperature of 80°C. If the reaction is carried out at lower temperature and interrupted before the complete conversion of butadiene has occurred, then 1,2-divinylcyclobutane (**4**) can be isolated in up to 40% yield.

$$
\left(P\!\!-\!\!O\!\!-\!\!\underset{}{\bigcirc}\!\!-\!\!\bigcirc \right)_3
$$

5

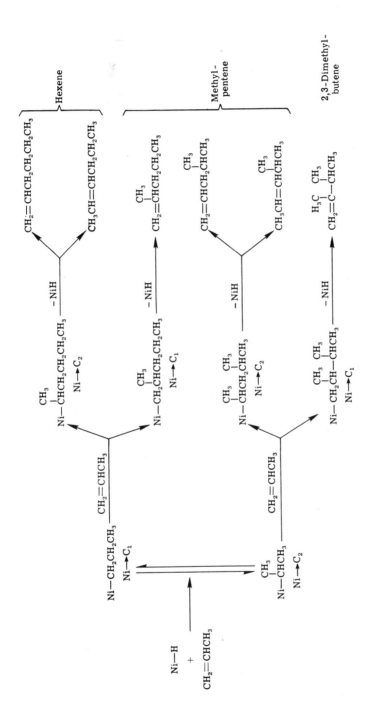

Scheme 6.5. Stepwise dimerization of propene (from Jolly and Wilke, 1975a, p. 23).

TABLE 6.11

Cyclodimerization of Butadiene[a,b]

Ligand	Rate (gm C_4H_6/gm Ni × hr)	Product distribution (%)			
		2	3	1	Misc.
$P(OC_6H_4\text{-}o\text{-}C_6H_5)_3$	780	96	3.1	0.2	0.2
$P(OC_6H_3\text{-}2\text{-}tert\text{-}C_5H_9, 4\text{-}CH_3)_3$	600	96	2.6	1.0	0.2
$P(OC_6H_3\text{-}2\text{-}CH_3, 4\text{-}tert\text{-}C_4H_9)_3$	315	91	6.8	1.5	—
$P(OC_6H_4\text{-}o\text{-}CH_3)_3$	400	92	5.7	1.4	0.6
$P(OC_6H_4\text{-}m\text{-}CH_3)_3$	110	74	8.1	10.8	7.1
$P(OC_6H_4\text{-}p\text{-}CH_3)_3$	70	73	8.8	9.7	8.2
$P(OC_6H_5)_3$	100	81	7.4	6.0	2.8
$P(C_6H_5)_3$	180	64	27	6.0	2.8
$P(cyclo\text{-}C_6H_{11})_3$	35	41	40	14	4.8

[a] Jolly and Wilke (1975a, p. 143).
[b] Ni/ligand 1:1; temp, 80°C.

2. Cyclotrimerization

Cyclodecatrienes are of industrial importance as precursors of cyclodo-decane, which is an intermediate in the production of 1,12-dodecanedioic acid and lauryllactam. The total production of cyclodecatrienes in the United States in 1975 was 20 million pounds.

A catalyst prepared from $TiCl_4$ and $AlCl_3$ (Ti/Al = 1:4.5) in an inert solvent rapidly converts butadiene to *trans,trans,cis*-1,5,9-cyclododeca-triene (**6**) in yields of over 80%. A still more active catalyst for this reac-tion was prepared from $(C_2H_5)_2AlH$ that was treated with $AlCl_3$ in ben-zene to which $TiCl_4$ was added. A suspension resulted, through which butadiene was passed at 40°C and atmospheric pressure (highly efficient cooling was necessary); cyclododecatriene was produced in 90–95% yield. It was calculated that a catalyst containing 1 gm of titanium can convert 800–1200 gm of butadiene per hour.

6

Another active catalyst for the cyclotrimerization of butadiene was pro-duced from nickel diacrylonitrile, $Ni(CH_2=CHCN)_2$. At 80°C in cyclooctadiene solvent under about 10 atm pressure, 96.5% of butadiene is converted to afford 80.4% of *trans,trans,trans*-1,5,9-cyclooctatriene (1) (Ono and Kihara, 1967).

V. HYDROCYANATION

In 1971 the Du Pont Company announced a commercial synthesis of adiponitride based on the selective addition of 2 mol of hydrogen cyanide to butadiene. In one variation of this process deduced from a series of patents, a zero-valent nickel catalyst, $Ni[P(OAr)_3]_4$, is effective. This catalyst can bring about a series of reactions which include the addition of HCN to butadiene, double-bond isomerization of the monoadduct, and hydrocyanation of 4-pentenenitrile (Parshall, 1978, p. 257).

$$CH_2=CH-CH=CH_2 \xrightarrow{\text{HCN}} CH_3CH=CHCH_2CN + CH_2=CHCHCN$$
$$\qquad\qquad\qquad\qquad\qquad\qquad\qquad\qquad\qquad\qquad\qquad\qquad\overset{|}{CH_3}$$

$$CH_3CH=CHCH_2CN \rightleftharpoons CH_2=CHCH_2CH_2CN$$

$$CH_2=CHCH_2CH_2CN + HCN \longrightarrow NCCH_2CH_2CH_2CH_2CN$$

Although the mechanisms of these reactions have not been described, it has been shown that HCN reacts with zero-valent nickel complexes to produce hydridonickel(II) cyanide $[L = PR_3, P(OR)_3; R = CH_3, C_2H_5, C_6H_5,$ etc.$]$.

$$NiL_4 + HCN \underset{+L}{\overset{-L}{\rightleftharpoons}} HNi(CN)L_3 \underset{+L}{\overset{-L}{\rightleftharpoons}} HNi(CN)L_2$$

Complexes such as the four-coordinated nickel hydride can coordinate an olefin and form an alkylnickel complex by insertion:

$$\qquad\qquad\qquad\qquad\qquad\qquad\qquad\overset{RCH=CH_2}{\overset{|}{}}$$
$$HNi(CN)L_2 + RCH=CH_2 \longrightarrow HNi(CN)L_2 \longrightarrow RCH_2CH_2Ni(CN)L_2$$

Reductive elimination of the alkyl and cyano ligands forms RCH_2CH_2CN and regenerates the catalyst, thus completing the catalytic cycle.

According to information published in 1971 the process consists of two steps. "In the first step butadiene is hydrocyanated to produce a mixture of unsaturated mononitriles which are separated, isomerized, and refined to a mixture of linear 3- and 4-pentenenitriles. In the second step, additional HCN is added to the mixed pentenenitriles and adiponitrile fraction separated from the mixture of dinitriles produced. Both steps use variations of the same catalyst system" (*Chem. Eng. News*, 1971).

VI. HETEROGENATION OF HOMOGENEOUS CATALYSTS

A. Introduction

Homogeneous transition metal organometallic catalysts can be heterogenized by linking them to a polymer. Haag and Whitehurst (1968) and Manassen (1969) are pioneers in this field of research. The heterogenized catalysts can be looked upon as consisting of the insoluble polymeric portion, which may be considered as catalyst support, and the catalytic portion, which projects into solution. Each of the components of the heterogenized catalyst must be tailored to the requirement of the catalytic reaction.

When homogeneous catalysts are linked to polymers it is possible to use those catalysts in batch-type reactions and then simply to decant the product and reuse the catalysts. It is also possible to use these catalysts in continuous-type systems. In general, polymer-supported transition metal organometallic compounds demonstrate catalytic behavior similar to the corresponding homogeneous complexes.

Effective heterogenized catalysts were obtained by complexing transition metal organometallic compounds with polymers containing pendant phosphine units. The polymeric material used in the preparation of the catalysts consist of either linear polystyrene or swellable cross-linked polymers composed of styrene units containing 1 or 2% divinylbenzene.

Examples of the synthesis of polymeric ligands with an enclosed phosphine group are presented in Scheme 6.6.

The polymeric analogues of benzyldiphenylphosphine (**I** and **II**) were

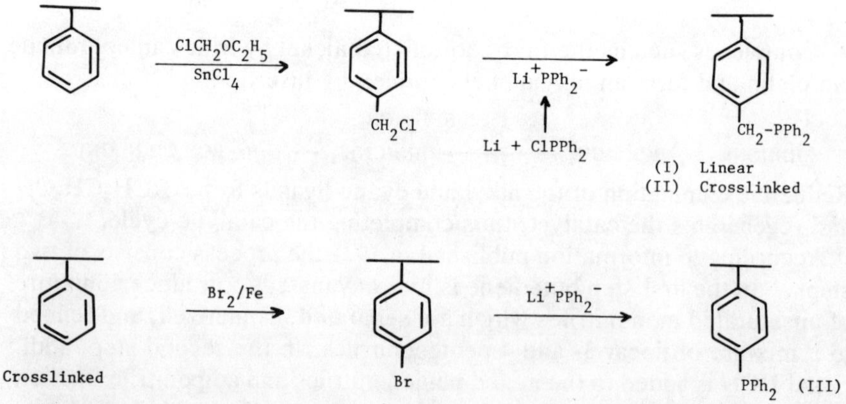

Scheme 6.6. Methods used in the preparation of polymeric ligands (from Evans *et al.* 1974).

prepared by reacting lithium diphenylphosphide (LiPPh$_2$) with the chloro-methylated polymers. The reaction proceeded nearly to completion.

Polymeric analogues of triphenylphosphine (**III**) were synthesized by reacting LiPPh$_2$ with brominated styrene/divinylbenzene copolymer.

Transition metal organometallic compounds were anchored to the two types of polymeric ligands to produce heterogenized catalysts (Scheme 6.7).

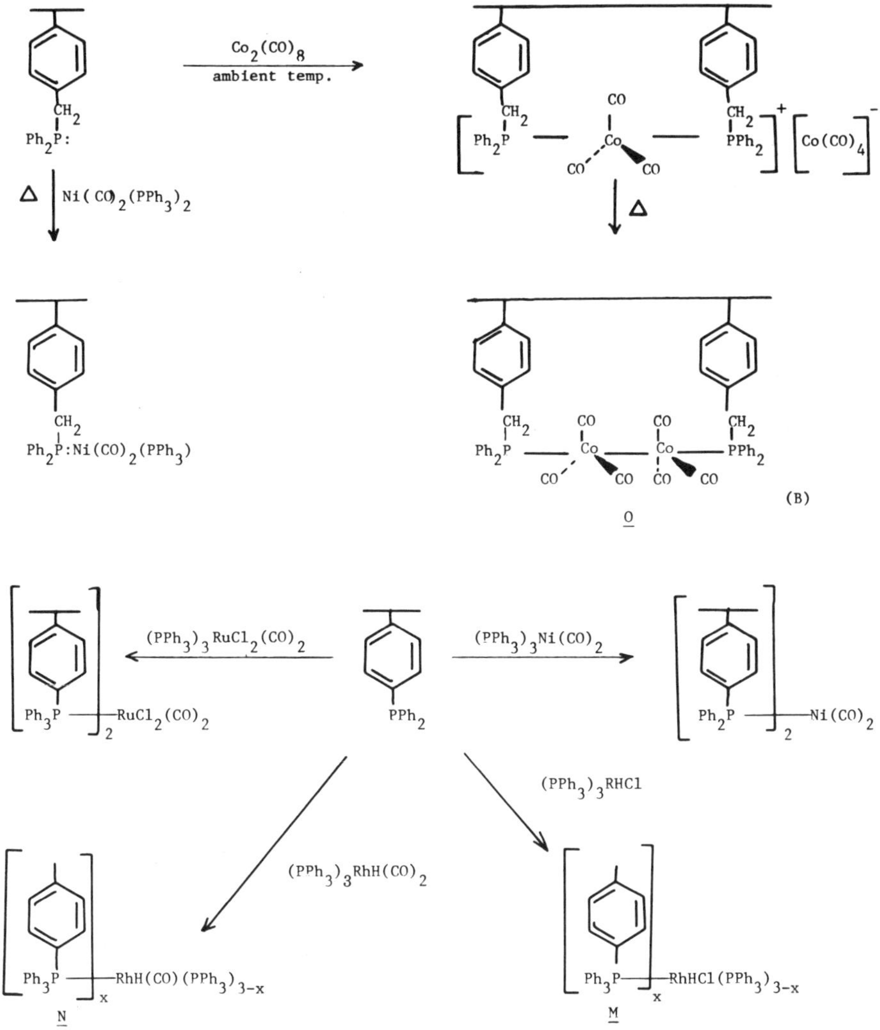

Scheme 6.7. Preparation of heterogenized transition metal organometallic catalysts.

B. Description of Anchoring Catalysts

1. To Diphenylphosphinated Cross-Linked Polymer

The diphenylphosphinated copolymer was swollen in a stirred toluene solution under a blanket of nitrogen. A toluene solution of the homogeneous catalyst to be anchored was added and the reaction carried out at 100°C for 24 hr. The solvent was decanted and the polymer beads extracted in a Soxhlet apparatus for 24 hr with benzene to remove any unanchored catalyst and free phenylphosphine from the copolymer ligand (Pittman, 1975).

2. Reaction of $Co_2(CO)_8$ with Polymeric Ligand

Cross-linked ligand of Type II (Scheme 6.6, 15 gm) was suspended in 200 ml toluene under a blanket of nitrogen. To this was added gradually with stirring a solution of 9.0 gm of $Co_2(CO)_8$ in 100 ml toluene and the stirring continued for 3 hr. The dark brown resin was collected by filtration and washed with 300 ml toluene and 200 ml petroleum ether. The material was then heated under nitrogen at 69°C for 27 hr. The product was again washed with 250 ml petroleum ether and dried under vacuum (Evans, 1974).

TABLE 6.12

Hydrogenation of 1,3-Butadiene Cyclooligomers by Homogeneous and Polymer Anchored $(PPh_3)_3RhCl$[a,b]

Run	Catalyst	mmol of catalyst	Olefin	mmol of olefins	Cyclo-alkane $(\%)$[c]
1	$(Ph_3P)_3RhCl$	0.100	3	23.6	100 (3a)
2	$(Ph_2P)_3RhCl$	0.100	2	24.1	100 (2a)
3	$(Ph_3P)_3RhCl$	0.100	1	27.8	100 (1a)
4	*M*	0.083	3	23.6	100 (3a)
5	Recycle *M*	0.083	3	23.6	100 (3a)
6	Recycle *M*	0.083	3	23.6	100 (3a)
7	*M*	0.083	2	24.1	100 (2a)
8	*M*	0.083	1	27.8	100 (1a)

[a] From Pittman *et al.* (1976).
[b] Reactions at 50°C, 29 atm, H_2 for 4 hr.
[c] The numbers in parentheses represent the following structures:

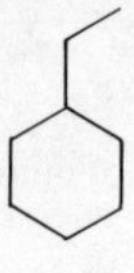

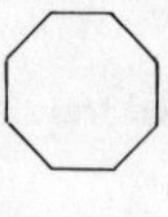

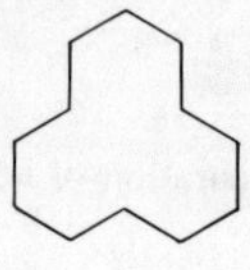

C. Reactions Using Heterogeneous Catalysts

There are some minor differences when a homogeneous catalyst is heterogenized. Here reactants must diffuse into a polymer matrix to reach an anchored catalyst site; reaction rates may thus be lowered because diffusion becomes a rate-limiting factor. Another related factor is that the concentration of reactants at the catalytic sites can be different for a heterogenized catalyst and a catalyst dissolved in solution. This is due to the differences in solvation between the bulk solution and the inside of a swollen polymer matrix.

1. Hydrogenation

A study of hydrogenation was made with olefins 1, 2, and 3, obtained from the oligomerization of butadiene (Section III,C); the catalysts were homogeneous and anchored [$(PPh_3)_3RhHCl$, M; Scheme 6.7]. The three olefins were quantitatively hydrogenated to the corresponding saturated hydrocarbons (1a to 3a, Table 6.12).

The rate of hydrogenation of 1,6-cyclooctadiene to 2a using heterogeneous catalyst was 0.8 times as fast as with $(PPh_3)_3RhCl$ catalyst under these conditions. No loss of Rh from heterogenized catalysts occurred even after repeated recycling.

2. Hydroformylations

Examples of hydroformylation reactions with anchored rhodium and cobalt catalyst are described.

a. Anchored $(PPh_3)_3RhH(CO)$ Catalyst. The title catalyst N, which was prepared according to Scheme 6.7, readily hydroformylates 1-pentene to a mixture of linear and branched-chain aldehydes (Table 6.13). This catalyst could be repeatedly recycled without drastic loss of activity, and rhodium was not lost upon recycling.

b. Anchored $Co_2(CO)_8$ Catalyst. The anchored dicobaltoctacarbonyl catalyst O was prepared as outlined in Scheme 6.7. Hydroformylation experiments with 1- and 2-pentene in the presence of this catalyst were carried out at a total pressure of approximately 68 atm at ($H_2/CO = 1/1$) for about 7 hr (Table 6.14).

At temperatures of 145–150°C, 1-pentene undergoes nearly complete conversion to a mixture of straight and branched-chain aldehydes. The ratio of linear to branched hexanals is of the order of 2 to 1, lower than that observed under somewhat different conditions in the presence of phosphine-modified cobalt catalysts (Table 6.8).

TABLE 6.13

Hydroformylation of Pentenes by Homogeneous and Anchored $(PPh_3)_3RhH(CO)$ Catalyst[a]

Run	Catalyst	Temp (°C)	Pressure (atm)	Time (hr)	Pentene consumed (%)	Total yield of aldehydes	Ratio of hexanal/2-methylpentanal
1	$(PPh_3)_3RhH(CO)$[b]	40	17	3.0	73.8	73.8	2.85
2	$(PPh_3)_3RhH(CO)$[b]	40	17	4.5	83.8	83.8	2.88
3	$(PPh_3)_3RhH(CO)$[b]	60	54	4.5	100[c]	92.9	2.69
1	N[h]	23	68	12.5	10.3	10.3	Only linear aldehyde
2[d]	N[h]	28	17	12.5	11.3	11.3	4.34
3	N[h]	40	34	12.5	68.8	68.8	4.26
4[e]	N[h]	53	68	12	56.5	56.5	4.44
5	N[h]	62	68	12.8	83.0	83.0	3.95
6[f]	N[h]	40	17	22.5	63.8	63.8	5.2
7[g]	N[h]	40	17	12.6	41.6	41.6	4.73
8[g]	N[h]	40	17	17	44.5	44.5	3.6
9[g]	N[h]	40	17	36.5	>80	>80	

[a] From Pittman et al. (1975).

[b] Concentration of catalyst was 1.8×10^{-2} mol l^{-1} and of the olefin was 1.53 mol l^{-1} in each run.

[c] About 5% pentane was found.

[d] This reaction employed catalyst that had previously been used in a 40°C, 17-hr hydroformylation.

[e] Run 4 was performed with catalyst that had previously been used fresh in run 5.

[f] Run 6 was performed with the catalyst that had just been isolated from run 1 (i.e., its third recycling).

[g] Runs 7–9 performed sequentially with the catalyst sample used in run 6. Note. It is active after six recycling procedures.

[h] Scheme 6.7.

TABLE 6.14

Hydroformylation of x-Pentenes Using Anchored Cobalt Catalyst[a,b]

Catalyst[c] (gm)	x-Pentane x (ml)	Benzene solvent (ml)	Temp (°C)	Pressure (Atm)	Products[d]			
					Aldehydes		Alcohols	
					Linear	Branched	Linear	Branched
1.0[e]	1 (4.0)	8.0	150	67	62.5	32.9	0.0	0.0
1.0[e]	1 (4.0)	8.0	147	68	65.2	29.4	0.0	0.0
1.0[e]	1 (4.0)	8.0	149	67	62.0	34.1	0.0	0.0
1.0	1 (4.0)	8.0	150	68	61.4	24.8	0.0	0.0
1.0[e]	1 (4.0)	8.0	172	68	26.8	14.9	28.1	18.4
1.0	1 (4.0)	8.0	164	68	35.1	25.8	22.0	15.5
1.0	1 (2.0)	4.0	147	27	50.2	40.2	0.0	0.0
1.0	2 (4.0)	8.0	144	66	63.8	30.8	0.0	0.0

[a] From Evans et al. (1974).
[b] Reactions carried out in a 150-ml stainless steel rocking autoclave using $H_2/CO = 1/1$. Duration of reaction was 6 to 7.5 hr.
[c] Catalyst O, Scheme 6.7.
[d] Mole percent based upon the amount of pentene used.
[e] Catalyst recycled from preceding reaction.

The anchored catalyst, which may be recovered by filtering the reaction mixture in air and washing the benzene, may be recycled in subsequent reaction with essentially no change in catalytic behavior.

The anchored catalyst O has been found to be active in the hydroformylation of 1-pentene at the relatively low pressure of 27 atm, much below that required for $Co_2(CO)_8$ and similar to results reported for $Co_2(CO)_6(PR_3)_2$ (Table 6.8).

REFERENCES

Candin, J. P., and Oldham, A. R. (1968). *Discuss. Faraday Soc.* **46**, 60.

Chem. Eng. News (1971). April 26, p. 30.

Cramer, R. (1965). *J. Am. Chem. Soc.* **87**, 4717.

* Davidson, P. J., Hignett, R. R., and Thompson, D. T. (1977). Homogeneous catalysis involving carbon monoxide. *In* "Catalysis," Specialist Periodical Reports, Vol. 1, pp. 369–411. Chem. Soc., London.

Evans, G. O. Pittman, C. U., Jr., McMillan, R., Beach, R. T., and Jones, R. (1974). *J. Organometal. Chem.* **67**, 295.

* Falbe, J. (1970). The hydroformylation reaction (oxo reaction/Roelen reaction. *In* "Carbon Monoxide in Organic Synthesis" (C. R. Adams, transl.), pp. 3–77. Springer-Verlag, Berlin and New York.

Forster, D. (1976). *J. Am. Chem. Soc.* **98**, 846.

Haag, W. O., and Whitehurst, D. D. (1968). Belgian Pat. **721**, 6868.

Heck, R. F. (1974). "Organotransition Metal Chemistry: A Mechanistic Approach," p. 218. Academic Press, New York.

* Heck, R. F., and Breslow, D. S. (1961). *J. Am. Chem. Soc.* **83**, 4023.

Hussey, A. S., and Takeuchi, Y. (1969). *J. Am. Chem. Soc.* **91**, 672.

Hussey, A. S., and Takeuchi, Y. (1970). *J. Org. Chem.* **35**, 643.

* James, B. R. (1973). Group VIII metal ions and complexes. Rhodium. *In* "Homogeneous Hydrogenation," pp. 198–287. Wiley, New York.

* Jolly, P. W., and Wilke, G. (1975a). The oligomerization of olefins and related reactions. *In* "The Organic Chemistry of Nickel," Vol. 2, pp. 1–54. Academic Press, New York.

* Jolly, P. W., and Wilke, G. (1975b). The oligomerization and cooligomerization of butadiene and substituted 1,3-dienes. *In* "The Organic Chemistry of Nickel," Vol. 2, pp. 133–340. Academic Press, New York.

Kagan, H. B., Langlois, N., and Dang, T. P. (1975). *J. Organomet. Chem.* **90**, 353.

Knowles, W. S., Sabacky, M. J., and Vineyard, B. D. (1972). *Chem. Commun.* p. 10.

Lai, R., and Lucciani, E. (1974). *Adv. Chem. Ser.* No. 132, p. 9.

Manassen, J. (1969). *Chim. Ind.* (Milan) **51**, 1058.

Moffat, A. J. (1970). *J. Catal.* **18**, 193.

* Morrison, J. D., Masler, W. F., and Neuberg, M. K. (1976). Asymmetric homogeneous hydrogenation. *Adv. Catal.* **25**, 81–124.

Ono, I., and Kihara, K. (1967). *Hydrocarbon Process.* **46**(8), 147.

Orchin, M., and Rupilius, W. (1972). On the mechanism of the oxo reaction. *Catal. Rev.* **6**, 85–131.

* Parshall, G. W. (1978). Industrial applications of homogeneous catalysis. *J. Mol. Catal.* **4**, 243–270.

Paulik, F. E., and Roth, J. F. (1968). *Chem. Commun.* p. 1578.

Pittman, C. U. Jr., Smith, L. R., and Hanes, R. M. (1975). *J. Am. Chem. Sec.* **97**, 1742.

* Pino, P., Piacenti, F., and Bianchi, M. (1977). Reactions of carbon monoxide and hydrogen with olefinic substrates. The hydroformylation (oxo) reaction. *In* "Organic Syntheses via Metal Carbonyls (I. Wender and P. Pino, eds.)," Vol. 2, pp. 44–231. Wiley, New York.

Roelen, O. (1938). German Patent 849548.

Slaugh, L. H., and Mullineaux, R. D. (1968). *J. Organomet. Chem.* **13**, 469.

* Tolman, C. A. (1972). The 16 and 18 electron rule in organometallic chemistry and homogenous catalysis. *Chem. Soc. Rev.* **1**, 337–353.

Wender, I., Metlin, S., Ergun, S., Sternberg, H. W., and Greenfield, H. (1956). *J. Am. Chem. Soc.* **78**, 5401.

* Wilke, G. (1963). Cylooligomerization of butadiene and transition metal π-complexes. *Angew. Chem., Int. Ed. Engl.* **2**, 105.

Young, J. F., Osborn, J. A., Jardine, F. H., and Wilkinson, G. (1965). *Chem. Commun.* (London) p. 131.

7

Metathesis of Unsaturated Hydrocarbons

I. INTRODUCTION

The discovery by Banks and Bailey (1964) that propene undergoes metathesis (disproportionation) to ethylene and butenes constituted a major development in catalytic conversion by hydrocarbons. This type of metathesis is catalyzed by transition metals and involves the interchange of alkylidene units between olefins.

$$
\begin{array}{ccc}
R^1CH=CHR^2 & & R^1CH \quad CHR^2 \\
+ & \rightleftharpoons & \| \quad + \| \\
R^1CH=CHR^2 & & R^1CH \quad CHR^2
\end{array}
$$

Olefin metathesis is essentially a thermoneutral catalytic rupture and re-formation of carbon–carbon double bonds and results in an equilibrium composition of reactants and products. All types of olefins may undergo this metathesis: ethylene, terminal and internal monoolefins, including those with tetrasubstituted double bonds, and aryl-substituted alkenes. Cyclic olefins, with the exception of cyclohexene, undergo metathesis, which results in the ultimate formation of macrocyclic polyenes.

II. CATALYSTS

A. Heterogeneous Catalysts

Both solid and soluble catalysts are used for this reaction. Solid catalysts are mainly transition metal oxides, carbonyls, or sulfides deposited on high surface area solid supports, such as alumina and silica. The most active catalysts contain molybdenum, tungsten, or rhenium.

The metals whose compounds provide the majority of heterogeneous catalysts are listed in Table 7.1. The metals located on the left-hand side of the second and third transition series are especially important. Cata-

264

TABLE 7.1

Promoter Compounds for Olefin Metathesis[a]

Oxides			Sulfides	Carbonyls
Mo	W	Re		Mo
V	Sn	Te	Mo	
Nb	Ta	La		W
Ru	Os	Ir	W	
Rh	Sr	Ba		Re

[a] From Rooney and Stewart (1977).

lysts produced from the oxides of molybdenum, tungsten, and rhenium are the most active.

A typical heterogeneous catalyst is prepared by impregnating alumina or silica support with an aqueous solution of ammonium paramolybdate or ammonium tungstate, drying, and calcining the product at 500°–600°C for 5 hr in the presence of a stream of dry air or nitrogen.

In general, the promoter oxide content of the catalyst is 1–15% by weight. The extent of side reaction, which lowers selectivity, can be reduced by treating the catalyst with metal ions. Double-bond migration is diminished by the addition of alkali metal ions. Coke formation is reduced by the incorporation of cobalt oxide. Highly efficient disproportionation of propene at 205°C was achieved with a catalyst containing by weight 3.4% Co, 11.0% MoO_3, and 85.6% Al_2O_3.

Metal oxide supported catalysts become deactivated through use. They can be regenerated by the use of a controlled amount of air to burn off the accumulated coke. The deposition of coke can be greatly reduced if the metathesis of propene is carried out in the liquid phase with hydrocarbon solvents.

The temperature at which metathesis is carried out depends on the promoter metal oxide used. A catalyst composed of 1–20% Re_2O_7/Al_2O_3 converts 1-butene to ethylene and 3-hexene with a selectivity of 95% at a temperature of 25°C. The same catalyst is suitable for cross-metathesis reactions carried out at about 100°C and atmospheric pressure. From ethylene and 4-methyl-2-pentene, 95% selectivity of propene and 3-methyl-1-butene is obtained.

$$CH_2{=}CH_2 + CH_3CH{=}C(CH_3)_2 \rightleftharpoons CH_2{=}CHCH_3 + CH_2{=}C(CH_3)_2$$

The activity of the catalyst can be enhanced by adding to the alkenes 1–4 mol % of a chelating olefin such as 1,5-hexadiene or dicyclopentadiene.

TABLE 7.2

Catalyst Components for Homogeneous Metathesis

Transition metal derivatives	Organometallic compounds or Lewis acids	Modifiers
WCl_6, WCl_5, WCl_4	RLi, RMgCl, R_2Mg	O_2, H_2O, ROH
$WOCl_4$, WBr_5, WF_6	R_3Al, R_2AlCl, $R_3Al_2Cl_3$	RCOOH, RCOOR, RCOR
$W(CO)_6$, $W(\pi\text{-allyl})_4$	$RAlCl_2$, $AlCl_3$, R_2Zn	RSH, RNH_2
$WCl_2(Py)_2(NO)_2$, $MoCl_5$	R_4Sn, R_2Hg	
$MoCl_2(Py)_2(NO)_2$		

[a] From Calderon (1977).

B. Homogeneous Catalysts

Homogeneous catalysts are composed either of salts of molybdenum or tungsten or their respective coordination compounds together with a co-catalyst which is an organometallic compound. Certain organic compounds such as alcohols, acids, and amines added in controlled amounts to the system, in addition to oxygen, improve their activity and selectivity. Table 7.2 lists the three classes of components that have been used to prepare homogeneous metathesis catalysts.

One of the procedures for the preparation of a homogeneous catalyst consists of treating a dilute solution of WCl_6 (0.05 M) in an aromatic solvent, benzene or chlorobenzene, with an equimolar amount of absolute ethanol under nitrogen atmosphere. The reaction is accompanied by the evolution of HCl. The moderately stable $WCl_5OC_2H_5$ thus formed decomposes to C_2H_5Cl and an orange precipitate, presumably $WOCl_4$. The organoaluminum component in a hydrocarbon solvent is handled separately. It is best to add the two catalyst components to the reactant mixture consecutively. Examples of typical catalyst systems for the homogeneous metathesis of 1- and 2-pentenes are given in Table 7.3.

III. MECHANISTIC ASPECTS AND REACTIONS

The metathesis of alkenes was originally assumed to proceed via quasicyclobutane intermediates, and this was compatible with the results obtained with ^{14}C-labeled alkenes. The metathesis of $[2\text{-}^{14}C]$propene in the presence of $MoO_3/CoO/Al_2O_3$ afforded ethylene with no radioactivity, whereas 2-butene showed a specific radioactivity twice as high as that of the starting alkene:

TABLE 7.3

Typical Catalyst Systems for Homogeneous Metathesis of Pentene[a]

Catalyst system	Molar ratio of components	Pentene	Alkene/ catalyst (mol/mol)	Temp (°C)	Reaction time	Solvent	Conversion[c] (%)	Selectivity[c] (%)
$WCl_6/C_2H_5AlCl_2/C_2H_5OH$	1:4:1	2-Pentene	10,000:1	Ambient	1–3 min	Benzene	49.9	99.6
$WCl_6/(C_2H_5)_3Al$	2:1	2-Pentene	270:1	−30	30 min	Chlorobenzene	51	96
$WCl_6/(n\text{-}C_4H_9)Li$	1:2	2-Pentene	50:1	Ambient	4 hr	Benzene	50	100
$WCl_6/(n\text{-}C_4H_9)Li/AlCl_3$	2:4:1	2-Pentene	50:1	Ambient	15 min	Benzene	48	93
$WCl_6/LiAlH_4$	1:1	2-Pentene	100:1	Ambient	15 min	Chlorobenzene	50	
$WCl_6/(n\text{-}C_3H_7)MgBr$	1:2	2-Pentene	50:1	Ambient	12 hr	Ether + benzene	31	
$W(C_5H_5N)_2Cl_4/C_2H_5AlCl_2/CO$	1:8	2-Pentene	400:1		5 min	Pentane + benzene	47	
$MoCl_2(NO)_2P[(C_6H_5)_3]_2/ (CH_3)_3Al_2Cl_3$	1:2	1-Pentene	200:1	0	50 min	Chlorobenzene	24[a]	95
$MoCl_5/(C_2H_5)_3Al/O_2$	1:2	2-Pentene	90:1	Ambient	2 hr	Chlorobenzene	14	71
$Na_2[(CO)_5Mo/Mo(CO)_5]/ N(n\text{-}C_4H_9)_4Cl/CH_3AlCl_2$	1:1:20	1-Pentene	350:1	Ambient	2 hr	Chlorobenzene	61[a]	
$ReCl_5/(n\text{-}C_4H_9)_4Sn$	2:3	2-Pentene	50:1	Ambient	24 hr	Chlorobenzene	41	84
$ReCl_5/(C_2H_5)_3Al/O_2$	1:4	2-Pentene	90:1	Ambient	2 hr	Chlorobenzene	49	100

[a] From Mol and Moulijn (1975).

[b] Equilibrium conversion about 50%.

[c] (Moles primary products/moles reactant consumed) × 100%.

[d] Conversion to octene; because ethene escaped from the system, the calculated equilibrium conversion may be exceeded.

$$CH_2={}^{14}CHCH_3 \rightleftharpoons \begin{bmatrix} CH_2 \cdots {}^{14}CHCH_3 \\ \vdots \qquad \vdots \\ CH_2 \cdots {}^{14}CHCH_3 \end{bmatrix} \rightleftharpoons \begin{matrix} CH_2 \\ \| \\ CH_2 \end{matrix} + \begin{matrix} {}^{14}CHCH_3 \\ \| \\ {}^{14}CHCH_3 \end{matrix}$$

Also, experiments with deuteriated alkenes showed that during the reaction no hydrogen exchange occurred between the alkylidene moieties. Thus, asymmetric ethane-1,1-d_2, $C_2H_2D_2$, was obtained from a mixture of C_2H_4 and C_2D_4 in the presence of Re_2O_7/Al_2O_3 catalyst:

$$\begin{matrix} CH_2=CH_2 \\ \\ CD_2=CD_2 \end{matrix} \rightleftharpoons \begin{bmatrix} CH_2 \cdots CH_2 \\ \vdots \qquad \vdots \\ CD_2 \cdots CD_2 \end{bmatrix} \rightleftharpoons \begin{matrix} CH_2 \\ \| \\ CD_2 \end{matrix} + \begin{matrix} CH_2 \\ \| \\ CD_2 \end{matrix}$$

The same kind of isotopic behavior takes place in homogeneous systems in which $WCl_6 \cdot EtAlCl_2 \cdot EtOH$ is the catalyst. Thus, from 2-butene and 2-butene-d_8 the only product was 2-butene-d_4.

It was realized that quasi-cyclobutane species could not be accepted as intermediates in the metathesis reaction. If they were, ethylene and cyclobutane would equilibrate in the presence of a catalyst for metathesis. A major advance in interpreting this reaction was the suggestion of a nonconcerted pairwise exchange of alkylidenes through the formation of metallocyclopentane. This proposal was based on the observation that tung-

$$\begin{matrix} CH_2 \\ \| \\ CHR \end{matrix} \rightarrow M \leftarrow \begin{matrix} CH_2 \\ \| \\ CHR \end{matrix} \rightleftharpoons \begin{matrix} H_2C-\!\!-CH_2 \\ | \qquad | \\ RHC \diagdown_{M} \diagup CHR \end{matrix} \rightleftharpoons \begin{matrix} H_2C-\!\!-CHR \\ | \qquad | \\ H_2C \diagdown_{M} \diagup CHR \end{matrix}$$

$$\Updownarrow$$

$$\begin{matrix} CHR \\ \| \\ CHR \end{matrix} \rightarrow M \leftarrow \begin{matrix} CH_2 \\ \| \\ CH_2 \end{matrix}$$

sten hexachloride reacts with 1,4-dilithiobutane to afford in benzene solution a quantitative yield of ethylene.

The presently accepted mechanism is that originally proposed in 1971 by Hérisson and Chauvin and involves the participation of alkylidene metals, metallocarbenes, which react with olefins to give intermediate metallocyclobutanes.

$$CHR^1=M-\!\!\begin{matrix} CHR^2 \\ \| \\ CHR^3 \end{matrix} \rightleftharpoons \begin{matrix} R^1HC-CHR^2 \\ | \qquad | \\ M-CHR^3 \end{matrix} \rightleftharpoons \begin{matrix} R^1HC=CHR^2 \\ \\ M=CHR^3 \end{matrix}$$

In support of the metallocyclobutane mechanism are the synthesis and reactivity of diphenylmethylenetungsten pentacarbonyl (**1**) with a variety of olefins. Upon heating of **1** with isobutylene at 100°C for 2.5 hr the following products were obtained:

$$\begin{matrix} H_5C_6 \\ \diagdown \\ H_5C_6 \diagup \end{matrix}C=W(CO)_5 + CH_2=\!\!\begin{matrix} C-CH_3 \\ | \\ CH_3 \end{matrix} \longrightarrow \begin{matrix} & CH_2 & \\ H_5C_6 \diagdown & \diagup\diagdown & \diagup CH_3 \\ & C-\!\!-C & \\ H_5C_6 \diagup & & \diagdown CH_3 \end{matrix} + CH_2=\!\!\begin{matrix} C \diagup^{C_6H_5} \\ \diagdown_{C_6H_5} \end{matrix} + W(CO)_6$$

1

Treatment of α-methoxystyrene with **1** at 32°C for 6 hr afforded metallocarbene **2** and 1,1-diphenylethylene as the main products.

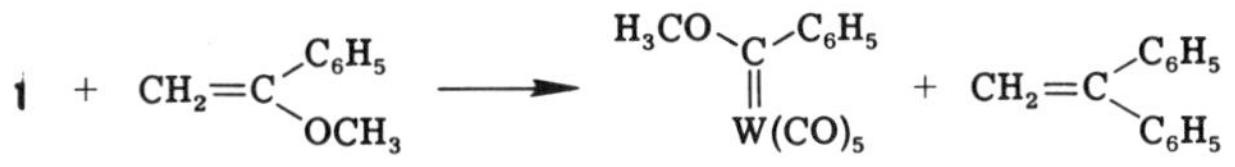

$$\mathbf{1} + CH_2=C\begin{smallmatrix}C_6H_5\\\\OCH_3\end{smallmatrix} \longrightarrow \begin{smallmatrix}H_3CO\\C\\\|\\W(CO)_5\end{smallmatrix}C_6H_5 + CH_2=C\begin{smallmatrix}C_6H_5\\\\C_6H_5\end{smallmatrix}$$

2

The mechanism proposed for this reaction, which accommodates the alkylidene transfer of the substrate to the diphenylcarbene and the cyclopropanation reaction, is presented in Scheme 7.1.

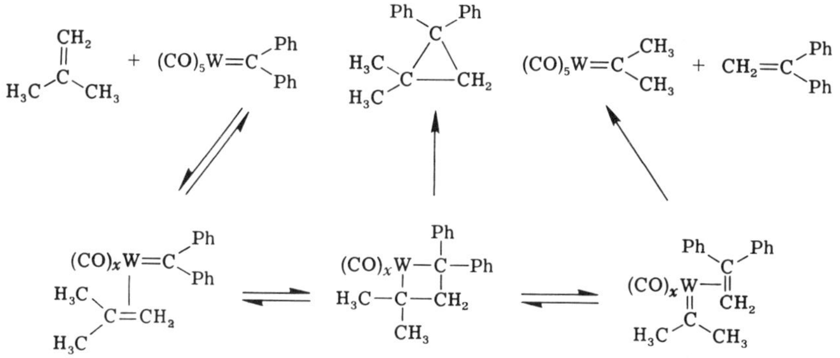

Ph = C_6H_5

Scheme 7.1. Mechanism of metathesis (from Casey and Burkhardt, 1974).

Cyclopropanes were formed from ethylene and metallocarbenes on heterogeneous metathesis catalysts. The presence of a metallocarbene was demonstrated during the reaction of ethyl acrylate (a carbene scavenger) in a metathesis system.

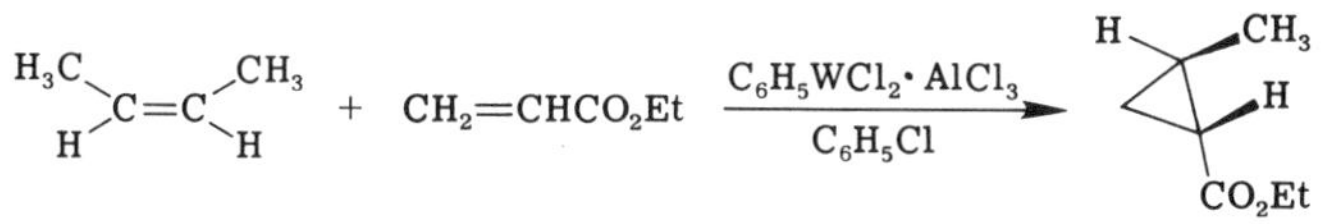

$$\begin{smallmatrix}H_3C\\\\H\end{smallmatrix}C=C\begin{smallmatrix}CH_3\\\\H\end{smallmatrix} + CH_2=CHCO_2Et \xrightarrow[C_6H_5Cl]{C_6H_5WCl_2\cdot AlCl_3}$$

The metallocarbene catalysts are generated by interaction of the organometallic cocatalysts with the transition metal derivatives (Table 7.2), according to the following scheme:

$$WCl_6 + (CH_3)_2Zn \xrightarrow{-ZnCl_2} Cl_4W\begin{smallmatrix}CH_3\\\\CH_3\end{smallmatrix} \rightleftharpoons Cl_4\overset{\underset{CH_3}{|}}{\overset{H}{W}}=CH_2 \longrightarrow Cl_4W=CH_2 + CH_4$$

The formation of metallocarbene species in systems that do not employ organometallics can be explained by a hydrogen transfer process between Mo and W complexes and olefins. The reaction steps involve π complexation, followed by a hydride shift producing a π-allylmetal hydride entity, which then rearranges to a metallocyclobutane through an attack of the hydride on the central atom of the π-allyl species.

$$
M \longleftarrow \begin{matrix} CHR \\ \| \\ CH \\ | \\ CH_2R \end{matrix} \quad \rightleftharpoons \quad HM \longleftarrow \begin{matrix} RHC \\ \diagdown \\ CH \\ \diagup \\ RHC \end{matrix} \quad \longrightarrow \quad M \begin{matrix} CHR \\ \diagdown \\ \diagup \\ CHR \end{matrix} CH_2
$$

$$
M{=}CHR \quad + \quad CH_2{=}CHR
$$

A. Alkenes

Examples of solid catalysts for the metathesis of propene are given in Table 7.4. The selectivity for the production of ethylene and butenes ranges between 82 and 100%, with conversion of propene between 18 and 44.8%.

The effect of substitution on the ease of participation of alkenes in metathesis is governed by steric factors: $CH_2{=} > RCH_2CH{=} > R_2CHCH{=} > R_2C{=}$. The length of the alkyl chain in olefins of the type $RCH{=}CH_2$ does not appear to be critical. The metathesis of alkenes substituted with bulky cycloalkyl, cycloalkenyl, or aryl groups occurs.

The metathesis of 1-alkenes having three to six carbon atoms over $Mo(CO)_6/Al_2O_3$ catalyst is presented in Table 7.5. Tetrasubstituted al-

TABLE 7.4

Examples of Solid Catalysts for the Metathesis of Propene[a]

Catalyst system	Temperature (°C)	Weight hourly space velocity (hr⁻¹)	Conversion[b] (%)	Selectivity[c] (%)	Ethene/butenes molar ratio[d]
$MoO_3/CoO/Al_2O_3$	163	8.5	42.9	94	0.93
$MoO_3/Cr_2O_3/Al_2O_3$	160	180[e]	36	97	0.87
MoO_3/SiO_2	538	3.5	28	95	1.26
$Mo(CO)_6/Al_2O_3$	71	1–2	25	97	0.76
WO_3/SiO_2	427	40	44.8	99	1.13
$WO_3/AlPO_4$	538	7.5	34	82	0.99
WS_2/SiO_2	538	2	18.3	100	1.25
Re_2O_7/Al_2O_3	25	6	30.2	100	0.98
$Re_2(CO)_{10}/Al_2O_3$	100	1600[e]	20.4	100	1.00

[a] From Mol and Moulijn (1975).
[b] Equilibrium conversion 42.3% at 25°C, 47.8% at 538°C.
[c] (Moles ethylene and butenes/moles propene consumed) × 100%.
[d] Theoretical: 1.00.
[e] Gas hourly space velocity.

TABLE 7.5

Metathesis of Various Olefins over Molybdenum Hexacarbonyl–Alumina Catalyst[a,b]

	Propene	1-Butene	1-Pentene	1-Hexene
Conversion (%)	25	10	60	54
Product distribution (mol %)				
Ethylene	42	8	2	5
Propene	Feed	34	21	13
Butenes	55	Feed	27	12
Pentenes	2	18	Feed	15
Hexenes	1	32	27	Feed
C_7 +		8	23	55
Butene distribution (%)				
1-Butene		84	33	52
2-Butene	100	16	67	48
Pentene distribution (%)				
1-Pentene		2		52
2-Pentene		98		40
Methylbutenes				8

[a] Banks and Bailey, 1964.
[b] Conditions: temp, 120°C; pressure, 32 atm; LHSV, 2.0.

kenes react with other alkenes, as shown for tetramethylethylene and ethylene; however, single tetrasubstituted alkenes such as $Me_2C=CEt_2$ do not give metathesis products.

$$(CH_3)_2C=C(CH_3)_2 \ + \ CH_2=CH_2 \ \rightleftharpoons \ 2 \ \substack{H_3C \\ H_3C} \!\!> \!\! C=CH_2$$

The heterometathesis of an internal olefin with a terminal olefin is much faster than the symmetric homometathesis of an internal olefin. Terminal olefins undergo rapid regenerative metathesis. If 1-pentene metathesizes with 1-pentent-d_{10}, the products formed contain 1-pentenes having two and eight deuteriums per molecule.

$$
\begin{array}{ccc}
CH_3CH_2CH_2CH=CH_2 & & CH_3CH_2CH_2CH \\
+ & \rightleftharpoons & \quad || \quad + CH_2CD_2 \\
CD_3CD_2CD_2CD=CD_2 & & CD_3CD_2CD_2CD
\end{array}
\rightleftharpoons
\begin{array}{c}
CH_3CH_2CH_2CH=CD_2 \\
+ \\
CD_3CD_2CD_2CD=CH_2
\end{array}
$$

Although the synthetic potential of alkene metathesis depends on the capacity to change the length of the carbon skeleton of hydrocarbons, this process appears not to be utilized commercially. In organic synthesis, olefin metathesis can provide an elegant and often cheaper route to certain compounds. To prepare cis-9-tricosene (**3**), the pheromone of the house-

$$
\begin{array}{ccc}
CH_3(CH_2)_7CH=CH_2 & & CH_3(CH_2)_7CH \quad CH_3(CH_2)_7CH \quad CH_3(CH_2)_{11}CH \quad CH_2 \\
+ & \rightleftharpoons & \quad || \qquad\quad + \qquad || \qquad\quad + \qquad || \qquad + \quad || \\
CH_3(CH_2)_{11}CH=CH_2 & & CH_3(CH_2)_{11}CH \quad CH_3(CH_2)_7CH \quad CH_3(CH_2)_{11}CH \quad CH_2
\end{array}
$$

3

fly, 1-decene and 1-tetradecene were subjected to a metathesis reaction. The cis and trans isomers of **3** were isolated in 26% yield, and separation gave pure cis product.

B. Cycloalkenes

The metathesis of cycloalkenes, with the exception of cyclohexene, is a general reaction occurring in homogeneous systems with transition metal complexes. Whereas marked stereospecificity in the metathesis of alkenes is rare, it is not uncommon in that of cycloalkenes. The microstructure of polypentenamers obtained via the metathesis of cyclopentene depends on the catalyst used (Fig. 7.1).

From the accepted mechanism of the metathesis reaction it stands to reason that once a cycloalkene monomer is incorporated into a high-molecular-weight chain, its vinylene moiety is no longer cyclic. Despite its linearity the pentamer repeat units maintain their stereospecificity when $WF_6 + C_2H_5AlCl_2$ is used as catalyst.

1. Cycloalkenes and Alkenes (Telomerization)

Alkenes act as chain scission agents in cycloalkene polymerization, and therefore they are used as molecular weight regulators. The metathesis between alkenes and cycloalkenes leads to copolymerization (telomerization). The results of the reaction of 2-pentene and cyclopentene in the presence of tungsten-based catalysts, $WOCl_4/Sn(n\text{-}C_4H_9)_4$, are given in Table 7.6, and the mechanism of telomerization is presented in Scheme 7.2.

The high-molecular-weight polymers (telomers) formed by the application of the metathesis reaction to cycloalkenes and alkenes display elasto-

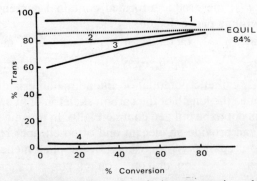

Fig. 7.1. Effect of catalyst and conversion of microstructure in cyclopentene polymerizations (from Calderon *et al.*, 1979). Key: 1, $WCl_6 + NCC_2H_4OH + C_2H_5AlCl_2$; 2, $WCl_6 + ClC_2H_4OH + C_2H_5AlCl_2$; 3, $WCl_6 + EtOH + C_2H_5AlCl_2$; 4, $WCl_6 + C_2H_5AlCl_2$.

TABLE 7.6

Product Distribution Obtained by the Metathesis of an Equimolar Mixture of Cyclopentene (C_5H_8) and 2-Pentene (C_5H_{10}) in the Presence of $WOCl_4/Sn(n-C_4H_9)_4$[a,b]

Reaction time (min)	C_5H_8	C_4H_8	C_5H_{10}	C_6H_{12}	C_9H_{16}	$C_{10}H_{18}$	$C_{11}H_{20}$	$C_{14}H_{24}$	$C_{15}H_{26}$	$C_{16}H_{28}$	$C_{19}H_{32}$	$C_{20}H_{34}$	$C_{21}H_{36}$	$C_{24}H_{40}$	$C_{25}H_{42}$	$C_{28}H_{44}$
0	58		58													
7	20	5.8	36	5.5	2.8	5.9	2.7	1.25	2.7	1.25	0.6	1.75	0.6			
						11.4			5.2			2.95				
14	19	10.4	17.4	9.8	4.5	10.1	4.6	2.0	3.8	1.9	0.8	1.7	0.8	0.4	0.9	0.4
						19.2			7.7			3.3			1.7	

[a] From Mol and Moulijn (1975).
[b] The numbers represent relative molar concentrations.

meric and plastic-like properties and have an application in the rubber industry.

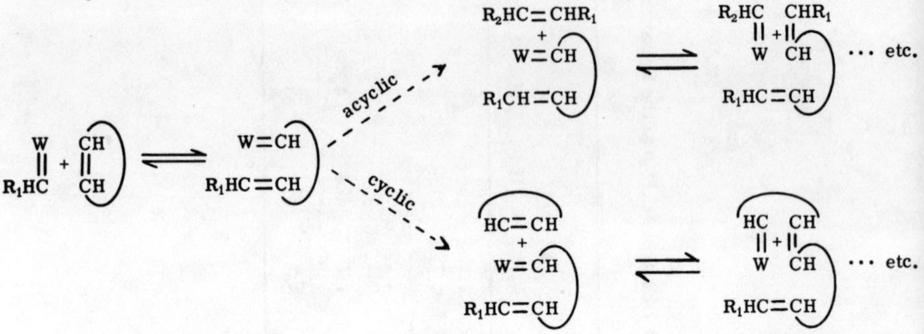

Scheme 7.2. Mechanism for the mutual metathesis of cyclic and acyclic alkenes (from Mol and Moulijn, 1975).

2. Large-Ring Cycloalkenes (Catenanes)

In the case of large-ring cycloalkenes, interlocked systems (catenanes) are reported to be formed. The catenanes are obtained by subjecting cyclododecene to metathesis in octane at ambient temperature using a $WCl_6/EtAlCl_2/EtOH$ catalyst in pentane. The formation of catenanes, as evidenced by mass spectroscopic analysis, was accounted for by assuming an intramolecular transalkylidenation of a twisted "strip" (Fig. 7.2). The presently accepted mechanism for metathesis reactions cannot be used to explain the formation of catenanes.

C. Alkynes

Alkynes undergo metathesis in heterogeneous systems. The reaction is analogous to the metathesis of alkenes inasmuch as transalkylidynation occurs.

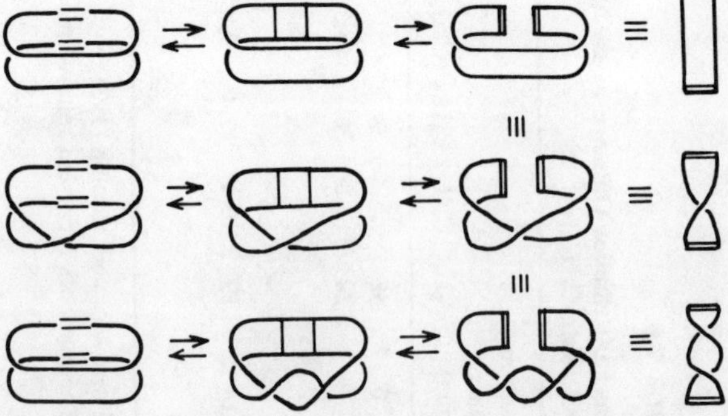

Fig. 7.2. Catenane formation (from Wolovsky, 1970).

$$H_3{}^{14}CC \equiv CCH_2CH_3$$
$$+$$
$$H_3{}^{14}CC \equiv CCH_2CH_3$$
$$\rightleftharpoons$$
$$H_3{}^{14}CC \equiv C^{14}CH_3$$
$$+$$
$$H_3CCH_2C \equiv CCH_2CH_3$$

1-Alkynes, as propyne, in the presence of WO_3/SiO_2 catalyst yield, in addition to metathesis products, a mixture of 1,2,4- and 1,3,5-trialkylbenzenes.

REFERENCES

* Banks, R. L., and Bailey, G. C. (1964). *Ind. Eng. Chem., Prod. Res. Dev.* **3**, 170.

* Calderon, N. (1977). The olefin metathesis reaction. *In* "Chemistry of Double Bonded Functional Groups," Supplement A (S. Patai, ed.), Part 2, pp. 913–964. Wiley, New York.

Calderon, N., Lawrence, J. P., and Ofstead, E. A. (1979). Olefin metathesis. *Adv. Organomet. Chem.* **17**, 449–492.

Casey, C. P., and Burkhardt, T. J. (1974). *J. Am. Chem. Soc.* **96**, 7808.

Hérrison, J. L., and Chauvin, Y. (1971). *Makrom. Chem.* **141**, 161.

* Mol, J. C., and Moulijn, A. (1975). The metathesis of unsaturated hydrocarbons catalyzed by transition metal compounds. *Adv. Catal.* **24**, 131–171.

* Rooney, J. J., and Stewart, A. (1977). Olefin metathesis. *In* Catalysis," Specialist Periodical Reports, Vol. 1, pp. 277–334. Chem. Soc., London.

Wolovsky, R. (1970). *J. Am. Chem. Soc.* **92**, 2132.

8

Synthesis of Liquid Hydrocarbons (Synthetic Fuels)

I. FROM CARBON MONOXIDE AND HYDROGEN

A. Introduction

In 1923 Franz Fischer and Hans Tropsch discovered that carbon monoxide and hydrogen react in the presence of iron, nickel, or cobalt catalysts to produce higher gaseous, liquid, and solid aliphatic hydrocarbons. The new reaction, known as the Fischer–Tropsch (F–T) or synthesis reaction, rapidly became the object of intensive studies, since in the late 1920's there was much concern, as there is today, with a possible shortage of natural petroleum sources.

The first measurable amounts of liquid product were obtained by recycling CO and H_2 over alkalized iron filings at 410°C and 115 atm. The products consisted entirely of oxygen-containing compounds. However, when the operation was carried out at 7 atm of pressure, hydrocarbons appeared in increasing but still small amounts.

Of the three metal catalysts mentioned above, cobalt has the greatest tendency to produce hydrocarbons with more than one carbon atom per molecule. Iron is less active. Nickel, especially at higher temperatures, shows high activity toward the formation of methane.

B. Catalysts

1. Cobalt Catalysts

A catalyst consisting of $CO/ThO_2/kieselguhr$ (100:18:100) became the so-called standard cobalt catalyst. It was prepared by precipitating a solution of cobalt nitrate and thorium nitrate with a solution of sodium carbonate containing dispersed kieselguhr. The cake thus obtained was washed until free of alkali, dried, and reduced at 365°C by a stream of hydrogen.

The average composition of the products obtained at atmospheric pressure with cobalt catalysts is given in Table 8.1. The composition varies with many factors. An increase in the age of the catalyst and in the temperature of reaction to above 175°C results in products of lower molecular weight. An increase in the amount of carbon monoxide and a decrease in the amount of hydrogen causes an increase in the percentage of olefins in the product.

The liquid products boiling within the C_5 to C_9 range obtained from a CO/H_2 ratio of 1:2 at atmospheric pressure, gaseous hourly space velocity (GHSV) of 100, and about 190°C consisted of 71.7 vol % paraffins, 19.2% olefins, and 9.1% residual (oxygenated material, etc.). The C_5 to C_8 paraffins (Table 8.2) consisted of about 85–95% of unbranched hydrocarbons; the remainder were composed almost exclusively of methylalkanes. The content of the branched hydrocarbons increased with molecular weight of the respective fractions.

The optimal pressure for cobalt-catalyzed synthesis with a CO/H_2 ratio of 1:2 is 5–20 atm at 175°–195°C. At pressures of 50 atm and higher there is a rapid deactivation of the catalyst (Table 8.3).

The catalyst was operated for a brief period at atmospheric pressure and 185°–190°C; then the temperature was decreased below the synthesis range, the pressure was increased to the desired value, and the temperature was increased to give the maximal yield of liquid and solid hydrocarbons. The temperature ranged from 175° to 185°C at 5 atm, 180° to 195°C at 15 atm, and 185° to 200°C at 150 atm. The rapid decline in activity of the catalyst at pressures of 50–150 atm was attributed to the removal of cobalt by the formation of cobalt carbonyl.

It is possible to increase the olefin content of the hydrocarbons by in-

TABLE 8.1

Reaction Products Obtained with Cobalt Catalysts[a]

Product	Boiling range (°C)	Amount (wt %)	Olefins (vol %)
Gasol (C_3 + C_4)	<30	4	50
Gasoline	30–200	62	30
Oil	>200	23	10
Paraffin in the oil	—	7	—
Paraffin obtained by extraction of the catalyst	—	4	—

[a] From Pichler (1952, p. 280).

TABLE 8.2

Composition of C_5 to C_8 Alkanes
from Co/ThO$_2$/kieselguhr-
Catalyzed Fischer–Tropsch
Reaction[a]

Component	Vol %
C_5	
Pentane	94.9
Isopentane	5.1
C_6	
Hexane	89.6
2-Methylpentane	5.8
3-Methylpentane	4.6
C_7	
Heptane	87.7
2-Methylhexane	4.6
3-Methylhexane	7.7
C_8	
Octane	84.5
2-Methylheptane	3.9
3-Methylheptane	7.2
4-Methylheptane	4.4

[a] From Friedel and Anderson
(1950).

TABLE 8.3

Products of Cobalt-Catalyzed Synthesis at Different Pressures[a,b]

Pressure (atm gauge)	Total hydrocarbons	C_1–C_4	Total	Gasoline, <200°C	Diesel oil, >200°C	Wax
			Hydrocarbon products (gm/m³)			
				Liquid plus solids		
0	155	38	117	69	38	10
1.5	181	50	134	73	43	15
5	183	33	150	39	51	60
15	178	33	145	39	36	70
50	159	21	138	47	37	54
150	144	31	104	43	34	27

[a] From Anderson (1956a, p. 100).
[b] Average of 4-week tests with standard Co/ThO$_2$/kieselguhr catalysts.

creasing the carbon monoxide concentration of the synthesis gas. At 7 atm and an H_2/CO ratio of 2:1, the olefin content of the gasoline is 20%, and with an H_2/CO ratio of 1:2 it is almost 70%.

2. Nickel Catalysts

Nickel catalysts with a behavior similar to that of the best cobalt catalysts consist of precipitated nickel–thoria on kieselguhr. A catalyst consisting of Ni/ThO$_2$/kieselguhr (100:18:100) was the best of this type and was obtained by treating an aqueous solution of metal nitrates with a solution of potassium carbonate as described for the corresponding cobalt-containing catalyst (Section I,B,1). The optimal reduction temperature was obtained with a flow of dry hydrogen at 450°C.

3. Iron Catalysts

The synthesis reaction using iron catalysts is more complicated than that using either cobalt or nickel catalysts. With cobalt and nickel catalysts there is little tendency for oxidation and carbonization of the reduced metal, whereas these reactions occur to various degrees with iron catalysts and may lead to less activity and/or disintegration of the catalyst. However, the availability and low cost of iron led to extensive research on iron catalysts.

An effective iron precipitation catalyst of the composition 75Fe(II)/25Fe(III)/20Cu/1K$_2$CO$_3$ was thus developed. The catalyst was prepared by precipitating at 70°C a solution containing the appropriate amounts of FeCl$_2$·4H$_2$O, FeCl$_3$·6H$_2$O, and CuCl$_2$·2H$_2$O with a solution containing Na$_2$CO$_3$. The precipitate was washed free of alkali and then alkalized with K$_2$CO$_3$ solution. The mixture was evaporated, dried at 105°C, and granulated. The catalyst was pretreated with synthesis gas, 2H$_2$/1CO, at 250°C. After 48 hr the paraffins that had collected on the catalyst were extracted, and pretreatment was continued for another 24 hr. After a second extraction, the catalyst was ready for use. The results obtained with this catalyst are summarized in Table 8.4.

A small-scale commercial plant, 5000 barrels/day, has been in operation for over 20 years at SASOL in South Africa, mainly to produce fuel from CO and H$_2$. Two types of reactors are used. One is a fixed-bed reactor containing iron precipitation catalyst which is operated at 200°–240°C and 25 atm and has a charge of H_2/CO = 1.7:1. The other reactor, containing powdered catalyst swept along by the gas stream (fluid bed), is operated at higher temperature, 320°–340°C, and with a H_2/CO ratio of 3:1 (Table 8.5). The higher temperatures of the powdered fused iron catalyst containing a small amount of potassium favor the formation of olefins and lower molecular weight products. A higher ratio of hydrogen to car-

TABLE 8.4

Experiments with Active Iron Precipitation Catalysts[a]

	Synthesis pressure	
	1 atm	10 atm
Temperature, °C	211	200
Space velocity	50	100
Yield, gm/m³ (1 state, no recycle)	135	137
Product distribution, wt %		
CH₄	2.4	3.2
C₂–C₄	10.1	9.0
Liquid products (below 180°C)	13.5	10.0
Liquid products and paraffins (above 180°C)	74.0	77.8
Extraction	Every 2–3 days	No extraction necessary

[a] From Pichler (1952, p. 287).

TABLE 8.5

Product Distribution from Fixed-Bed and Fluid-Bed Synthesis on Iron Catalysts[a]

	Process I			
	Fixed-bed		Fluid-bed	
Temperature (°C)	220–240		320–340	
Pressure (atm)	26		22	
H₂/CO ratio in feed gas	1.7:1		3:1	
	Total (wt %)	Olefins (wt %)	Total (wt %)	Olefins (wt %)
Primary products				
C₁	7.8	—	13.1	—
C₂	3.2	23	10.2	43
C₃	6.1	64	16.2	79
C₄	4.9	51	13.2	76
C₅–C₁₁	24.8	50	33.4	70
C₁₂–C₂₀	14.7	40	5.1	60
>C₂₀	36.2	~15	—	—
Alcohols, ketones	2.3	—	7.8	—
Acids	—	—	1.0	—

[a] Frohning and Cornils (1974).

bon monoxide, $H_2/CO = 6:1$, improves the yield of liquid hydrocarbons. A new commercial SASOL plant is scheduled to produce 50,000 barrels/stream day of products from carbon monoxide and hydrogen.

C. Mechanism

Various mechanisms have been proposed to explain the conversion of carbon monoxide and hydrogen to higher hydrocarbons. The "carbide" theory, the first of the theories proposed, assumed that carbon monoxide was chemisorbed on the metal and hydrogenated to produce chemisorbed carbon (carbide) and water; the carbide was then hydrogenated to methylene groups. Chain growth was accomplished by the attachment of adjacent methylene groups to each other, with the subsequent hydrogenolysis of the remaining carbon–metal bonds. The shortcoming of the carbide theory is that it does not explain the formation of oxygen-containing products that accompanies the production of hydrocarbons.

The participation of oxygenated intermediates in the synthesis of hydrocarbons was subsequently proposed. It was assumed that the adsorbed carbon monoxide is hydrogenated to hydroxylated species, and the latter are linked by condensation with the removal of water and then hydrogenolyzed away from the metal at one of the carbon atoms, thus allowing for chain growth. Further hydrogenolysis leads to alcohols and hydrocarbons (Scheme 8.1). The proposed mechanism made an attempt to explain the fact that side chains longer than methyl are not usually observed in the synthesis reaction. Random methyl branching was also explained by assuming that a single surface site is involved in a hydrogen transfer reaction by which chain branching can occur (LeRoux, 1972).

$$
\begin{array}{ccc}
\underset{M}{\overset{R}{\underset{|}{\overset{|}{CH_2}}}-CH_2} & \longrightarrow & \underset{M}{\overset{R}{\underset{\diagup}{H_2C}}-CH_2} & \longrightarrow & \underset{M}{\overset{R\ H\ CH_3}{\underset{|}{C}}}
\end{array}
$$

More recently a modified carbide theory was revived on the basis of new experimental data with ^{14}C-labeled compounds, spectroscopic studies, and the careful analysis of products. It was also observed that the insertion of alkenes in the growing chains occurs. The product obtained from the addition of $^{14}CH_2{=}CH_2$ to the synthesis gas in the presence of cobalt catalyst showed considerable radioactivity, and the molar radioactivity of the product was found to increase with the chain length. The incorporation of propene also resulted in the production of methyl branching. Incorporation of higher olefins including α-alkenes does not occur frequently. It was also observed that the terminal olefinic carbon in α-alkenes can be split to give methane. Radioactive methane was also found

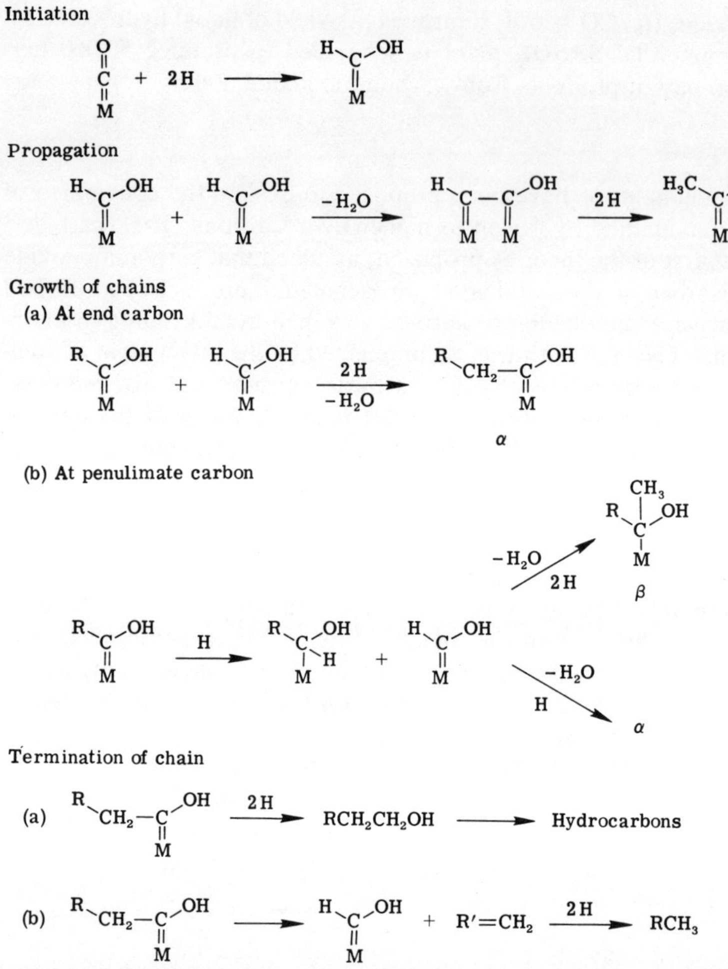
Initiation
Propagation
etc.
Growth of chains
(a) At end carbon
(b) At penulimate carbon
Termination of chain
Hydrocarbons

Scheme 8.1. Mechanism involving oxygenated intermediates (M = metal) (from Storch *et al.*, 1951).

in experiments in which the central carbon atom in propene was labeled with radioactive carbon.

The addition of [1-^{14}C]hexadecene to synthesis gas showed that product with fewer than 16 carbon atoms contained radioactivity that increased with chain length. This was ascribed to an *in situ* metathesis (see Chapter 7).

It is now the prevailing opinion that the Fischer–Tropsch reaction actually occurs through the interaction of the carbon deposited on the surface of the catalyst, or carbide, with carbon monoxide and hydrogen of the synthesis gas. Experiments conducted with nickel in the absence of hydrogen at 150°C showed that disproportionation of carbon monoxide takes place with the formation of carbon dioxide and the accumulation of carbon on the surface of the catalyst:

$$CO \xrightarrow{\text{metal}} CO_{ads} \longrightarrow C_{ads} + O_{ads} \text{ (initiation)}$$
$$CO_{ads} + O_{ads} \longrightarrow CO_2 + \text{metal}$$

This dissociation of CO appears to be a prerequisite for the Fischer–Tropsch synthesis.

Under the conditions of the synthesis reaction, labeled carbides undergo hydrogenation to form labeled methane. It could be envisaged that partial hydrogenation of the carbide would result in the formation of a carbene and a metal methyl compound, respectively. Various proposals were suggested to explain the propagation steps, for example:

1. Carbon monoxide insertion involving an adsorbed carbon monoxide and metal methyl (alkyl) groups, followed by reaction with hydrogen:

$$\underset{M}{\overset{R}{|}} + \underset{M}{\overset{CO}{\|}} \longrightarrow \underset{M}{\overset{R\diagdown C=O}{|}} \xrightarrow{2\,H} \underset{M}{\overset{R\diagdown CH-OH}{|}} \xrightarrow[-H_2O]{2\,H} \underset{M}{\overset{R\diagdown CH_2}{|}}$$

2. Methylidene (carbene) reaction with an alkyl group attached to the same metal atom:

$$\underset{M}{\overset{H_2C\diagdown \diagup CH_2R}{}} \longrightarrow \underset{M}{\overset{H_2C\text{---}CH_2R}{}}$$

The above type of intermediate was proposed earlier in connection with reforming (Chapter 1, Section VI,B). The basis for this mechanism is the observation that oxygen-free species are incorporated at a high rate in the growing chain. This suggests that in the Fischer–Tropsch synthesis carbon–oxygen bond rupture precedes carbon–carbon bond formation Biloen and Sachtler, 1981).

The assumed existence of carbene explains the formation of radioactive molecules with fewer than 16 carbon atoms from the addition of $^{14}CH_2{=}CHC_{14}H_{29}$ to the synthesis gas:

$$\underset{M}{\overset{CH_2}{\|}} + \underset{CH-R}{\overset{^{14}CH_2}{\|}} \longrightarrow \overset{CH_2{=}^{14}CH_2}{\underset{M{=}CH-R}{+}}$$

The termination step is assumed to occur through a chain transfer reaction:

$$\underset{M}{\overset{RCH_2CH_2}{|}} \rightleftharpoons RCH{=}CH_2 + \underset{M}{\overset{H}{|}}$$

This also explains the insertion of α-alkenes into the growing chain of a Fischer–Tropsch synthesis.

II. FROM METHANOL AND DIMETHYL ETHER

A. Introduction

A process for the manufacture of gasoline from methanol and/or dimethyl ether was announced by Mobil Corporation (Meisel *et al.*, 1976). This process is based on the catalytic conversion of methanol to hydrocarbons over zeolites. The hydrocarbons produced by this process, C_4 to C_{10}, are predominantly in the gasoline boiling range. Unlike the Fischer–Tropsch process, the methanol process produces very little light gas and higher molecular weight hydrocarbons; in addition, it produces aromatic hydrocarbons.

B. Catalyst

The preparation of the ZSM-5 class zeolites that are used in the methanol process is described in a series of patents. The following is one of the preparations of ZSM-5 zeolite given by Argauer *et al.* (1972). SiO_2 (22.9 g) was partially dissolved in 100 ml, 2.8 N tetrapropylammonium hydroxide by heating to a temperature of about 100°C. There was then added a mixture of 3.9 g $NaAlO_2$ (composed by weight of 42.0% Al_2O_3, 30.9% Na_2O, 27.1% H_2O) dissolved in 53.8 ml H_2O. The mixture was heated at 150°C for six days. The resultant solid product was cooled to room temperature, filtered, washed with 1 liter H_2O, dried at 100°C, and calcined at 435° for 16 hr.

The approximate structure of one member of the zeolite class catalysts is given in Fig. 8.1. The pore openings are both at the top and the side of the cylinders. These openings are intermediate in dimension between the wide-pore faujasites (9–10 Å) used in catalytic cracking, and the narrow-

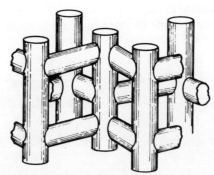

Fig. 8.1. Approximate model of pore structure of zeolites ZSM-5 (from Meisel *et al.*, 1976).

pore zeolites (5 Å), which admit only unbranched alkanes (see Chapter 1, Section VI,C). This property leads to the diffusion constraints on substituted benzenes of greater than 10 carbon atoms and is responsible for the sharp cutoff in the product distribution.

The zeolites used in this process have the capacity to function for long periods of time with negligible coke formation relative to the amount of hydrocarbon processed.

C. Experimental Data

The experimental results obtained from a flow-type reactor containing up to 6 cm³ of catalyst are given in Table 8.6. The experiments were performed by passing methanol over the catalyst at 371°C and LHSV of 1080, 108, and 1. The results at the lowest contact time, LHSV of 1080, show that the main reaction is the conversion of methanol to dimethyl ether. The ether-forming reaction can thus be regarded as the first step in the overall sequence of the conversion of methanol to hydrocarbons. The

TABLE 8.6

Effect of Space Velocity on Methanol Conversion and Hydrocarbon Distribution at 371°C[a]

	LHSV[b]		
	1080	108	1
Product distribution (wt %)			
Water	8.9	33.0	56.0
Methanol	67.4	21.4	0.0
Dimethyl ether	23.5	31.0	0.0
Hydrocarbons	0.2	14.6	44.0
Conversion, MeOH + MeOMe (wt %)	9.1	47.5	100.0
Hydrocarbon distribution (wt %)			
Methane	1.5	1.1	1.1
Ethane	—	0.1	0.6
Ethylene	18.1	12.4	0.5
Propane	2.0	2.5	16.2
Propene	48.2	26.7	1.0
i-Butane	13.8	6.5	18.7
n-Butane	—	1.3	5.6
Butenes	11.9	15.8	1.3
C_5 + aliphatic compounds	4.4	27.0	14.0
Aromatic compounds	—	6.6	41.1

[a] From Chang and Silvestri (1977).
[b] Volume of liquid methanol/volume of catalyst/hour.

small amounts of hydrocarbons formed at the lowest contact time consist mainly of C_2 to C_4 alkenes. As the contact time increases to LHSV of 1, arenes and paraffins become the predominant hydrocarbons.

The zeolite shape selectivity manifests not only by a relatively low boiling range of products, but also in the isomer distribution of the aromatic fraction (Table 8.7). The normalized isomer distribution of xylenes found in the product corresponds essentially to the thermodynamic equilibrium. However, in tri- and tetramethylbenzenes certain isomers, namely, 1,3,5-trimethylbenzene and 1,2,3,5- and 1,2,4,5-tetramethylbenzenes, fall significantly short of their equilibrium concentrations. These results can be attributed to lower diffusivity of these isomers due to steric constraints.

The effect of temperature on product distribution is presented in Fig. 8.2. The data were obtained at moderately low space velocity. At 260°C, the main reaction is the conversion of methanol to dimethyl ether. The hydrocarbons produced are mainly C_2 to C_5 alkenes. Between 340° and 375°C conversion of $CH_3OH/(CH_3)_2O$ approaches completion with the formation of substantial amounts of aromatic hydrocarbons. With a fur-

TABLE 8.7

Aromatic Hydrocarbon Distribution from Methanol[a]

	Normalized distribution (wt %)	Normalized isomer distributions	Equilibrium distributions (371°C)
Benzene	4.1		
Toluene	25.6		
Ethylbenzene	1.9		
Xylenes			
o-	9.0	21.5	23.8
m-	22.8	54.6	52.7
p-	10.0	23.9	23.5
Trimethylbenzenes			
1,2,3-	0.9	6.4	7.8
1,2,4-	11.1	78.7	66.0
1,3,5-	2.1	14.9	26.2
Ethyltoluenes			
o-	0.7		
m- + p-	4.1		
Isopropylbenzene	0.2		
Tetramethylbenzenes			
1,2,3,4-	0.4	9.3	16.0
1,2,3,5-	1.9	44.2	50.6
1,2,4,5-	2.0	46.5	33.4

[a] From Chang and Silvestri (1977).

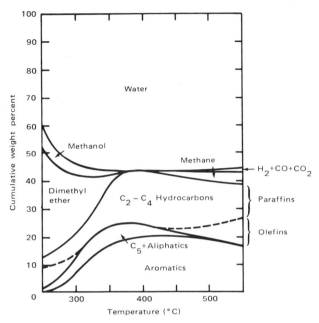

Fig. 8.2. Zeolite-catalyzed methanol conversion; yield structure vs temperature (LHSV 0.6–0.7) (from Chang and Silvestri, 1977).

ther increase in temperature only second-order changes in product distribution occur.

D. Mechanism

The reaction path for the conversion of methanol to hydrocarbons over ZSM-5 class zeolites involves the following sequence:

$$2\,CH_3OH \underset{+H_2O}{\overset{-H_2O}{\rightleftharpoons}} (CH_3)_2O \overset{-H_2O}{\longrightarrow} \text{Light alkenes} \begin{cases} \text{alkanes} \\ \text{arenes} \end{cases}$$

The initial reversible step of dimethyl ether formation is fast and essentially at equilibrium under the range of conditions of the hydrocarbon synthesis reaction.

The basic question to be answered is: What is the nature of the process whereby methanol and dimethyl ether may undergo conversion to olefins? It is suggested that the primary mechanism of methanol–dimethyl ether conversion to olefins is a concerted bimolecular reaction between methylene (carbene) donor and acceptor, methanol or its ether. The methylene is labilized by the zeolite crystal field. Subsequent protolysis affords ethylene:

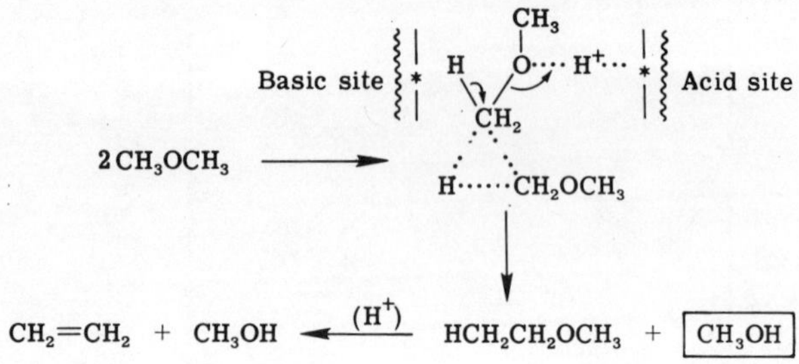

$$2\,CH_3OCH_3 \longrightarrow$$

$$CH_2{=}CH_2 + CH_3OH \xleftarrow{(H^+)} HCH_2CH_2OCH_3 + \boxed{CH_3OH}$$

The methylene insertion can be continued to form higher alkenes (Chang and Silvestri, 1977):

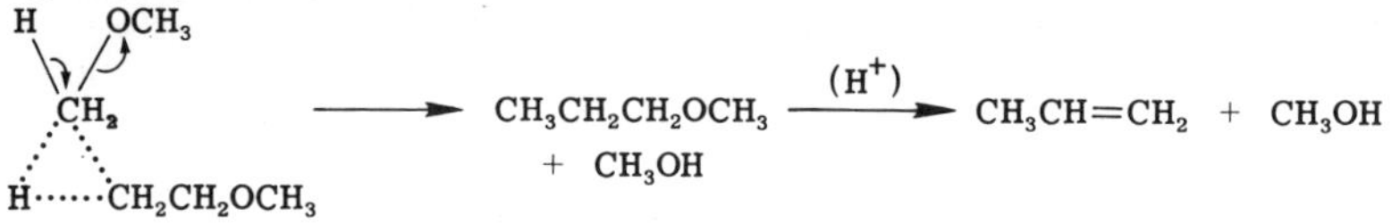

$$\longrightarrow \quad CH_3CH_2CH_2OCH_3 \xrightarrow{(H^+)} CH_3CH{=}CH_2 + CH_3OH$$
$$+ CH_3OH$$

The addition of methylene to the double bond of the initially formed olefins may also occur.

The fact that xylene distribution is equilibrium-controlled indicates that a reversible isomerization of xylenes occurs in the presence of the zeolites. This type of isomerization is usually acid-catalyzed, and therefore the zeolites must have relatively strong acid sites to accomplish this isomerization. The presence of isobutane and arenes is reminiscent of an acid-catalyzed conjunct polymerization of ethylene in the presence of phosphoric acid catalyst in which over 15% of the ethylene converted was composed of isobutane and a substantial amount of aromatic hydrocarbons were formed (see Chapter 1, Section III,C,6).

A similar type of conjunct polymerization may also be considered to occur in the synthesis of alkanes and arenes over zeolite catalyst, with limitations usually imposed by the molecular-shape-controlled action of zeolites.

Methanol Synthesis

a. Introduction. The synthesis of methanol from carbon monoxide and hydrogen was developed in the early 1920's in Germany. Before that time methanol was produced by the destructive distillation of wood wastes, and thus it was given the trivial name "wood alcohol."

The synthesis of methanol has assumed greater importance in recent years as new fuel uses for it have been discovered, such as gasoline blending, the production of synthetic fuels, and the production of methyl *tert*-butyl ether as an octane booster. It is projected that the global market for methanol will increase from 2.8×10^9 gallons in 1980 (1.1×10^9 in the United States) to 5.1×10^9 gallons in 1985 (1.5×10^9 in the United States) (*Chem. Eng. News*, 1980).

Until recently the main starting material for the synthesis of methanol was methane, which is converted to CO and H_2 at pressures of up to 30 atm and temperatures of up to 850°C. A suitable catalyst was found to be composed of 32% NiO, 14% CaO, and 54% Al_2O_3. The supports were previously calcined to 1500°C (Hydrocarbon Reforming Catalysts, 1970). More recently there has been a shift in non-oil-exporting countries toward the use of coal as starting material for the manufacture of CO and H_2.

$$CH_4 + H_2O \rightleftharpoons CO + 3H_2$$

The synthesis of methanol is an exothermic reaction. The theoretical yield of methanol depends on the pressure and temperature (Table 8.8).

$$CO + 2H_2 \rightleftharpoons CH_3OH \qquad \Delta H_{350°C} = -24.5 \text{ kcal}$$

Two main types of catalysts are employed for the synthesis of methanol. One is based on zinc oxide and is used with chromia as promoter. The second type comprises catalysts containing copper and zinc oxide together with stabilizers such as aluminum and chromium oxides. Copper-containing catalysts have higher activities and can be used at lower temperatures, where the thermodynamics of the methanol synthesis are more favorable. These catalysts have replaced the earlier industrial zinc oxide–chromia catalysts.

b. Zinc Oxide–Chromia Catalysts. An active zinc oxide–chromia catalyst was obtained with zinc oxide prepared from $Zn(NO_3)_2$ by precipitation with aqueous ammonia followed by heating with the precipitated zinc

TABLE 8.8

Theoretical and Practical Yields of Methanol as a Function of Temperature and Pressure[a,b]

Pressure (atm)	300°C		325°C		350°C		375°C		400°C	
	x	y	x	y	x	y	x	y	x	y
100	83.9	63.6	76.7	52.3	67.2	40.5	54.8	28.8	44.4	21.3
200	90.5	76.1	86.3	67.6	80.5	57.9	73.0	47.4	65.8	39.1
300	92.8	80.4	89.6	74.2	82.5	66.2	80.3	57.5	71.6	49.5
400	94.2	84.1	91.6	78.4	85.5	71.3	84.0	63.5	74.6	56.6

[a] From Dolgov (1963).
[b] x = theoretical yield of methanol; y = practical yield of methanol.

hydroxide to about 500°C. Chromium oxide was introduced by impregnating ZnO with an aqueous solution of H_2CrO_4. A catalyst consisting of a Zn/Cr ratio of 8:1 was reported to have high catalytic activity.

Methanol synthesis in the presence of zinc oxide–chromia catalyst was carried out at 320°–400°C, 200–300 atm, with a CO/H_2 ratio of 1:1.5 to 1:2.5. About 1.2–1.6 volumes of methanol were formed per volume of catalyst every hour.

c. **Copper–Zinc Oxide Catalysts.** Currently employed industrial low-pressure (20–100 atm) catalysts consist of $Cu/ZnO/Cr_2O_3$ or $Cu/ZnO/Al_2O_3$. In the commercial synthesis of methanol over copper-based catalysts it is desirable for the synthesis gas to contain a small amount of carbon dioxide, approximately 6 vol %. Without the presence of CO_2 in the gas the catalyst rapidly and irreversibly deactivates, although the selectivity to methanol formation is maintained.

The reaction conditions and variations in composition of the catalysts are given in Table 8.9. Methanol synthesis on copper-containing catalysts is carried out at lower temperatures than that on non-copper-containing catalysts, and therefore from thermodynamic considerations the use of lower pressures is possible. With copper catalysts temperatures of 225°–275°C and pressures of 50–100 atm are employed. Although both types of catalysts show high selectivity toward methanol formation, the selectivity

TABLE 8.9

Synthesis of Methanol in the Presence of Copper–Zinc-Containing Catalysts[a]

Catalyst (wt %)	Reactants[b]	Temp (°C)	Pressure (atm)	GHSV[c]	Yield[d] (kg/liter × hr)
$CuO/ZnO/Al_2O_3$					
12:62:25	2	230	200	10,000	3.29
12:62:25	2	230	100	10,000	2.09
24:38:38	2	226	50	12,000	0.7
60:22:8	1	250	50	40,000	0.5
60:17:17	1	275	70	200[e]	4.75
$CuO/ZnO/Cr_2O_3$					
40:10:50	1	260	100	10,000	0.48
40:40:20	2	250	40	6,000	0.26
60:30:10	1	250	100	9,800	2.28

[a] From Herman et al. (1979).
[b] 1, $H_2 + CO + CO_2$; 2, $H_2 + CO + CO_2 + CH_4$.
[c] Gaseous hourly space velocity.
[d] Kilograms of methanol/volume of catalyst × hour.
[e] Molar hourly space velocity.

with copper-containing catalysts may reach over 99%. This high selectivity is possibly due to the lower temperature of operation. In the presence of both types of catalysts and with a temperature above 250°C, a parallel formation of methane is observed.

The laboratory preparation of copper–zinc oxide catalysts consists of the coprecipitation of basic salts of copper and zinc from 1 M nitrate solution by the dropwise addition of 1 M sodium carbonate at ~90°C until a pH of about 7.0 is obtained. The precipitate is collected, washed with water, and dried to 110°C. The material is then calcined up to 350°C and reduced with 2% H_2 in N_2 at 250°C.

Catalysts containing oxides of Cu/Zn/Al and Cu/Zn/Cr are prepared and treated by the same procedure. The third component is then added by coprecipitation from a nitrate solution. The activity of the catalysts thus prepared depends on the catalyst composition (Table 8.10).

An active low-pressure methanol catalyst requires the simultaneous presence of copper and zinc oxide, and the effects of alumina and chromia must be of secondary importance. Optical measurement of the binary 30% Cu/70% ZnO catalyst, which has the optimal catalytic activity, revealed a new strong spectrum, to which the Cu(I) species dissolved in ZnO was assigned. As the copper concentration decreases from 30 to 5%, the ab-

TABLE 8.10

Activity of Copper–Zinc Catalysts as a Function of
Their Composition[a,b]

Catalyst composition, $CuO/Zn/M_2O_3$ (wt %)	Carbon conversion (%)	Yield[c]
0:100:0	0	0
10:90:0	1.0	0.02
20:80:0	10.2	0.24
30:70:0	51.1	1.35
40:60:0	9.6	0.18
50:50:0	11.3	0.20
67:33:0	21.8	0.41
100:0:0	0	0
60:30:10[d]	17.0	0.58
60:30:10[e]	47.0	1.01

[a] From Herman et al. (1979).
[b] The experiments were carried out in a flow system at 250°C, 75 atm, GHSV 5000, with $H_2/CO/CO_2$ = 70:24:6.
[c] Weight of methanol/weight of catalyst × hour.
[d] M_2O_3 = Al_2O_3; pressure, 100 atm.
[e] M_2O_3 = Cr_2O_3; pressure, 100 atm.

sorption intensity also decreases. The dissolution of Cu(I) in ZnO is favored by the fact that Cu(I) is isoelectric with Zn(II) and it assumes, like ZnO, a tetrahedral coordination.

It was proposed that the initial step of methanol synthesis is the chemisorption and activation of CO on the Cu(I) centers and of hydrogen on the surrounding ZnO surface. The hydrogenation of CO occurs by a series of steps, one of which causes the hydrogenation of the oxygen end of CO and another the hydrogenolysis of the Cu–C bond (Scheme 8.2).

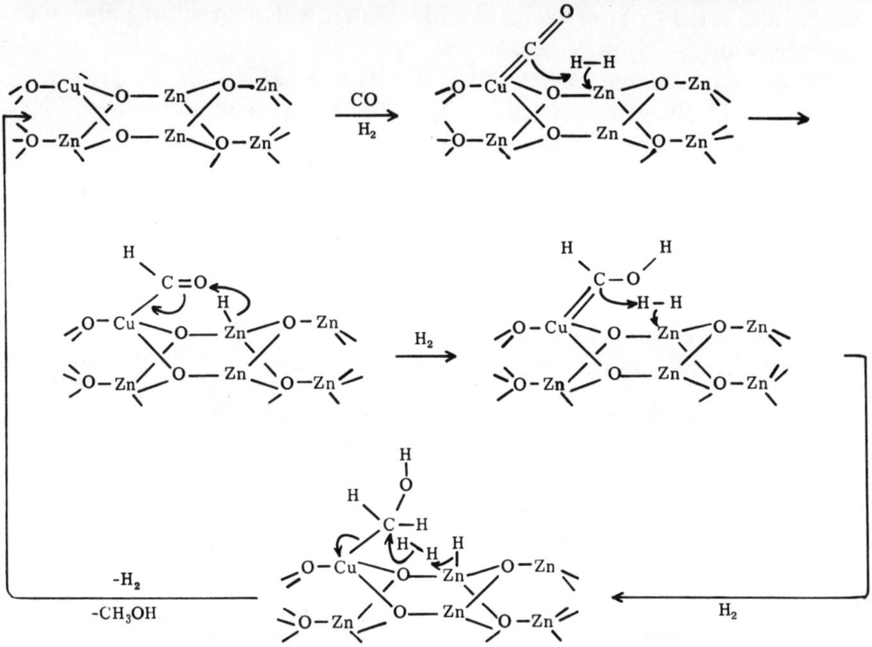

Scheme 8.2. Reduction of carbon monoxide to methanol over copper–zinc oxide catalysts (from Herman *et al.*, 1979).

REFERENCES

* Anderson, R. B. (1956a). Catalysts for Fischer–Tropsch synthesis. *In* "Catalysis" (P. H. Emmett, ed.), Vol. 4, pp. 29–256. Reinhold, New York.
* Anderson, R. B. (1956b). Kinetics and reaction mechanism of the Fischer–Tropsch synthesis. *In* "Catalysis" (P. H. Emmett, ed.), Vol. 4, pp. 257–372. Reinhold, New York.
Argauer, R. J., and Landolt, G. R. (1972). U. S. Patent 3,702,886.
Biloen, P., and Sachtler, W. M. H. (1981). Mechanisms of Hydrocarbon Synthesis over Fischer–Tropsch Catalysts. *Adv. Catal.* **30,** in press.
Chang, C. D., and Silvestri, A. J. (1977). *J. Catal.* **47,** 249.
Chem. Eng. News (1980). April 7, p. 16.

Denny, P. J., and Whan, D. A. (1979). The heterogeneously catalyzed hydrogenation of carbon monoxide. *In* "Catalysis," Specialist Periodical Reports, Vol. 2, pp. 46–86. Chem. Soc., London.

Dolgov, B. N. (1963). "Die Katalyse in der Organischen Chemie," p. 679. VEB Dtsch. Verlag Wiss., Berlin.

Friedel, R. A., and Anderson, R. B. (1950). *J. Am. Chem. Soc.* **72,** 1212.

Frohning, C. D., and Cornils, B. (1974). *Hydrocarbon Process.* **53**(11), 143.

* Herman, R. G., Klier, K., Simmons, G. W., Finn, B. P., Bulko, J. B. and Kobylinski, T. P. (1979). Catalytic synthesis of methanol from CO/H_2. I. Phase composition, electronic properties, and activities of the $Cu/ZnO/M_2O_3$ catalysts. *J. Catal.* **56,** 407.

Hydrocarbon Reforming Catalysts. (1970). *In* "Catalyst Handbook," pp. 64–96. Springer-Verlag, Berlin and New York.

LeRoux, J. H. (1972). *J. Appl. Chem. Biotechnol.* **22,** 719.

Meisel, S. L., McCullogh, J. P., Lechthaler, C. H., and Weisz, P. B. (1976). *Chemtech* **6,** 86.

Oblad, A. G. (1976). Catalytic liquefaction of coal and refining of products. *Catal. Rev. Sci. Eng.* **14,** 83–95.

* Pichler, H. (1952). Twenty-five years of synthesis of gasoline by catalytic conversion of carbon monoxide and hydrogen. *Adv. Catal.* **4,** 271–342.

Storch, H. H., Golumbic, H., and Anderson, R. B. (1951). "Fischer–Tropsch and Related Syntheses." Wiley, New York.

Vannice, M. A. (1976). The catalytic synthesis of hydrocarbons from carbon monoxide and hydrogen. *Catal. Rev. Sci. Eng.* **14,** 153–191.

Wender, I. (1976). Catalytic synthesis of chemicals from coal. *Catal. Rev. Sci. Eng.* **14,** 97–129.

Index

A

Acetaldehyde, ethylene oxidation, 217–219
Acetic acid
 production, 213, 214
 methanol carbonylation, 247
Acetone, production from cumene, 229–230
Acetylenes, see Alkynes
Acid-catalyzed reactions, see also Superacids
 acid strength, 5, 6
 hydrocarbon, 1–122
Acrolein, production from propene, 221
Acrylonitrile, production from propene, 221–223
Adipic acid, production, 214, 215
Adiponitrile, production, 255
Alkanes
 alkylation, acid-catalyzed, 50–57
 superacid catalyst, 98, 99
 aromatization of unbranched, 195
 cracking over silica–alumina, 83–85
 dehydrogenation, 185–187
 hydrocracking, 111–114
 isomerization, Friedel–Crafts catalyst, 12–18
 octane numbers, 101
 oxidation, 213, 214
 reaction with benzene, dehydroalkylation, 68

Alkenes, see also Olefins
 alkylations, 50–57
 cracking over silica–alumina, 83–85
 dimerization, 248–252
 hydroformylation, 243
 hydrogenation, homogeneous, 233, 235
 metathesis, 270–272
 oxidation, 216–226
 oxidative dehydrogenation, 223–226
 polymerization, acid-catalyzed, 32–43, 49
Alkenylation, alkylbenzenes, base-catalyzed, 147–150
Alkylation
 alkane, superacid catalyst, 98, 99
 of aromatic hydrocarbons, 59–79
  mechanism, 60, 61
  orientation, 61–63
  relative reactivity, 63
  steric effect, 63
 benzene, carbonium ion transfer, 67–69
 of cycloalkanes, 57, 58
 with cyclopropanes and cyclobutanes, 77–79
 mechanism, 54–57
 of saturated hydrocarbons, acid-catalyzed, 50–57
Alkylbenzenes
 alkenylation, base-catalyzed, 147–150
 aralkylation, base-catalyzed, 150–152
 dehydrogenation, 208–211